ADVANCES IN CHEMICAL PHYSICS

VOLUME LIII

Advances in
CHEMICAL PHYSICS

EDITED BY

I. PRIGOGINE

University of Brussels
Brussels, Belgium
and
University of Texas
Austin, Texas

AND

STUART A. RICE

Department of Chemistry
and
The James Franck Institute
The University of Chicago
Chicago, Illinois

VOLUME LIII

AN INTERSCIENCE® PUBLICATION
JOHN WILEY & SONS
NEW YORK · CHICHESTER · BRISBANE · TORONTO · SINGAPORE

Library of Congress Cataloging in Publication Data:
Library of Congress Catalog Card Number: 58-9935

ISBN 0-471-89569-5

Printed in the United States of America

10 9 8 7 6 5 4 3 2 1

CONTRIBUTORS TO VOLUME LIII

STEVEN A. ADELMAN, Purdue University, Department of Chemistry, West Lafayette, Indiana

MICHAEL A. COLLINS, Australian National University, Research School of Chemistry, Canberra, Australia

P. A. EGELSTAFF, University of Guelph, Physics Department, Guelph, Ontario, Canada

PETER PFEIFER, Fakultät für Chemie, Universität Bielefeld, D-4800 Bielefeld, Federal Republic of Germany

INTRODUCTION

Few of us can any longer keep up with the flood of scientific literature, even in specialized subfields. Any attempt to do more, and be broadly educated with respect to a large domain of science, has the appearance of tilting at windmills. Yet the synthesis of ideas drawn from different subjects into new, powerful, general concepts is as valuable as ever, and the desire to remain educated persists in all scientists. This series, *Advances in Chemical Physics*, is devoted to helping the reader obtain general information about a wide variety of topics in chemical physics, which field we interpret very broadly. Our intent is to have experts present comprehensive analyses of subjects of interest and to encourage the expression of individual points of view. We hope that this approach to the presentation of an overview of a subject will both stimulate new research and serve as a personalized learning text for beginners in a field.

ILYA PRIGOGINE

STUART A. RICE

CONTENTS

ADVANCES IN CHEMICAL PHYSICS

VOLUME LIII

NEW EXPERIMENTAL STUDIES OF
THE STRUCTURE OF FLUIDS

P. A. EGELSTAFF

Physics Department
University of Guelph
Guelph, Ontario, N1G 2W1, Canada

CONTENTS

I. INTRODUCTION

The study of fluids is a frustrating field, because the underlying ideas have been known for some time but the practical means through which high-quality experiments can be done are not available. Progress is made through a number of inadequate tests, and advances often result from improvements on old methods. The structure of fluids is a good example, since its theoretical importance has been recognized for a long time, whereas the ability to either calculate it or measure it accurately is comparatively recent. Even now, good studies of structure as a function of state parameters are rare. Many papers on fluids in general and their structure in particular are written with

1

a "this closes the book" conclusion, though in fact only the first page of the book has been turned. In this review we shall open more "books" than we shall close, even though the first experiments on fluid structure are more than 50 years old and the field has received a great deal of attention since then.

The conventional framework through which fluid structure is discussed is a hierarchy of correlation functions starting with the pair correlation function $g(r)$. Classically this is related to the probability of finding two atoms at the separation $\mathbf{r}$ (at a given time) if the space origin is located at one of them. If the system is in thermal equilibrium, the choice of the given time and origin is unimportant. Because this function extends over the range $0 < r < \infty$ and varies significantly with density and temperature, and because small changes in it are usually important, it has never been studied properly. Moreover, quantum corrections are generally ignored because other sources of error are larger. In the case of molecular fluids there are several pair correlation functions corresponding to the number of possible combinations of different atoms in pairs. This description is applicable to all classes (rigid, flexible, or dissociating) of molecules, but most work has been concerned with rigid molecules. However, because the whole set of pair correlation functions does not contain complete information on angular correlations, it is not as useful theoretically as the single pair correlation function for an atomic fluid (in this paper we shall distinguish between molecular and atomic fluids).

The next member of the hierarchy is the three-body correlation function, and this is related to the probability that three atoms (at a given time) are separated by $\mathbf{r}$ from 1 to 2 and $\mathbf{s}$ from 1 to 3 with the origin located on 1. The n-body correlation functions are defined in a similar way, and it is generally believed that their importance to structure decreases as n increases. A satisfactory theory cannot be developed using the pair correlation alone and at least some data on the three-body function are required as well. Although the description of fluid structure through n-body correlation functions is independent of any ideas about underlying forces, it is usual to invoke another hierarchy, namely, that of n-body forces. The most important member is the pair force and most theories of fluids involve a discussion in terms of this force alone. However for real fluids the three-body force is usually significant (e.g., produces effects $\sim 10\%$), and in some cases other forces are required.

In the case of atomic fluids some theories represent the potential energy as a sum of pair interactions, while others include the higher-order terms. Very few treatments go beyond the pair interactions in the molecular case, because the additional complexity cannot be justified. Thus realistic pair potentials are normally employed for atomic fluids, whereas effective pair potentials are often used for molecular fluids (i.e., a pair potential adjusted to give data similar to what would be obtained if the correct combination of forces were used). Where possible we shall use a true pair potential and discuss the difference between pair-theory predictions and experimental data in terms of many-body effects. However, because of uncertainties in pair potentials (especially for molecules), we shall look for situations in which it appears that the experimental results can be explained only through a many-body potential. To observe the effects of different forces and to study the changes in structure as a function of state, which depend on the type of force present, requires experimental data of high quality. Such a need implies the measurement of the shape and absolute magnitude of a smooth function to a precision ~ 1% in a reasonably short time. This is no easy task and requires special techniques, to be discussed later in this article.

There are similarities between the pair correlation functions for different fluids or for the same fluid in different states. These similarities arise because the structure is influenced by "excluded-volume effects," that is, because atoms and molecules do not overlap. To extract the remaining and interesting part of the structure requires precise data, so that relatively small differences caused by attractive forces or anisotropic forces may be distinguished. Much of the published experimental data does not meet this requirement, and (as indicated above) a task facing the experimentalist is to improve techniques to the point where they are able to make a significant contribution to this field. In the next section the techniques will be reviewed briefly and conclusions as to which methods offer the best prospects will be drawn.

In subsequent sections we discuss the experimental study of atomic fluids, homonuclear diatomic molecular fluids, and then more complex cases. In this review, because of space limitations, only the static structure will be considered. However, it should be remembered that the van Hove space and time-dependent correlation functions can be measured accurately now and provide new and important information on the dynamics at the atomic or molecular level. Usually static structure is regarded as the starting point for a discussion of dynamics, and so it is appropriate to confine our discussion to that.

II. EXPERIMENTAL METHODS

A. X-Ray and γ-Ray Diffraction

The precision required from an experimental method depends on the state of knowledge in the field. In the case of fluid structure the early experiments in the 1930s were made in order to verify the principal features of $g(r)$. After this period the requirement changed rapidly to one of high precision as outlined in the introduction. But the existing technique (X-rays) was incapable of meeting this need, even after many years of technical improvements which led to highly sophisticated experimental programs. The reasons for this will be described briefly.

For all diffraction experiments the layout is essentially similar to the neutron-scattering case shown in Fig. 1. On passing through the sample the radiation is scattered one or more times, and corrections are made to $I(\theta)$ for the multiple scattering to yield an effective single-scattering function. The sample material attenuates the beam, and this reduces $I(\theta)$, so that corrections are needed for "absorption." The latter consists of both scattering and true absorption, and when (as for X-rays) the true absorption is large, the multiple-scattering effect is reduced. A number of other experimental points are listed in the box on Fig. 1, and after these have been covered the experimentalist obtains a differential cross section ($d\sigma/d\Omega$) on a relative scale. The next step is to convert $d\sigma/d\Omega$ to a structure factor $S(q)$, which is simply related to $g(r)$ by the equation (for one kind of atom)

$$S(q) = 1 + \rho \int \exp(i\mathbf{q}\cdot\mathbf{r})[g(r)-1]\, d\mathbf{r} \tag{2.1a}$$

where ρ is the atomic number density, $q = (4\pi \sin \frac{1}{2}\theta)/\lambda$, and λ is the wavelength of the radiation. If we call an experiment that measures $S(q)$ directly

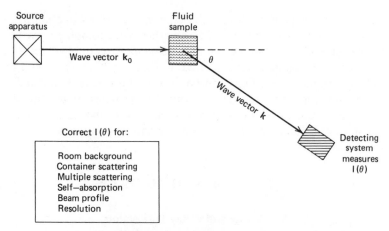

Fig. 1. Layout of a typical neutron diffraction or scattering experiment. The source apparatus generates either a monochromatic beam in the conventional experiment, or a pulsed "white" beam in the time-of-flight experiment. Six corrections, listed in the box, are applied to $I(\theta)$ to convert it to a relative differential cross section.

a "diffraction" experiment, then this name does not apply to the experiments reviewed here, because they are "scattering experiments" in which the differential cross section $d\sigma/d\Omega$ is measured and subsequently converted to $S(q)$ in a computer. However, these two names are used interchangeably in the literature.

For X-rays the steps between $d\sigma/d\Omega$ and $S(q)$ involve adjustments of varying sizes for (i) form factors, which are particularly difficult for molecules, (ii) polarization corrections, (iii) fluorescence, (iv) Compton scattering, (v) absolute normalization, and—if the data are to be better than 1%—(vi) Born-approximation corrections. The inherent inaccuracy of X-ray data (due to this list) is several percent: this is in addition to any experimental errors. Unfortunately this means that X-ray data cannot be used for the objectives discussed in this article, which require data good to ~1%. The lack of appreciation of this fact has delayed the development of this field for many years. To overcome these problems some workers have imposed theoretical requirements on their observations (i.e., adjusted their observations to fit the theory). A typical example is provided by the technique of forcing the Fourier transform to vanish at low r to simulate atomic overlap. Apart from the fact that this is usually done on the wrong function (i.e., the deduced nuclear function rather than the measured electron function), the errors discovered this way only highlight the inherent imprecision of the method.

Electron-diffraction data suffer from several of the problems associated with X-ray data and the experimental technique is restricted by the high absorption coefficients and large Born-approximation corrections, which make it necessary to use high energies. In addition, the correction for the Coulomb potential involves multiplying by θ^4, which introduces a relative factor of ~10^8 (if q is varied from 0.2 to 20 Å^{-1}), and this magnifies small defects which could otherwise be ignored.

If γ-rays, having an energy many times that of X-rays, are used, the effects of the above six X-ray corrections are reduced. In this case scattering through small angles is observed (with small polarization corrections) due to the short wavelength. Because of the high energy, the photoelectric cross section of the sample is reduced, thus reducing absorption and increasing penetrability. Fluorescent scattering is more easily rejected, and the Compton scattering at high angles can be used for absolute normalization. From the experimental point of view, radioactive sources are very stable, and a wide choice of detectors are available. These advantages have been exploited by crystallographers for some years, but have yet to be exploited for the study of fluids. The largest uncertainty remaining is then the form factor, and it may be preferable to use γ-ray diffraction to measure electron–electron correlation functions and—by comparison with accurate neutron data—to determine the form factors experimentally. In this way the behavior of electrons in the fluid or in molecules may be studied.

The scattered flux in Fig. 1 is a flux integrated over all energy changes ΔE that do not alter the quantum state of the scatterer (i.e., the nucleus for neutrons or the atom for X-rays and γ-rays). The scattering vector, $\mathbf{q}$, ought to be independent of ΔE, but is not, due to the need to conserve energy and momentum. However, if

$$\Delta E \ll E_0 \qquad\qquad (2.1b)$$

where $E_0 = c(\sqrt{m^2c^2 + \hbar^2k_0^2}) - mc^2$ is the energy of a photon (electron or neutron) of mass m, the result in Eq. (2.1a) is obtained. It can be seen that diffraction data do not represent "elastic" scattering, as often claimed, but total scattering integrated over all the energy transfers generated by the allowed modes of motion of the atoms in the fluid. For this reason it is useful to understand the dynamics of the fluid when conducting diffraction experiments. Similarly it can be seen that the Fourier transform of the true elastic-scattering amplitude is merely the (stationary) center of mass of the fluid. Equation (2.1b) is satisfied easily for X-rays and γ-rays, but for slow neutrons a correction is usually required and will be discussed below.

In discussing the Compton scattering of X-rays or γ-rays it is necessary to consider corrections of this kind too. The usual method is to apply a kinematic correction (known as the Breit–Dirac correction) by multiplying the Born-approximation cross section by a factor $k/k_0 \simeq 1 - (2\hbar q \sin\frac{1}{2}\theta)/mc$, where m is the electron mass (if higher-order terms are needed in this expansion a relativistic correction will be needed). However, this does not allow for the new momentum transfer, $q\sqrt{k/k_0}$, and entries in tables of Compton cross sections should be made at that value. For accurate work with γ-rays these corrections are worthwhile.

Some experimentalists try to avoid these questions by using an energy-sensitive detector to separate Compton scattering from the remainder. Unless small values of q are discarded or the resolution of the detector is very good, this technique requires corrections for detector resolution which (in the opinion of this author) may generate greater uncertainties than the original problem. The Fourier transformation of high-precision γ-ray experiments will yield the electron–electron correlation function for a fluid of atoms without excited states if the Compton scattering has been subtracted, or for atoms with excited states if it has been included.

B. Slow-Neutron Diffraction

The low flux of neutrons from nuclear reactors (usually much less than from an X-ray source) and the primitive techniques used during the development of neutron diffraction prevented its inherent precision from being exploited. Of the six adjustments required to X-ray data, numbers (i) to (iv) do not occur with neutrons, and (v) and (vi) generate much smaller errors. For these reasons the inherent precision of slow-neutron experiments is about an order of magnitude better than for X-rays, and is a few tenths of a percent. Although still far from this limit, neutron data with a precision of ~1% can be obtained today. Data of this kind will be discussed in later sections. However, to obtain 1% precision by any technique requires proper care and attention to many experimental matters, for example, proper definition of the state conditions and their uniformity over a large sample, accurate determination of the beam profile and position, stable flux and counting equipment, and redundancy in data collection. Only a few published experiments meet all these requirements.

In the case of slow neutron data Eq. (2.1b) is not normally satisfied and an "inelasticity" or "Placzek" correction is needed. The size of (and error caused by) this correction varies with E_0 and θ. If the experimentalist specifies an acceptable error, then the minimum E_0 for a given small θ and the maximum θ for a given E_0 will be specified automatically. Because of the importance of this topic, a critical review of inelasticity corrections is given in Appendix A. For atomic or diatomic molecular fluids this correction has been derived by Egelstaff and Soper[1] with sufficient precision that $S(q)$ correct to 1% may be obtained normally. Their formulas are easily generalized to other molecules. Earlier treatments[2,3] involve approximations leading to larger errors except in limiting circumstances. The usual Placzek series is a power series in M^{-1}, where M is the molecular mass. It is truncated at M^{-1} because this term may be calculated without knowing the detailed molecular dynamics. The minimum acceptable value of E_0 (for a given small θ) is such that the M^{-1} term gives a reasonable estimate for the size of the correction, and the maximum acceptable θ (for a given E_0) is such that the error on the correction is a reasonably small fraction of the amplitude of $|S(q)-1|$. If other methods of correction are used, similar principles are applied. It is useful, therefore, for the experimentalist to manipulate E_0 and θ. For molecules containing light atoms (e.g., HCl or D_2O) these corrections are not well known and are usually calculated from models. As a result the errors are larger in these cases. The only solution would be to employ fast neutrons (e.g., 30 eV) at *low* angles where Eq. (2.1b) is satisfied even for light elements. In this case $d\sigma/d\Omega$ is equal to $S(q)$ times the square of the scattering length. Unfortunately the flux at 30 eV is too small, with present sources, for good quality experiments.

To illustrate these points we show in Fig. 2 the Placzek correction for krypton at 297 K, and a conventional experiment with fixed k_0 ($=2\pi/\lambda$). Curves for two possible values of λ are plotted, and from (Fig. 2a) it will

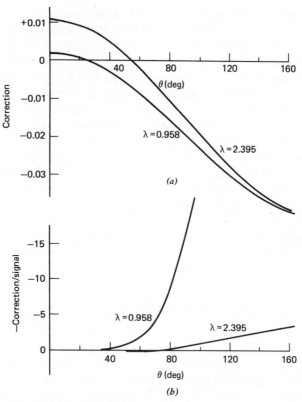

Fig. 2. (a) The Placzek or inelasticity correction for krypton at 297 K calculated for two wavelengths λ and counter efficiencies often used by the author. (b) The ratio of the Placzek correction to $|S(q)-1|_{max}$ for krypton at 297 K and $\rho = 6.2 \times 10^{27}$ atoms/m³. Here $|S(q)-1|_{max}$ is called the signal and is the envelope of $|S(q)-1|$ (i.e., a smooth function drawn through the points of maximum amplitude of $|S(q)-1|$).

beseen that for the shorter wavelength the absolute value of the correction is lower at low angles,[‡] becoming the larger of the two values for angles greater than 45°. The size of the correction ranges from +1% to −3% of the asymptotic level of $S(q)$ over the range of angles normally used (3–120°). In Fig. 2b the size of the correction relative to the maximum amplitude of $|S(q)-1|$ is plotted. For the longer wavelength the correction/signal ratio

[‡]In some papers it is implied that short wavelengths are superior at all angles, whereas the correct statement is that at a given q the correction decreases as λ is decreased. Figure 2 shows that the *largest* absolute and relative correction occurs for the shortest wavelength unless θ is restricted.

rises to a value of about unity at 110°, while for the shorter wavelength this value is reached at 55°. The steep rise in the ratio at larger angles illustrates the need to use low angles with short wavelengths. Although the numerical values in Fig. 2 apply to krypton at the wavelengths used by the author, the general trends of these curves are common to all fluids and wavelengths. In the krypton experiment reviewed in Section III the longer wavelength was used out to an angle of 110°, and the shorter wavelength was not used, as it would extend the q-range from 4.2 to 6.1 Å^{-1} only, while doubling the experimental effort and providing little new information (see Fig. 4). In general longer wavelengths are preferred, as the wavelength resolution and the angular resolution are better, the container scattering is less serious, and frequently the intensity is higher. The Placzek series is an asymptotic expansion, and for convergence the terms must decrease rapidly. Krypton is an example for which the M^{-2} term has been calculated, and the ratio (M^{-2} term)/(M^{-1} term) is $+0.010$ at 0° and -0.011 at 90° for $\lambda = 2.395$ Å and $T = 297$ K.

The conventional technique used with reactor neutron beams (Fig. 1) is to keep k_0 fixed and to vary θ. From the above remarks it is clear that the choice of k_0 is relatively unimportant in determining the quality of the results, although small k_0 is preferred. Usually it is chosen to correspond with the q-range of greatest interest (as $q_{max} = 2k_0$). An alternative technique is to keep θ fixed and to vary k_0. Because of the wide range of k_0-values available with pulsed neutron sources, this method is used frequently with them. Neutrons are selected according to their flight time from source to detector, which leads to a relationship between the different wave numbers:

$$\frac{a}{k_0} + \frac{1}{k} = \frac{a+1}{k_e} \tag{2.2a}$$

where a is the ratio of distances (from source to sample divided by sample to detector) and k_e is the wave number for elastic scattering. Because of this relationship and because k_e is readily measured, it is convenient to express quantities in terms of k_e. The conventional experiment corresponds to $k_0 = k_e$ or $a \to \infty$, and so this formalism can be used for either that case or for time-of-flight experiments.

The measured cross section is (using the notation of Fig. 1):

$$\frac{d\sigma}{d\Omega} = K \int_{-\infty}^{+\infty} \phi(k_0) \varepsilon(k) \frac{k}{k_0} S(q_\omega, \omega) F(\omega) \, d\omega \tag{2.2b}$$

where K is a normalization constant, $\phi(k_0)$ is the incident flux at wave number k_0, $\varepsilon(k)$ is the detector efficiency, $S(q, \omega)$ is the van Hove dynamic structure factor, and $\hbar q_\omega$ is the momentum transfer and is a function of the energy transfer $\Delta E = \hbar\omega$. Because of the relationship between k, k_0, and k_e, the dynamic structure factor is multiplied by a sampling factor $F(\omega)$, where

$$F(\omega) = \frac{a+1}{a + (k_0/k)^3} \tag{2.3}$$

When Eq. (2.1b) is satisfied, we have $k \simeq k_0 \simeq k_e$ and Eq. (2.2) reduces to

$$\frac{d\sigma}{d\Omega} \simeq K' \int_{-\infty}^{+\infty} S(Q, \omega)\, d\omega = K'S(q) \tag{2.4}$$

where $S(q)$ is the desired structure factor and K' is another constant. Some examples of $F(\omega)$ and of q/q_e (at $\theta = 180°$) as a function of $\hbar\omega/E_e$ for different a are shown in Fig. 3. It is seen that if the ratio $\hbar\omega/E_e$ is allowed to vary widely both functions deviate substantially from the values which lead to Eq. (2.4). For high angles the intensity often spreads over a range of ω, but for low angles (especially with large k_0) this difficulty can be avoided. Equation (2.4) is the leading term in an expansion of Eq. (2.2b) about $q = q_e$. At low angles ω is small and this expansion (see Appendix A) might be expected to work, but some caution is needed. The reason can be seen by expanding q/q_e in terms of $\hbar\omega/E_e$, namely,

$$\left(\frac{q}{q_e}\right)^2 = 1 - \frac{a-1}{a+1}\frac{\hbar\omega}{2E_e}$$

$$+ \left[\frac{1}{2(1-\cos\theta)} - \frac{1}{2} + \frac{3a}{(a+1)^2}\right]\left(\frac{\hbar\omega}{2E_e}\right)^2 + \cdots \tag{2.5}$$

The third term will be small when θ is small provided $(\hbar\omega/2E_e)^2 \ll \theta^2 \simeq q_e^2/k_e^2$, that is, provided $m\omega^2 \ll 2E_e q_e^2$, where m is the neutron mass. When ω^2 is replaced by the second moment $\int \omega^2 S(q_e\omega)\, d\omega = kTq^2/M$ for a system of point masses M, a useful condition is obtained: $E_e \gg mkT/2M$. However, the restriction on ω^2 must be considered in the general case.

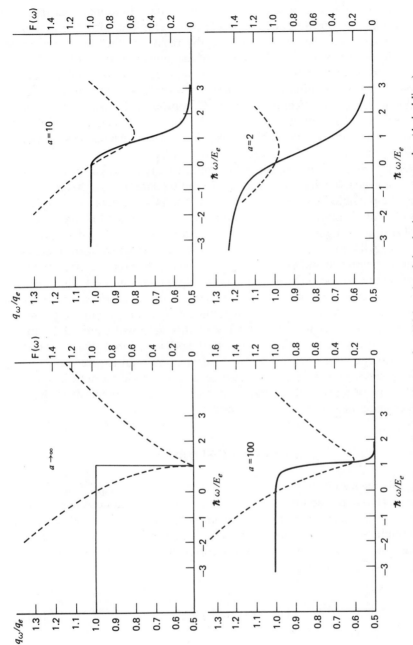

Fig. 3. The real momentum transfer $\hbar q_\omega$ for $\theta = 180°$ in units of the elastic momentum $\hbar q_e$ (dashed lines), and the sampling factor $F(\omega)$ from Eq. 2.3 (full lines), plotted as a function of the ratio of energy transfer to elastic energy $(\hbar \omega / E_e)$ for several values of the distance ratio a. Note: $q_\omega / q_e = |\mathbf{k}_0 - \mathbf{k}| / (2k_e \sin \frac{1}{2}\theta)$.

When E_0 is varied over a range which includes the vibrational level spacing or the dissociation energy of a molecule, some additional structure is seen in the cross section. This arises because the region of the steep drop in $F(\omega)$—Fig. 3—become a significant part of the integrand in Eq. (2.2b). For the case of metal hydrides this effect is large and is well known,[4] and another example is the dissociation of molecules. For diffraction experiments it is desirable to avoid this effect by choosing E_e appropriately, although it may be an effect of interest in its own right for the study of dissociation energies. In most experiments it is desirable to keep E_e less than or much larger than the vibrational energies of molecules (see Appendix A).

It will be realized from the foregoing that a variety of slow-neutron techniques are available which are capable of yielding data of high precision. It will not be possible in a number of cases to obtain good data over the full range of q of theoretical interest; nevertheless, it is better for the experimentalist to obtain good data over a restricted range than poor data over a wider range. Useful high-quality experiments are designed with a background of theoretical and experimental knowledge; they require great attention to detail and careful execution. Inevitably this means weak data must be discarded and experiments repeated. Because of the time and effort required to do this, the majority of the published work has not met these requirements, and it seems likely that low-quality data will continue to appear. Therefore some screening of published work is required, which itself is time consuming, and often needs more information than is given, as most authors overestimate the quality of their results. The reviewer must be quite critical in assessing such data, and must apply the same standards to theoretical as he applies to experimental results.

III. ATOMIC FLUIDS

The pair theory of atomic fluids interrelates the pair potential $u(r)$ and the pair correlation function $g(r)$, and given one there is a unique numerical function for the other. In approximate theories this relationship is simplified. For example, the Percus–Yevick (PY) approximation is $g(r) = c(r) + y(r)$, and the Fourier transform of the direct correlation function $c(r)$, denoted by $\hat{c}(q)$, is defined by

$$\rho\hat{c}(q) = 1 - [S(q)]^{-1} \tag{3.1}$$

where $y(r) = g(r)\exp\beta u(r)$ with $\beta = (kT)^{-1}$. In the general case the expansion of $c(r)$ in a power series in the density ρ is

$$c(r) = [f(r)+1] + \rho f(r) \int f(s) f(|\mathbf{r}-\mathbf{s}|) \, d\mathbf{s}$$

$$+ \rho [f(r)+1] \int [f(s)+1][f(|r-s|)+1]\{\exp[-\beta u_3(r,s)]-1\} \, d\mathbf{s}$$

$$+ O(\rho^2) \tag{3.2}$$

where $f(r) = \exp[-\beta u(r)]-1$. The first two terms of this series are given correctly by the PY equation, but the term involving the three-body potential $u_3(\mathbf{r},\mathbf{s})$ is not included in either the PY approximation or the pair theory. If this term is significant, it may be studied by comparing calculated and measured values of $c(r)$ as a function of density, assuming that $u(r)$ is known. This is not easy, because the measurable quantity is $\rho \hat{c}(q)$, which, as Eq. (3.1) shows, vanishes as $\rho \to 0$. Also, to test the published data on $u(r)$ it is desirable to isolate the leading term in Eq. (3.2) by working with $\rho \sim 0.2 \times 10^{27}$ atoms/m^3.

Experiments on gases are difficult because $|S(q)-1|$ is small relative to 1, and because containers for high pressures usually produce large background signals. This problem is made worse if corrosive gases are to be measured (e.g., HCl or Cl$_2$) which require containers of materials having a large neutron cross section, or if expensive isotopically separated samples are used which are available in small quantities only. In these respects, however, argon and krypton are favorable and relatively good experiments are possible. For example, the ratio of sample intensity to container plus background intensity was about 6 for the experiments on krypton ($\rho = 2.88 \times 10^{27}$ atoms/m^3) reviewed in Section III.B.

A. Argon

An attempt to measure $f(r)$ for argon was made by Fredrikze et al.[5] using neutron diffraction at one state with $T = 141.3$ K and $\rho = 2.52 \times 10^{27}$ atoms/m^3. Karnicky et al.,[6] using X-rays, did five experiments on the 173-K isotherm at densities of 1.24, 2.01, 3.14 (twice), and 4.70×10^{27} atoms/m^3. The lowest density was used to establish the form factor for argon. They calculated the terms $O(\rho)$ in Eq. (3.2) using the Axilrod–Teller (AT) triple-dipole term for $u_3(\mathbf{r},\mathbf{s})$ and an initial estimate for $u(r)$. The object of both experiments was to improve the quality of the data so that the leading term $f(r)$ could be extrated from the Percus–Yevick equation. Eventually these data were in reasonable agreement with $u(r)$ derived by the conventional method (i.e., by finding the parameters in an empirical formula though fitting to experimental data). There was the possibility of a discrepancy in the region 5–6 Å for both experiments, but this was about the size of the experimental error. Although this discrepancy was small in magnitude, it may spread over a substantial range of r, so that its cumulative effect might be significant. By making comparisons in Fourier space, discrepancies of this kind are more visible, as low values of q are sensitive to areas in real space. In addition, experimental errors are smaller if experimental (i.e., Fourier) space is used. Nevertheless these two experiments demonstrated the possibility of exploiting the virial series for $g(r)$ as a means of studying $u(r)$.

Liquid argon was studied with neutrons by Yarnell et al.[7] using a wavelength of 0.978 Å and an angular range of $3 < \theta < 90°$. Their state was on the coexistence curve at 85 K, and the density was $\rho = 21.3 \times 10^{27}$ atoms/m^3. For this state the Placzek correction was about twice the size of $|S(q) - 1|_{max}$ at their highest angle. Although this was a carefully executed experiment and the corrections were carried out (and described) fully, these data are fairly insensitive to the details of the pair potential. This may be understood by considering the correlation between r-space effects and q-space effects. As an illustration of this correlation we show in Fig. 4 (due to Ram[8]) the PY structure factors at $T = 297$ K and $\rho = 6 \times 10^{27}$ atoms/m^3 for the Lennard–Jones[9] and Barker et al.[10] krypton potentials. The small shift between the two potentials at the position where $u(r) = 0$ (Fig. 4a) causes the oscillations of $S(q)$ for $q > 1.5$ Å^{-1} to be slightly out of phase in Fig. 4b. However, the major differences occur for $q < 1.0$ Å^{-1} and are related to the differences of $u(r)$ in the range $4 < r < 10$ Å. As this is the range over which both $u(r)$ and $u_3(\mathbf{r}, \mathbf{s})$ show important features, experimental measurements for $q < 1.0$ Å^{-1} need to be carried out carefully. However in the liquid case the intensity in this region is very small (about 3% of Fig. 4b as $q \rightarrow 0$), and although the data of Yarnell et al were larger than the computer simulation data (using an LJ potential) for $q < 1.0$ Å^{-1}, their errors were such that no significance could be attached to this difference. In fact, if the point where $u(r) = 0$ is chosen appropriately, it appears from Fig. 4 that either potential would yield a reasonable fit to the liquid data. This conclusion is, of course, in accord with the results of perturbation theory. Improved measurements on the liquid for $q < 1$ Å^{-1} would be desirable to observe the difference between predictions based on different $u(r)$'s.

A comparison of the data taken by Yarnell et al.[8] with the data reported in the next section illustrates the technical improvements that have taken place in recent years. For the gas experiments significantly greater intensities were obtained than in this liquid-argon experiment, even though a smaller fraction of the beam was scattered, and in addition the ratio of sample intensity to container plus background intensity was twice as large as that for liquid argon.

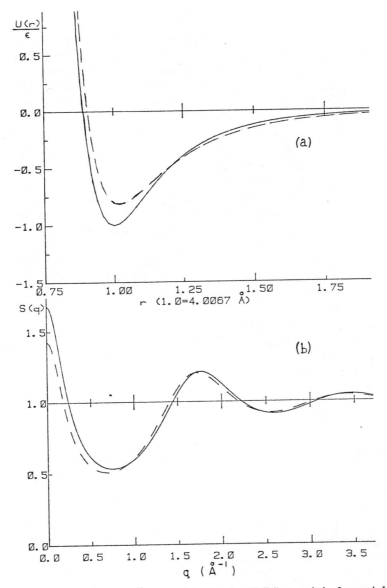

Fig. 4. (a) The Barker et al.[10] potential for krypton (full line) and the Lennard–Jones potential deduced from the properties of the solid[9] (dashed line). (b) The structure factors of krypton gas at 297 K and $\rho = 6 \times 10^{27}$ atoms/m³ calculated from the PY equation using the two potentials shown in (a).[8]

B. Krypton

The next noble gas, krypton, is a better experimental subject than argon, partly because its higher mass means that the quantum terms (which contribute at high q) are smaller and the inelasticity correction is smaller too. In addition, room temperature (rather than $\simeq 180$ K) is a reasonable choice, and the neutron cross sections of natural krypton are more favorable than those of natural argon. Teitsma and Egelstaff[11] made an extensive study of krypton along the room-temperature isotherm, $T \simeq 1.5\varepsilon/k$. This is a convenient temperature at which to study effects due to the bowl of the potential while working at some distance from the critical point. To cover in detail the region of q over which the principal effects occur ($q < 4$ Å^{-1}) they used a wavelength of 2.395 Å (see Fig. 2), and Fig. 5 shows their data for seven of the seventeen experiments carried out. The circles and lines are the data, while the crosses are results based on Monte Carlo[12] and virial-expansion calculations. It can be seen that for $\rho = 0.258 \times 10^{27}$ atoms/m^3, the maximum value of $|S(q)-1|$ is about 0.02, and the excellent agreement in this case shows that the experimental precision is 1% or better. With increasing density the magnitude of $|S(q)-1|$ increases, and differences between the calculated and measured results increase also. This difference indicates the contribution of the term involving $u_3(\mathbf{r},\mathbf{s})$ in Eq. (3.2). Teitsma and Egelstaff calculated $\hat{c}(q)$ from equation (3.1) and plotted it as a function of ρ. They determined the intercept and initial slope at $\rho \to 0$ from the straight lines obtained. These quantities give respectively

$$-2B(q,T) = \int \{\exp[-\beta u(r)] - 1\}\exp(i\mathbf{q}\cdot\mathbf{r}) \tag{3.3}$$

$$-3C_2(q,T) = \iint [f(r)+1][f(s)+1][f(|\mathbf{r}-\mathbf{s}|)-1]$$
$$\times \{\exp[-\beta u_3(\mathbf{r},\mathbf{s})] - 1\}\exp(i\mathbf{q}\cdot\mathbf{r})\,d\mathbf{r}\,d\mathbf{s} \tag{3.4}$$

In Figs. 6 and 7 (from Refs. 12 and 11, respectively) we show these data compared with the results obtained from a recent[13] pair potential and the AT expression for u_3 in Eq. (3.4).

In the case of Fig. 6 the agreement is excellent, except in the neighborhood of $q \simeq 0.7$ Å^{-1}. By assuming the theoretical form for $B(q,T)$ in the range $q > 4$ Å^{-1} the data may be extended to $q = 20$ Å^{-1} and then Fourier transformed, and $u(r)$ computed. This shows a small deviation from the published potentials[10, 13] in the range $6 < r < 12$ Å similar to that discussed above. One explanation for this discrepancy is that these potentials are obtained by fitting models to experimental data to find the model parameters,

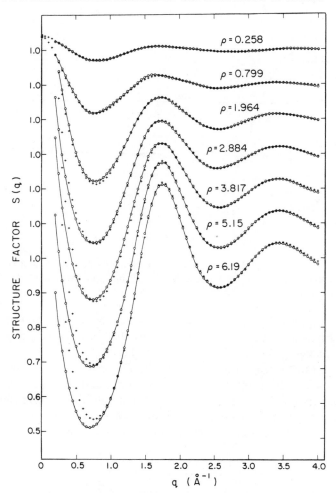

Fig. 5. Structure factor for krypton gas at $T = 297$ K and densities shown (in units of 10^{27} atoms/m^3). The experimental results[11] are shown by the circles and lines, while the theoretical calculations using a published pair potential[10] are shown by the crosses. (Reproduced from ref. 11 by permission of American Physical Society.)

whereas in this method $u(r)$ is calculated from Eq. (3.3). At large r the model takes the form

$$u(r) \sim \sum_{n=3,4,\ldots} \frac{C_{2n}}{r^{2n}}, \tag{3.5}$$

which is a valid series as $r \to \infty$. However for r in the above range it is possi-

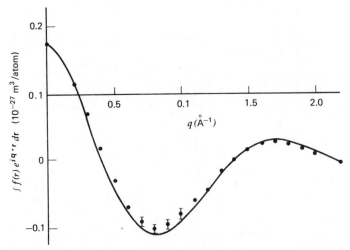

Fig. 6.   The Fourier transformation of $f(r) = \exp[-u(r)]-1$ derived from the data of Fig. 5 (solid circles) and calculated from a published $u(r)$[12] (full line). (Reproduced from ref. 12 by permission of American Physical Society.)

ble that Eq. (3.5) is not an accurate representation of $u(r)$, and the (scattering) experimental data cannot be fitted by this series. The large size of the virtual excited states should be taken into account.

In Fig. 7 the measured coefficient $3C_2(q, T)$—Eq. (3.4)—is compared with that calculated from the AT triple dipole term, and the agreement is poor. On Fourier-transforming these data it is apparent that discrepancies occur again for $r < 10$ Å. In many papers the AT term is used over this range even though it is a valid approximation only as $r \to \infty$. Figure 7 makes it clear

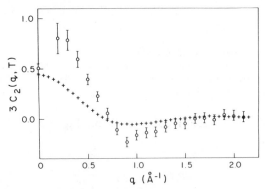

Fig. 7.   The function $3C_2(q, T)$—Eq. (3.4)—derived from the data of Fig. 5 (open circles) and calculated using the Axilrod–Teller triple-dipole potential (crosses). (Reproduced from ref. 11 by permission of American Physical Society.)

that this approximation will give data of the right magnitude but wrong in detail. The Fourier components at which large discrepancies occur are evident from this figure. If better theoretical models for $u(r)$ and $u_3(\mathbf{r},\mathbf{s})$ are prepared, they should be compared with the Fourier-space data of Figs. 6 and 7, since r-space results depend upon some assumption about the high-q form of $S(q)$ or $\hat{c}(q)$. While this could be measured using shorter-wavelength radiation, it is difficult to obtain accurate results, as $|S(q)-1|$ is small (see Fig. 2b), and it would seem to be more reliable to transform the theoretical results.

At densities greater than 6×10^{27} atoms/m^3 the higher terms in the virial series matter and calculated data may be obtained by computer simulation only. Egelstaff et al.[12] show that the contribution to $g(r)$ or $S(q)$ from terms involving $u_3(\mathbf{r},\mathbf{s})$ decreases with increasing density if the AT term is used. This has been confirmed by Ram,[8] who used the effective potential of Sinha et al.[14] with $u(r)$ from Barker et al.[10] and $u_3(\mathbf{r},\mathbf{s})$ as the sum of triple-dipole and dipole–dipole–quadrupole terms. His Monte Carlo results for $\rho = 14 \times 10^{27}$ atoms/m^3 (or 80% of the triple-point liquid density) are given in Fig. 8 for krypton at 297 K, and show an observable effect only at small q. Ram

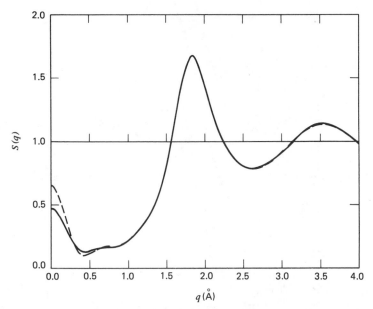

Fig. 8.   Monte Carlo simulations of structure factors for krypton gas at 297 K and $\rho = 14 \times 10^{27}$ atoms/m^3. The full line is simulated with the Barker et al.[10] pair potential, and the dashed line is a simulation using an effective potential[14] obtained by adding triple-dipole and double-dipole–quadrupole terms to the former potential.[8]

used a 500-particle system for the simulation, and this gives unreliable data for $q < 0.5$ Å^{-1}. It is important to perform simulations with much larger systems if good predictions in this field are to be made.

In Fig. 9 we show three examples of $S(q)$ at higher densities taken from the experimental results of Egelstaff et al.[15] who measured 12 states in krypton from 6 to 14×10^{27} atoms/m^3. The full lines in this figure are simulations[12] using the Barker et al.[10] potential. It can be seen that, in contrast to Fig. 8, the real many-body forces cause $S(q)$ to deviate from the pair force simulation. The major discrepancy occurs for $q < 1.5$ Å^{-1}, with smaller effects at higher q. These discrepancies could be related to an effective potential with state-dependent parameters, but an adequate theory is not available.

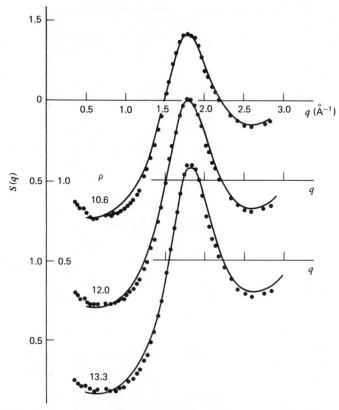

Fig. 9.   Observed structure factors for dense krypton gas at $T = 297$ K and $\rho = (10.6, 12.0,$ and $13.3) \times 10^{27}$ atoms/m^3 are shown by the solid circles.[15] Monte Carlo simulations using the pair potential[12] are shown by the full lines.

The detailed and precise measurements of $S(q)$ along one isotherm have produced a wealth of useful information on forces and structure in krypton gas, and similar data on another isotherm would help to make the interpretation more definite. Although the experiments on krypton may possibly be the most accurate ones carried out yet, they are not at the limit of present techniques. The results for $\rho = 2.88$ are stated[11] to be significantly better than those at the other densities, and therefore with additional precautions data might be determined to 0.5%. Because the effects being studied are small, this additional precision would be worthwhile. Nevertheless the major limitation, for both low- and high-density results, now lies with the theory.

IV. HOMONUCLEAR DIATOMIC FLUIDS

These are the simplest fluids after atomic fluids, because of the simple molecular structure and because there is only one pair correlation function, $g(r)$. However, most of the complications arising with molecular fluids arise in this case as well. If we work at a temperature small compared to the vibrational frequency times $\hbar/k$, the molecule may be treated as rigid. The pair potential depends on the separation (r_{12}) between the centers of two molecules and upon their angular orientations ω_1 and ω_2 relative to an intermolecular axis. This pair potential $u(r_{12}, \omega_1, \omega_2)$ can be used in a computer simulation to find $g(r)$, but, unlike the atomic case, $g(r)$ is not uniquely related to the pair potentials. There are a number of published potentials for (e.g.) nitrogen and chlorine, but they are not as reliable as the noble-gas potentials. Any study of the structure of a molecular fluid reduces to some extent to a study of the pair potential.

There are two methods by which intermolecular potentials are obtained. The most widely used method is to invent an analytical formula containing a number of constants and which represents the main physical processes occurring, and then to find the constants by fitting predicted results to experimental data. If we wish to obtain a pair potential, it is necessary to fit to low-density data only; otherwise an effective potential (including many-body forces) is obtained. This method seems to work fairly well for noble gases, but as discussed above is not wholly satisfactory. However, intermolecular potentials are functions of several variables, and the validity of this procedure is then more doubtful. An alternative is to calculate the intermolecular energy surface using the best available quantum-mechanical techniques and find the parameters in the formula by fitting to this surface. This gives a so-called "ab initio" potential which is approximately a true pair potential. For molecular fluids in which some configurations are improbable this method may give a more satisfactory potential, and in some cases has been found to fit the experimental data better than the "fitted potential." Consequently the older idea that the fitted potentials would serve as a model for the ab initio potentials seems to be outmoded. A more recent procedure is to use an ab initio potential as a first approximation to the pair potential and to adjust one or two constants to fit low-density data while maintaining the fit to the potential-energy surface. Then it is possible to study many-body forces through the differences between the predictions of the pair theory and experimental data. For example, in the case of nitrogen, Mulder et al.[16a] have obtained an ab initio potential for large r and Berns et al.[16b] for small. This ab initio potential for nitrogen has not been used in the pair-theory simulations so far, and simpler potentials (e.g., site–site) are normally used.

The expansion of the general pair correlation function for a rigid molecule fluid, $g(r_{12}, \omega_1, \omega_2)$, is

$$g(r_{12}, \omega_1, \omega_2) = f(12) + 1 + \frac{\rho}{4\pi}[f(12)+1]$$
$$\times \left\{ \int f(13)f(23)\, d3 + \int [f(13)+1][f(23)+1] \right.$$
$$\times [\exp(-\beta u_3(1,2,3)-1]\, d3 \right\}$$
$$+ O(\rho^2) \tag{4.1}$$

where $f(12) = \exp[-\beta u(r_{12}, \omega_1\omega_2)]-1$, and so on. The relationship[17] between $g(r)$ and the general function is

$$g(r) = \left\langle g\left(\mathbf{r} - \frac{\mathbf{d}_1}{2} + \frac{\mathbf{d}_2}{2}, \omega_1, \omega_2\right)\right\rangle_{\omega_1\omega_2} \tag{4.2}$$

and that between $S(q)$ and $g(r_{12}, \omega_1, \omega_2)$ is $S(q) = 1 + j_0(qd) + D(q)$, where

$$D(q) = \rho \int e^{i\mathbf{q}\cdot\mathbf{r}_{12}}\left\langle[g(r_{12}, \omega_1, \omega_2)-1]F(\omega_1, q)F^*(\omega_2, q)\right\rangle_{\omega_1\omega_2} d\mathbf{r}_{12}$$

with

$$F(\omega_1, q) = \tfrac{1}{2}\sum e^{i\mathbf{q}\cdot\mathbf{d}_1/2}$$

and where $\mathbf{d}$ is the internuclear vector and ρ is the atomic number density. By combining Eqs. (4.1) and (4.3) a virial expansion for $S(q)$ is obtained, but it is clear that this series is not as useful as the analogous series [Eqs. (3.1) and (3.2)] for atomic fluids, because of the average over ω_1 and ω_2. In this paper we write $S(q) = f_1(q) + D(q)$ and normalize so that $S(q) \to 1$ as $q \to \infty$ and the single-molecule structure factor $f_1(q) \to 1$ as well.

For the case of atomic fluids it was advantageous to make comparisons between theoretical predictions and experimental data in Fourier space, and eventually this may be the case for molecular fluids too. However, our knowledge of molecular pair potentials is weak and the discrepancies are probably larger, and therefore it is worthwhile to make comparisons in real space in the hope they may help explain the physical reason for the weakness. If care is taken with the truncation problem of Fourier transformations, meaningful comparisons may be made with limited data (Appendix B). This is important in many cases because computer-simulation data are usually truncated badly in r-space, while experimental data are truncated in q-space.

A.   Nitrogen

A number of measurements have been made on nitrogen gas and are listed in Table I. Sullivan and Egelstaff[18,19] studied several states with good precision, but only to $q = 4$ Å^{-1}. Powles et al.[20] studied one state with lower precision, but to high $q = 22$ Å^{-1}. However, after the single-molecule term had been subtracted their data on the intermolecular interference function $D(q)$ extended to 4 Å^{-1} as well. Both groups made comparisons with $D(q)$ calculated from Eqs. (4.1) and (4.3), using models for $u(r_{12}, \omega_1, \omega_2)$ but putting $u_3 = 0$. Sullivan and Egelstaff[19] included an approximate correction for the term $O(\rho)$ in (4.1), while Powles et al. did not. Comparisons were also made in r-space, especially in Ref. 19. In this work the first two terms of the virial series were calculated to large r, then Fourier transformed and truncated in q-space at the same point as the experimental data. Finally, after transforming back to r-space, an intercomparison of predictions and experi-

TABLE I
Structure Measurements on N$_2$

ρ (10^{27} molecules/m^3)	T (K)	Reference
2.38	160	19
4.07	160	19
5.58	160	19
3.60	200	18, 19
4.69	200	18, 19
2.54	294	20
3.41	296	18, 19
4.73	296	18, 19
17.3 (liquid)	77	24

mental data was possible on a comparable basis (Appendix B). The data presented in Figs. 10a, 11, and 13 were obtained in this way.

Models used in these comparisons included the site-site and site–site quadrupole models of Cheung and Powles[21] and the Raich–Mills and Raich–Gillis[22] and MSKM[23] potential models. The differences observed by Sullivan and Egelstaff were in general less than those observed by Powles et al. They concluded that none of these potential models gave a satisfactory fit to the whole (r, ρ, T) dependence of their data. Of the models considered the Raich–Gillis and site–site were most successful; a comparison for one case is shown in Fig. 10. The site–site model gives the right peak position but wrong amplitude at this temperature. At room temperature the amplitude was in better agreement, which indicates the potential is unsatisfac-

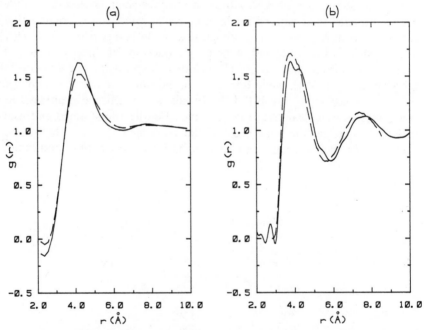

Fig. 10. (a) Pair correlation function $g(r)$ for nitrogen gas at 160 K and $\rho = 4.07 \times 10^{27}$ atoms/m³: full line, experimental result[19]; dashed line, virial-expansion calculation using site–site potential. (b) $g(r)$ for liquid nitrogen at 77 K and $\rho = 17.3 \times 10^{27}$ atoms/m³: full line, Fourier transform of data in Fig. 25; dashed line, computer simulation using site–site potential.[27] If a quadrupole term is added to the site–site potential, the theoretical curves are modified in a minor way only.

tory. Because of the significance of these small differences, work of higher precision would be useful in testing better potential models. Nitrogen is a favorable experimental case, and this improvement should be possible.

The present models were adjusted to fit liquid or solid data available at the time of publication. Recently new structural data on liquid nitrogen have been published,[24, 25] and it would be useful to compare these with the pair-theory predictions. However, it is necessary to rework the published data[26] (see Appendix A and Fig. 25 below) before the quality of the measurements becomes evident in $S(q)$ or $g(r)$. In Fig. 10b the Fourier transformation of the reworked data of Clarke et al.[24] are shown compared with a computer simulation using the site–site potential (Haile[27]). In contrast to the gas (Fig. 10a), these curves are not in phase, which from the classical point of view may indicate that the repulsive wall of $u(r)$ is too steep. But Egelstaff[28] has estimated the size of the quantum corrections to liquid nitrogen from a hard-sphere model and concluded that they are about as important as for liquid neon. If the simulation is modified by this model quantum correction, the phase difference largely disappears and the predicted and measured amplitudes are more consistent with the gas data of Fig.

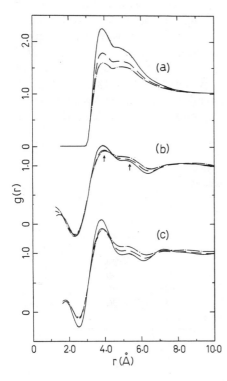

Fig. 11. Pair correlation functions for chlorine. (a) $\rho \to 0$ term in virial expansion of $g(r)$ for $T = 296$ K (full line), 358 K (dashed line), and 399 K (dash–dot line); calculated using a site–site potential.[31] (b) Experimental results for states a, b, and c of Table II. Notation as in (a). Arrows mark the two peaks discussed in the text. (c) Calculations using the interaction-site approximation (ISA) and a site–site potential for states a, b, and c of Table II. Notation as in (a). The ISA (or ISM) was proposed by D. Chandler and H. C. Anderson [*J. Chem. Phys.* **57**, 1930 (1972)].

10a. Calculations which include quantum corrections and use a correct pair potential are needed before progress can be made. Egelstaff[26] points out that the most recent X-ray and neutron data are in disagreement as published. This discrepancy can be removed if in the interpretation of the X-ray data a model of the electron distribution in the molecule is used in which about 5% of the electrons are moved to the center of the bond. Because this gives larger effects than are observed in Fig. 10, the X-ray data cannot be used for studies of potential models at this time.

B. Chlorine

Chlorine is an interesting molecular liquid because the molecule is more elongated than nitrogen and therefore orientational effects may be more evident. It is also more flexible than nitrogen. Three states of liquid chlorine and three states of chlorine gas have been studied by Sullivan and Egelstaff[29, 30] and are listed in Table II. They found that the principal peak of $g(r)$ has two maxima due to the molecular shape. Figure 11b shows the three liquid results. Also shown, in Fig. 11a, are curves corresponding to the $\rho \to 0$ predictions [Eqs. (4.1) and (4.2)] using the site–site potential of Singer et al.,[31] and in Fig. 11c, the predictions of the interaction-site approximation (ISA). It is clear that the relative size of the two maxima is a function of density and temperature and reflects changes in the orientational population. For example because the order of the peak heights is reversed in going from the $\rho \to 0$ curves to either set of data at the experimental density, some open configurations (e.g. end-to-end or T) must be excluded in favor of compact configurations (e.g. diagonal or parallel) with increasing density. A similar effect must occur also with decreasing temperature because the intermolecular well depths favor[29] that variation, and this can be seen in all

TABLE II
Structure Measurements on Cl_2

State[a]	ρ (10^{27} molecules/m^3)	T (K)	Reference
a	11.91	296	29
b	10.07	358	29
c	8.24	399	29
	1.54	398	30
	1.06	398	30
	0.73	398	30

[a]See Fig. 11.

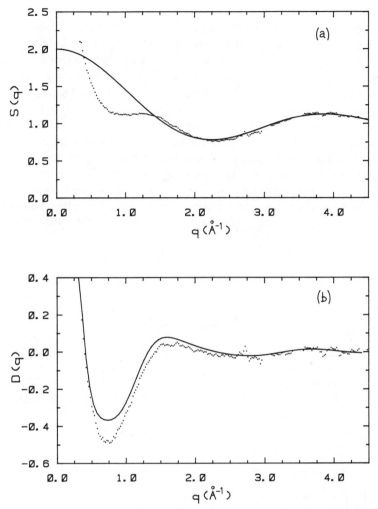

Fig. 12. (a) The structure factor for chlorine gas at $T = 398$ K and $\rho = 1.54 \times 10^{27}$ molecules/m³: dots indicate experimental data,[30] and the gaps occur at Bragg reflections from the apparatus. The full line is the single-molecule structure factor, $f_1(q)$, using the published bond length $d = 2.00$ Å.[29] (b) The dots show the difference between the two curves of (a), and the full line shows the virial expansion (two terms) using a site–site potential. These two results, after Fourier transformation of the part shown here, are drawn in r-space in Fig. 13a.

the families of curves. The experimental peaks in b appear to be broader than the predicted peaks in c. While this may be due to the ISA, it is possible that it is due to the potential, and a more detailed study of the liquid structure would then require computer simulations and an improved potential model.

Figure 12a shows the observed structure factor for chlorine gas (for $T = 398$ K and $\rho = 1.54 \times 10^{27}$ molecule/m³) and the structure factor $f_1(q)$ for a single molecule. These functions coincide approximately for $q > 3$ Å^{-1} if the correct bond length d has been chosen (the relationship between d and r_e is explained in Ref. 20). This comparison could provide a method of determining the bond length in a dense system and of observing its dependence on state parameters. Among the gas states listed in Table II, a variation of d of $\simeq 2\%$ is within experimental error. We have to associate any density-dependent change in d with the intermolecular field overlapping the intramolecular field. A proper theory is not yet available, but it is clear that this effect

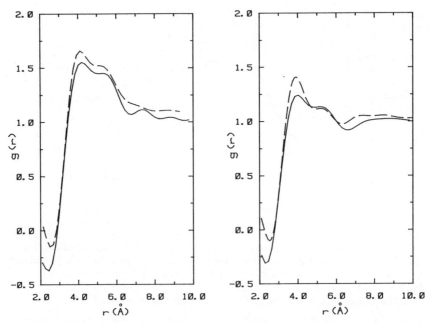

Fig. 13. (a) Pair correlation function $g(r)$ for chlorine gas at $T = 398$ K and $\rho = 1.54 \times 10^{27}$ atoms/m³. Full line, experimental result;[30] dashed line, virial-expansion calculation using site–site potential. (b) $g(r)$ for liquid chlorine at $T = 398$ K and $\rho = 8.24 \times 10^{27}$ atoms/m³. Full line, experimental result; dashed line, interaction-site approximation using site–site potential.

must be substantially larger in chlorine than in nitrogen for example.[19] Fig. 12b shows $S(q) - f_1(q) = D(q)$ compared to model calculations.

For the 398-K isotherm, data are available for both sides of the coexistence curve, and are shown in Fig. 13. The discrepancy for the low-density gas (Fig. 12b) is removed partly by the transform in Fig. 13a. The liquid $g(r)$ in Fig. 13b was calculated using the ISA, and part of its discrepancy may be due to that approximation. This comparison is an important test of the pair theory, but is somewhat inconclusive with present knowledge. Again more theoretical work is needed to improve the chlorine potential and to produce computer simulations for the states in Table II, and the gas experiments need to be improved.

V. HYDROGENOUS FLUIDS

It is appreciated widely that hydrogen is seen in neutron-diffraction experiments and that neutron-scattering amplitudes are different for H and D. In principle this enables experiments to be designed for molecular fluids made of $H_n A_m$ molecules, which will separate the three partial pair correlation functions $g_{HH}(r)$, $g_{HA}(r)$, and $g_{AA}(r)$ for atoms on *different* molecules. However, in practice there are several obstacles which have to be surmounted, and this section begins with a brief review of them.

In addition to the normal difficulties, discussed in previous sections, samples which contain hydrogen have a large incoherent scattering cross section, which unfortunately dominates the differential cross section. The coherent scattering cross section may be only $\simeq 5\%$ of the total for H_2O, or $\simeq 15\%$ for HCl, and therefore it is necessary to make measurements of very high precision in order to extract $S(q) - 1$. This high precision applies, of course, to all aspects of the conduct of the experiment, to data reduction and corrections, and to the statistical counting fluctuations, which may be made small by long counting times. For a good diffraction experiment on a fluid the sample and container should together scatter $\leqslant 20\%$ of the incident beam, in order that a reliable estimate of the multiple scattering can be made. Consequently it is important to use sample thicknesses which vary with the H:D ratio to compensate for the fact that the total scattering cross section for H is $\simeq 80$ barns, while for D it is $\simeq 7$ barns. Multiple scattering calculations have been reported for slabs by Vineyard,[32] for cylinders by Blech and Averbach,[33] and for several sample shapes by Sears.[34] These calculations are not accurate enough for hydrogenous samples, because high precision is needed due to the relatively small coherent cross section. Soper and Egelstaff[35] have used a more precise numerical technique that is satisfactory for samples of restricted thickness. Attenuation or absorption corrections are easier to handle, but in carrying them out the proper geometry of the sample, container, and detector must be included and the beam profile needs to be measured and taken into account.[35]

Finally, the inelasticity corrections for H and D are larger than for other elements (e.g. compare Fig. 18 with Fig. 2). A number of models have been used for estimating these corrections but none are wholly satisfactory (see Appendix A). For these reasons experiments involving H:D isotopic variations are less satisfactory than for other elements, and special methods are desirable to overcome some of the problems. Soper and Egelstaff[36] suggested a triple-sample difference method in which the H–H pair correlation is extracted without the need to know the self-terms, the inelasticity corrections, most of the single-molecule terms, and the other two correlation functions. Their method can be applied only to the A–A correlation function for an element in which the scattering amplitudes (b_A) of two isotopes of A are very different. There are a small number of suitable elements; hydrogen is one, as $b_H = -0.378 \times 10^{-14}$ m while $b_D = +0.667 \times 10^{-14}$ m. Because of this limitation the H–H data obtained by this method have to be used with conventional isotopic data to find the other two correlation functions.

In this paper the cross sections are normalized in such a way that $S_{\alpha\beta}(q) \to 1$ as $q \to \infty$, where $S_{\alpha\beta}(q)$ is a partial structure factor for atom α on one molecule and atom β on another: that is, for all $S(q)$'s the self-terms are normalized to 1. To do this we follow Placzek and work with the product $(\sum_{i=j} b_{ij}^2)^{-1} d\sigma/d\Omega$, where $b_{ij}^2 = b_i^c b_j^c + b_i^n b_j^n \delta_{ij}$, and with the reduction formulae in appendix A (b_i^c and b_i^n are the coherent and incoherent scattering amplitudes respectively for atom i). Here (i, j) are any atoms in the fluid, and $S_{\alpha\beta}(q) - 1 = \int_{i \neq j} S_{ij}(q, \omega) d\omega - f_{1ij}(q)$. The advantage of this procedure is that the different $S_{\alpha\beta}$'s or S_{ij}'s may be intercompared readily. In many papers a normalization factor $(\sum_i b_i^c)^{-2}$ is used, but this has the disadvantage that $S_{ij}(q, \omega)$ or $S_{\alpha\beta}(q)$ for $q \to \infty$ varies widely between the different partial structure factors. An extreme case is provided by the observed structure factors for H_2O and D_2O, which in Placzek's normalization approach 1, but in the alternative notation differ by a factor of 66.

A. Hydrogen Chloride

Heteronuclear diatomic molecules usually have a dipole moment, and often (e.g., HF) have strongly directional forces. Hydrogen chloride is an interesting case for these reasons, and in addition ^{35}Cl has a large coherent scattering amplitude and reasonably large mass, so that some of the experimental problems are not as difficult as for (say) water. This molecule also has a large quadrupole moment, so that the local orientational correlations are due to the mutual effect of several factors. It is a good case in which to use the isotopic-substitution technique (except for the fact that doubly isotopic samples such as $D^{35}Cl$ or $D^{37}Cl$ are hard to obtain), and it has been studied widely by other techniques.

Soper and Egelstaff[36,37] made measurements to 4.2 Å^{-1} with five isotopically different samples of hydrogen chloride, including three different H:D ratios with natural chlorine as well as samples of $H^{35}Cl$ and $H^{37}Cl$. They studied two states, one gaseous (2.91×10^{27} molecules/m^3 at 82.5°C) and one liquid (13.8×10^{27} molecules/m^3 at 24°C). Powles et al.[38] measured DCl liquid at 20°C to a large $q \simeq 22$ Å^{-1}. Whereas Soper and Egelstaff claimed that $D(q)$ behaved normally and the oscillations had almost died away at 4.2 Å^{-1}, Powles et al., in liquid DCl, observed a smaller principal peak, as well as an anomalous bump between 3 and 4 Å^{-1} and a smaller bump at 5 Å^{-1}. Over this anomalous region the nickel-alloy container used by Soper and Egelstaff had a flat response at a level somewhat less than that of the sample alone, whereas the vitreous-silica container of Powles et al. had an oscillatory response at a level twice as great as the sample alone. It is not clear whether the discrepancy is related to the difference in containers, and further measurements would be desirable. Since there is a difference in the measured cross sections of e.g., $\simeq 15\%$ at 3.5 Å^{-1} greater than the calculated difference, it should be possible to resolve this question.

Powles et al.[39] also reported results on liquid DBr at -40°C, observing a similar although smaller effect in a similar container. They Fourier-transformed their data on liquid DCl and DBr and identified three regions in r-space: (i) an intermolecular contribution, (ii) a "hydrogen bond" contribution, (iii) the "packing signal" or conventional liquid-structure region. They concluded that about half the molecules in DCl are hydrogen-bonded, and less in DBr. The intramolecular peak was distorted by these effects, and the mean squared amplitude of the intramolecular distance was found to be abnormally large.

Soper and Egelstaff reduced their five isotopic results to three partial correlation functions for the two states assuming the equivalence of H and D, and several authors have compared these results with computer-simulated data. Votava and Ahlrichs[40] derived an ab initio potential for HCl, and in Fig. 14 we show their computer simulation for the gas compared with the experimental results.[37] The experimental errors in this figure arise from the inconsistencies between the five isotopic experiments. In order to compare different isotopic results it is important that the states should be essentially the same, but Soper and Egelstaff[37] in their gas experiment did not maintain adequate control over the states and were forced to make adjustments for density differences later. It seems likely that the inconsistencies arose for this reason and the consequent errors overwhelmed other sources of error

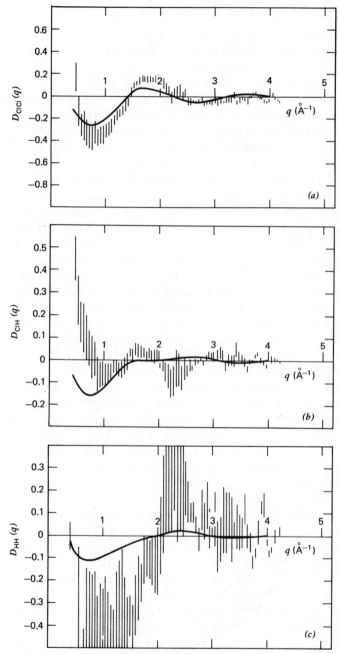

Fig. 14. Partial structure factors for HCl gas[37] at $T = 83°$ C and $\rho = 2.91 \times 10^{27}$ molecules/ m^3. The experimental values are shown as vertical bars (which are large due to inconsistencies between the five isotopically different samples), and the computer simulations of Votava and Ahlrichs[40] are shown as the full lines. (a) is the Cl–Cl, (b) the Cl–H, and (c) the H–H struc- ture factor, respectively. [$D_{\alpha\beta}(q) = S_{\alpha\beta}(q) - 1$.]

(this problem did not occur for their liquid data). Nevertheless, some discrepancies can be seen: the best data are for $S_{ClCl}(q)$, and here the amplitude of the theoretical curve is less than that of the experimental. Recently,[41] Votava and co-workers have revised their potential, and it is possible that this discrepancy will be reduced. McDonald et al.[42] and Klein et al.[43] have simulated the structure of liquid HCl using site–site potentials, and the comparison of their "best" case with the partial correlation functions of Soper and Egelstaff[36] for the liquid is given in Fig. 15. This model assumed that a specific attraction existed between the Cl and H, which had the effect of stabilizing the bent dimer rather than a cyclic arrangement. They took this to represent a crude form of hydrogen bond. The most recent ab initio potential of Votava et al.[41] gives an improvement in the Cl–Cl peak height (Fig. 15), but the position and height of the H–H peak are worsened. This could imply either that improvements are needed to these potentials or that many-body contributions affect the H–H correlation more than the Cl–Cl. In the latter case, these contributions would be equivalent to performing calculations of g_{HH} at a lower temperature in order to increase its peak amplitude.

The experiments discussed above are the first ones to separate the three partial structure factors for a heteronuclear diatomic system containing hydrogen. It seems likely that, especially if new isotopic samples of $D^{35}Cl$ and $D^{37}Cl$ can be obtained, the techniques could be improved and better precision achieved. The comparisons in Figs. 14 and 15 are encouraging and suggest that with realistic improvements in both the experimental and theoretical data a much better understanding of fluid HCl will be achieved.

B. Water

The possibility of finding the three partial structure factors of water has excited considerable interest for several years. To avoid the difficulties associated with the proposed H : D ratio in neutron experiments, Kálmán et al.[44] made electron-diffraction experiments and combined their data with the results of Narten and Levy[45] for X-rays on H_2O and those of Narten[46] for neutrons on D_2O. These three types of radiation have different scattering amplitudes, and so, after correcting the electron-diffraction data in the usual way (Section II), Kálmán et al. were able to separate the three structure factors. Their results are shown in Fig. 16, compared with a computer simulation by Lie et al.[47] using an ab initio pair potential (MCY[48]). The curves are similar, showing that both experimental and theoretical techniques are converging on the same goal. However, there are discrepancies in the position and height of peaks and in the depth of valleys, and some of these may be associated with either the essential imprecision of the diffraction data or small differences in raw data due to the different instruments and techniques

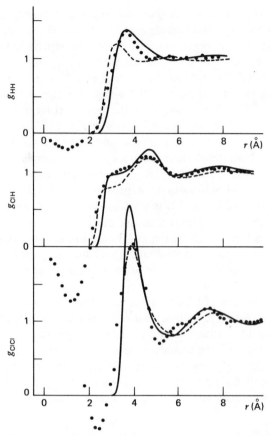

Fig. 15. The experimental partial pair correlation functions for liquid HCl (37) at 297 K and $\rho = 13.8 \times 10^{27}$ molecules/m^3 are shown as solid circles. The full line is the computer simulation (model c) of Klein and McDonald,[43] and the dashed line is the simulation of Ahlrichs and co-workers[41] using an ab initio potential.

employed. Therefore several authors have worked to improve the conventional three-neutron-data experiment using mixtures of H and D.

The nature of this experiment is illustrated in Fig. 17. This shows the structure factors for liquid H_2O, D_2O, and a hypothetical liquid HDO calculated from the computer-simulation results of Lie et al.[47] The striking differences arise because the coherent-scattering amplitudes of H and D have opposite signs, and if the structure factors for these liquids could be mea-

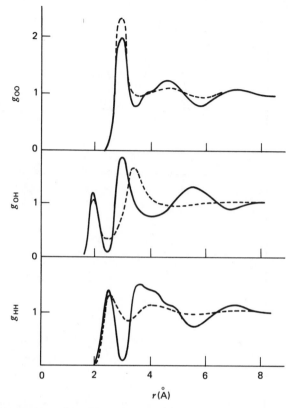

Fig. 16.   The experimental partial structure factors for water, at 297 K and $\rho = 33.4 \times 10^{27}$ molecules/m^3 (or 1 g/cm^3) obtained by Kálmán et al.[44] from electron, X-ray, and neutron diffraction data are shown by the full lines. The computer simulations by Lie et al.[47] are shown by the dashed lines; these are for the same state with the MCY potential.

sured properly they would enable excellent results for $S_{OO}(q)$, $S_{HH}(q)$ and $S_{OH}(q)$ to be obtained (apart from the error involved in assuming that H and D behave in the same way in water). However, because of the large incoherent cross section of hydrogen, the observed cross section (for liquid H_2O and samples containing more than 50% H_2O) will be about 50 times greater than the small part of it proportional to $|S(q)-1|_{max}$ over most of the range to be covered. Thus to measure $S(q)-1$ the cross-section measurements and their analysis must be good to $\sim 0.1\%$, or significantly better than has been achieved in the cases of krypton and nitrogen.

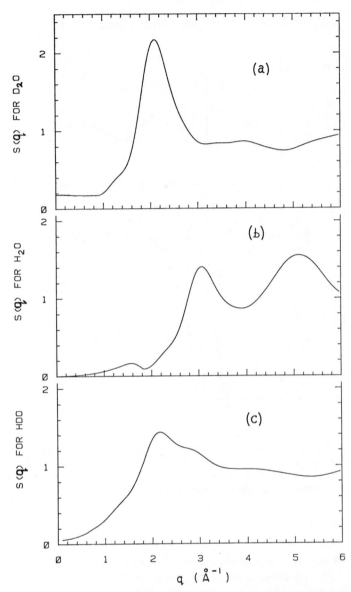

Fig. 17. Calculated structure factors for (a) liquid D_2O, (b) liquid H_2O, and (c) the hypothetical liquid HDO. The partial structure factors for O–O, O–H, and H–H, correlations were
evaluated from the computer-simulation data of Lie et al.[47] and were smoothed for $q < 1\ \mathring{A}^{-1}$
to remove Fourier oscillations. The major peak in (a) occurs roughly at $2\pi/$(intermolecular
spacing), and because H has a negative scattering length, it is inverted in (b). For the same reason the "peaks" in (b) are "troughs" in (a). We have

$$S_{D_2O} = 0.28 D_{OO} + 1.26 D_{OH} + 1.45 D_{HH} + f_1(D_2O)$$

$$S_{H_2O} = 0.55 D_{OO} - 1.41 D_{OH} + 0.91 D_{HH} + f_1(H_2O)$$

$$S_{HDO} = 0.37 D_{OO} + 0.37 D_{OH} + 0.09 D_{HH} + f_1(HDO)$$

To examine this problem it is useful to review an exploratory experiment by Powles et al.,[49] who used long counting times with samples of liquid H_2O and a H_2O-D_2O mixture, and showed that reasonable diffraction patterns could be obtained. In order to subtract the incoherent and multiple scattering from the H_2O sample they adopted an experimental method. The mixture of H_2O and D_2O was adjusted so that the scattering amplitude for hydrogen was zero. Thus the coherent scattering was related to the oxygen–oxygen correlation function only, and after taking $S_{OO}(q)$ from X-ray data,[45] they were able to subtract the coherent part of the mixture. The resulting curve (which includes the inelasticity correction) was scaled by a normalization factor and then subtracted from the normalized H_2O cross section to give $S(q)$ for H_2O. This subtraction was done on a point-by-point basis, although smoother data could have been obtained by fitting an analytical curve to the incoherent scattering results before subtracting them. It was evident to Powles et al.[49] that subtracting a smooth curve from data with small statistical fluctuations would generate a relatively smooth function which looked like $S(q)$ but which nevertheless contained significant errors. They discussed several sources of these errors, and in particular pointed out that the magnitude of the interference scattering was much less than the inelasticity correction, so that large errors may have occurred there. To illustrate this point we show in Fig. 18 the ratio of this correction (calculated by Soper and Silver[50] using Rahman's[51] free-molecule theory) to the maxima in $|S(q)-1|$ for H_2O calculated from the data of Fig. 17. The large ratios in this figure indicate the extra experimental precision needed and the high quality required of the theoretical calculations.

To improve the theoretical data Rovere et al.[52] reworked the free-molecule calculation and made numerical predictions. Unfortunately, the data given in their paper have to be corrected, as they do not have the right absolute cross section (1.9 barns per steradian in place of 1.55) and disagree with the numerical results from Rahman's theory.[51] In the case of the self-term for H_2O (which has the largest correction), other methods are available from neutron-thermalization work[53] that can be used to improve these calculations by including some properties of the dense fluid. Even so, the precision is not likely to meet the requirement shown in Fig. 18. Also, the best multiple-scattering and attenuation correction calculations, which have been discussed above (Ref. 35 and section II), for an H_2O sample of the maximum thickness (1 mm) are good to 1% of the H_2O cross section, rather than the required 0.1%. Limited progress has been made in the ten years since this experiment, and therefore it is to be expected that experiments[†] similar to

[†]A recent paper, W. E. Thiessen and A. H. Narten, *J. Chem. Phys.* **77**, 2656, (1982), reports such an experiment but the precautions described here were ignored. The real errors on their data may be several orders of magnitude larger than those quoted.

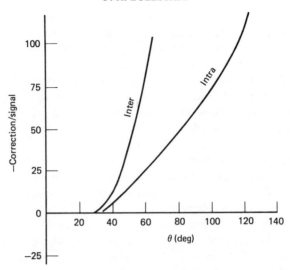

Fig. 18.   The ratio of the inelasticity correction for light water (H_2O) at 297 K and 33.4×10^{27} molecules/m^3 to the useful signal as a function of the scattering angle θ for neutrons with $\lambda = 0.9$ Å. The useful signal is defined as ·$|S(q)-1|_{max}$, namely the envelope of $|S(q)-1|$. "Inter" refers to the intermolecular contribution to $S(q)-1$, while "Intra" refers to the intramolecular contribution.

those of Powles et al.,[49] but with better statistics and several isotopic ratios, will yield data of quality similar to those obtained already by Kálmán et al.[44]

For these reasons Soper and Silver[50] used the pulsed-neutron-source method (Section II) at an angle of 40° and the triple-sample difference technique[36] to obtain $g_{HH}(r)$. This method minimizes the errors, and moreover the inelasticity correction in the pulsed-source method is nearly independent of energy and so produces less distortion. But it cannot yield the other two functions in the same way, although $g_{HH}(r)$ may be used with the conventional isotope experiment to improve it and as a good test of the theory. Soper and Silver's result at 25°C is shown in Fig. 19, compared with the computer simulation at -31°C of Impey et al.[54] using the MCY potential (it will be shown later that this is the best comparison to make). In common with other workers, they assumed the equivalence of H and D in the structure. Their experimental data differ from those shown in Fig. 16, and the fit between the prediction and experimental data is better than in Fig. 16, except for the nearest-neighbor peak. Soper and Silver's peak is smaller and broader than the predicted peak. It is believed that this difference is due to zero-point

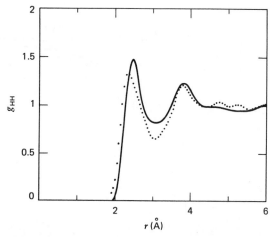

Fig. 19. The partial pair correlation function $g_{HH}(r)$ for water at 297 K and 33.4×10^{27} molecules/m³. The solid circles are the experimental data of Soper and Silver,[50] and the full line is a computer simulation[54] at 242 K and the same density using the MCY potential. The data of Fig. 23 suggests that the experimental results should be compared with a simulation at 207 K, and in this case the agreement would be improved in the trough at 3 Å, but at 2.4 Å the calculated peak would be taller and sharper.

libational motion of the molecule, which is not included in the classical simulation. If so, the amplitude of zero point motion is $\simeq 0.3$ Å, which is reasonable in comparison with the shape of the MCY potential near the hydrogen-bond configuration. The most useful additional information is obtained by eliminating $g_{HH}(r)$ from the heavy-water diffraction pattern. This gives the combination $g_{OH}(r) + 0.22 \, g_{OO}(r)$, which allows the discrepancy in the OH function in Fig. 16 to be checked. Soper and Silver's data are much closer to the computer simulation than are the data of Kálmán et al. It is encouraging that substantial progress has been made on a long-standing problem, and that these techniques may be improved further.

The predictions of the pair theory using the ab initio (MCY) potential of Matsuoka et al.[48] and eleven other potentials have been compared with an extensive range of experimental data by Reimers et al.,[55] and the agreement for the MCY case is better than for some "fitted" potentials, although not as good as for the best of the latter. This suggests that it is a reasonable pair potential, and if so, it might be expected that the second virial coefficient of steam would be given correctly.[56] These data are shown in Fig. 20 together with the predictions obtained for the Watts[57] potential, which was fitted to

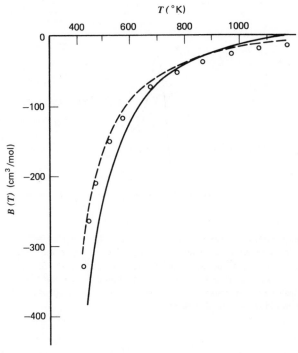

Fig. 20.   The second virial coefficient $B(T)$ of the pressure for steam. The circles are experimental results; the full line and dashed line are calculations using the MCY[48] and Watts[57] potentials respectively [from Table 5 of Ref. 55, where it is pointed out that the calculated $B(T)$'s are sensitive to small changes in the geometric parameters of the H_2O molecule].

these data. Two modifications to the calculations are needed before a good fit with the MCY potential can be obtained. To fit the lower-temperature part of a correction for the long-range contribution to $B(T)$ is required; that is, a long-range tail must be added to the potential (probably an r^{-6} term). This term would make little difference to the correlation functions in Fig. 16. In the higher-temperature part the calculations give a value greater than experiment because they are for rigid rather than flexible molecules. In the absence of this correction the MCY potential will not predict all properties correctly for $T \geqslant 800$ K, and for the liquid-structure work this correction will modify the structure factor at high q. For the comparisons we make this region is relatively unimportant, and so we can use the MCY-potential simulated correlation functions as a reasonable approximation to the predictions of a pair theory. Differences between these predictions and experimental structure data can then be considered in terms of many-body potentials.

For this purpose we consider the temperature dependence of the partial correlation functions of water. Also because it is probable that many-body effects are large for water, we shall start by reinterpreting the published X-ray data.[45] X-rays preferentially observe $g_{OO}(r)$, with some contribution from $g_{OH}(r)$, and it is useful to extract the temperature dependence of $g_{OO}(r)$. Narten[46] made the approximation

$$S_x(q) \simeq f_1(q)S_{OO}(q), \qquad (5.1)$$

where $S_x(q)$ is the measured X-ray structure factor and $f_1(q)$ is the molecular form factor for X-rays. This approximation is valid for perfect spheres with an oxygen at the center; and for real water there are significant corrections to it, which may be evaluated from a model. For example, Egelstaff and Root[58] point out that the independent-atom model is an improvement over Eq. (5.1), and deduce the peak height of $g_{OO}(r)$ as a function of temperature at $\rho = 33.4 \times 10^{27}$ molecules/m^3 (i.e., 1 g/cm^3). Their results are shown in Fig. 21, compared with the computer-simulated data[47,54] based on the MCY potential. The experimental data[45] were taken along the coexistence curve, and from 4 to 25°C the density change is 0.3%, while from 4 to 50°C it is 1.2%, which entailed another small correction to constant density. However, it is clear that the experimental temperature dependence is much less than predicted. The other parts of $g_{OO}(r)$ show a significantly larger temperature dependence,[58] and so we have the unusual situation in which

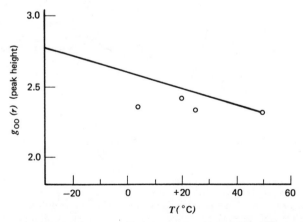

Fig. 21.   Peak height of $g_{OO}(r)$ on the isochore at $\rho = 33.4 \times 10^{27}$ molecules/m^3 (i.e., 1 g/cm^3). The full line is derived from computer simulations using the MCY potential, and the circles are experimental values deduced from X-ray data (see text).

the nearest-neighbor peak height (and the appropriately defined coordination number[58]) show less temperature dependence than the part due to more distant neighbors.

Water is a convenient system for measuring the isochoric temperature derivative (ITD) of any property as the liquid is passed through the density maximum at one atmosphere. Egelstaff et al.[59] exploited this property in measuring the ITD of the neutron structure factor of D_2O at 11°C. Figure 22 shows the Fourier transform of their data compared with the computer simulation of Impey et al.[54] In this case it is clear that the sum of the nearest-neighbor peaks in $g_{OH}(r)$ and $g_{HH}(r)$ at $\simeq 2$ Å increase with decrease in temperature at a rate greater than predicted. This is opposite to the behavior of the nearest-neighbor $g_{OO}(r)$ peak and so is difficult to understand without introducing many-body forces. Qualitatively the oxygen-atom distances remain nearly fixed while the molecules rotate into the hydrogen-bond configurations. An approximate quantitative discussion of the magnitude of many-body potential effects on $g_{OH}(r)$ and $g_{HH}(r)$ may be given in the following way. In Fig. 23a we show the observed neutron structure factor for D_2O at 25°C compared with the prediction of Lie et al.[47] The difference of these two curves is shown as a full line in Fig. 23b, where it is compared with the measured ITD for a temperature difference of 90°. The

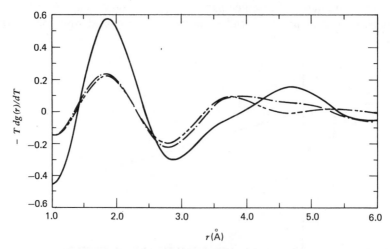

Fig. 22.   The isochoric temperature derivative of the "neutron pair correlation function" (i.e., the combination of partial correlation functions observed in neutron diffraction by liquid D_2O). The full line shows the experimental result,[59] and the dash–dot lines show two results deduced from the computer simulations[54] using the MCY potential. The peak at $\simeq 2$ Å is the combined effect produced by the nearest-neighbor OH and HH peaks, although more heavily weighted by the OH peak. (Reproduced from ref. 59 by permission of American Physical Society.)

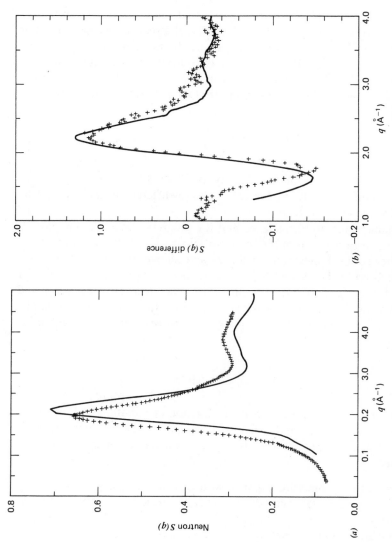

Fig. 23. (a) The neutron structure factor for heavy water at $T = 297$ K and $\rho = 33.4 \times 10^{27}$ molecules/m^3 (or 1 g/cm^3), from experiment[58] (crosses) and from computer simulation[47] (full line). (b) The difference between the two curves in (a) is shown as the full line, and the measured isochoric temperature derivative[58] for heavy water is shown by the crosses for a temperature difference ΔT of 90 K.

excellent agreement in this figure suggests that the many-body effects may be represented by a pair potential calculation at a reduced temperature. For 25°C this would be $\simeq -85$°C, which is why the low-temperature comparison was made in Fig. 19 (-31°C is the lowest temperature available). Egelstaff and Root[58] combine the data in Figs. 22 and 23 into an approximate formula for an effective potential (for OH and HH correlations):

$$u_{eff}(r) \simeq \left(1.3 + 0.7\frac{\Delta T}{T}\right)u_{MCY}(r), \qquad (5.2)$$

where $\Delta T = 297 - T$, and the density is 33.4×10^{27} molecules/m^3 (or 1 g/cm^3). Unfortunately, as Fig. 21 shows, a different formula would be needed for the O–O correlations.

In summary, the problems faced and the progress made over the past 10 years in the measurement of the three partial structure factors of water have been reviewed. Unconventional methods were shown to be needed, and one has been tried by Soper and Silver[50] that shows considerable promise. In addition it was reported that the temperature dependences of the neutron and X-ray structure factors near the density maximum (at 1 atm) contain significant information about the role of many-body forces in nearest-neighbor interactions.

VI. DISCUSSION

The object of modern work on the diffraction of radiation by fluids is to obtain precise, accurate, and meaningful data that may be compared with the results of advanced theories. In order to do this three things must be considered:

1. The ultimate precision possible with each technique;
2. The experimental methods that may be exploited to obtain precise results for the cross sections;
3. The correction or interpretation of the cross section so as to derive meaningful correlation functions.

The first part of this review was devoted to a discussion of these matters, and the later parts to the presentation of material that in general confirmed the conclusions drawn from this discussion. We have concentrated on the simpler fluids, that is, atomic and diatomic molecular fluids. Krypton and nitrogen are good subjects for experimental work, and the structural data on them are the best available at present. They are also good cases for comparison between theory and experiment, and further improvements can be foreseen. The other fluids reviewed present substantial experimental problems, and the data are not good enough. They can be improved only through long and carefully executed experimental programs. Nevertheless, enough data of reasonable quality have been obtained on the structure of fluid, HCl, and H_2O to test present theories and to raise conjectures about pair and many-body forces. It was shown that quantum effects were relevant in two cases, which deserve further exploration.

A lot of work has been published on a polyatomic fluids, but this work is not as meaningful, for several reasons. In particular, the partial correlation functions were not obtained, only one state was studied, and the interpretation of the cross section (see Appendix A) was oversimplified. In addition the intermolecular potentials are not as well known, and frequently ad hoc models or crude potentials (e.g. fused spheres) must be used in the interpretation. Nevertheless, by varying the chemical composition of the molecules, and by comparing X-ray and neutron data, the systematic behavior of complicated molecular systems may be investigated in a preliminary way. For example, for tetrahedral molecules (AB_4) a great deal of work has been published in recent years (e.g., Narten,[61] van Tricht,[62] Clarke et al.,[63] Gibson and Dore,[64] Granada et al.,[65] and Habenschuss et al.[66]). In none of the neutron-diffraction papers have the inelasticity corrections been made correctly (see Appendix A), so that the reduction of $d\sigma/d\Omega$ should be reworked in all cases. For the molecules containing only heavy nuclei this will not be important. The X-ray papers report data on polyatomic molecules which cannot be interpreted precisely with our present knowledge of electron–electron correlation functions. Nevertheless it has been confirmed that strong intermolecular correlations exist in the liquid state, and some ideas on the frequency of particular orientational states have been tested. In several papers, particularly those on neutron diffraction, the intramolecular bond lengths have been investigated and reported to be close to the values measured for the vapor; but this conclusion should be reviewed after $d\sigma/d\Omega$ has been reworked. There is insufficient space to cover these and other polyatomic fluids in this review. However with the improvements in ab initio potentials and computer simulation, it is likely that new predictions will be made during the next few years and then tested through better diffraction experiments on these fluids.

In selecting material for this review, priority has been given to cases in which the role of intermolecular forces could be investigated. The case of rare-gas fluids is the most obvious, as the pair potential is relatively well known from other work.[11, 14] The first question is how well the pair theory with this potential explains the structure of a dense fluid, since following the successes of the perturbation version of pair theory there is a feeling that the pair theory is a very good theory for liquids. The experiments on krypton along a single isotherm from the dilute gas to 80% of the triple-point density have demonstrated that both the three-body and higher potentials are important in accurate calculations of the structure factor. Work on proper theoretical estimates of these terms, as well as approximate models for them, is required for further progress. The question of how reliable the pair theory is for real liquids will not be settled until these matters are resolved.

In the case of molecular fluids there is a large range of diffraction experiments which could be done; although a few have been completed with the required precision, most of them have not been attempted. The difficulties of interpretation are more severe than for atomic fluids, but several examples were given to illustrate that this is no longer a limitation with modern experimental and theoretical techniques. Also, the precision required of molecular experiments is very great, as we are trying to extract information about multivariate functions from a function of a single variable. If a heteronuclear molecule is used, several partial correlation functions must be determined, and this lengthens the experimental work but can be very informative.

It is one of the ironies of this subject that reasonable ab initio potentials for HCl and H_2O have been obtained and used before similar work is completed on N_2 or Cl_2. Possibly this is due to the fact that the former interact more strongly than the latter, so that less attention needs to be paid to weak long range forces. In the latter two cases the importance of precise measurements of $S(q)$ as a function of ρ and T has been demonstrated, and the present limitation lies with the theoretical models. None of the tentative conclusions presented here about fluid N_2 or Cl_2 can be accepted until computer simulations have been carried out with good ab initio pair potentials to confirm or modify present theoretical results. Future developments with any molecular fluid now depend on obtaining a good pair potential from (approximately) first principles, adjusting it to low-density macroscopic data, and then generating good predictions for the partial correlation functions and comparing them with high-quality measurements to check the pair potential and to determine the contribution of many-body forces as a function of state parameters. This is no small task, but it seems to be one that is possible with modern techniques. The first attempts of this kind appear to be those involving the hydrogenous liquids, HCl and H_2O; perhaps because they represent a major scientific challenge. The quality of the ab initio potentials is encouraging, and the simulations are reasonable good, although the size of the system chosen is often too small for satisfactory comparisons with scattering data. In any case the principal features of the diffraction patterns have been obtained from first principles. The discrepancies are significant and were discussed in terms of possible many-body contributions. From an unusual state dependence or from the behavior of the discrepancies for different partial correlation functions, it appears to be possible to argue that there are some effects that need many-body forces to explain them (even when the pair potential is not known accurately). Clearly more theoretical work is needed on this topic to define such effects in a quantitative way.

Enough material has been reviewed here to attempt an initial assessment of the overall structural effect produced by many-body forces. The principal modifications to pair-theory predictions are found for $q < 2q_0$ (and mainly for $q < q_0$), where $\hbar q_0$ is the momentum transfer at the principal maximum in $S(q)$. From a set of low-density measurements along an isotherm, the three-body contribution to $S(q)$ may be extracted. It can then be used as a vehicle for testing models or theories of three-body forces in molecular as well as atomic fluids. It would be pleasant if the effect of the higher-order forces decreased as the density increased, but the available experimental evidence suggests otherwise. In this situation it would be convenient if the observed effects could be classified in terms of some simple equivalent effect in the pair-theory data. If successful, this procedure might suggest a mechanism behind the effect.

In the cases of HCl and H_2O it was suggested that the higher-order many-body forces produce an effect similar to that of reducing the temperature in the pair theory. For krypton it appears to be simpler to discuss this effect in terms of an effective pair potential deduced from $S(q)$ itself—a procedure that has a celebrated history. But in all three cases the many-body effect seems to be associated with more structure, that is, with sharper peaks in $S(q)$ or $g(r)$. A lot more work is needed before this approach to the question of many-body forces is properly developed and tested, and before a particular solution can be accepted. Nevertheless, the outlook is more promising today than it has been for many years, because both the experimental and theoretical techniques have developed to the point where they can focus jointly on these problems.

Perhaps this review has shown that experiment can move ahead of theory. To make proper contact with experiment the theoretician must start with an ab initio potential, make a minimum adjustment to fit low-density data, perform computer simulations with large enough systems to predict $S(q)$ for $q > 0.1\text{Å}^{-1}$, include three-body potential terms in a reasonable way, specify how many-body potential effects are to be handled, include quantum effects and terms depending on intramolecular vibrations in all of the above, and finally develop an analytical theory for $g(r)$. A computer simulation with a few hundred classical rigid molecules interacting with an approximate pair potential is not the world of real experiments, though it gives a qualitative approximation to it. An analytical theory that omits important diagrams and that uses an approximate pair potential, the classical limit, and rigid molecules is not close enough to real experiments to be useful.

An optimistic attitude has been adopted in this review, namely, that potentials obtained from (inexact) "first principles" are worth using and improving, and that experiments of high quality can be done in spite of enormous difficulties. This optimistic attitude may be criticized, because it can only be an attitude until it is proven (or disproven) by experience. Consequently it has not been discussed in depth, and the reader is left to be pessimistic or optimistic as he or she chooses.

Recent theoretical improvements have both helped and stimulated the new experiments discussed in this review. It is to be hoped that the new experimental data will in turn suggest new approaches to the theorist. The structure of fluids is still a frustrating field, but it is possible that a goal such as the experimental and theoretical determination of structural correlation functions to ~ 1% may eventually become a reasonable possibility.

Acknowledgments

It is a pleasure to acknowledge the assistance of Mr. J. D. Sullivan and Mr. J. H. Root in the preparation of Figs. 10 and 25 and of Fig. 17 respectively, and of Dr. J. Ram, Dr. J. M. Haile, Dr. R. Alrichs, Dr. A. K. Soper, and Dr. R. N. Silver, who very kindly provided copies of their work prior to publication. Without their help some key questions could not have been addressed. Dr. M. L. Klein is thanked for tables of the computer simulations in Ref. 54. And the Natural Sciences and Engineering Research Council of Canada are thanked for their financial support of this research.

APPENDIX A: A CRITICAL REVIEW OF THE INELASTICITY CORRECTIONS IN NEUTRON DIFFRACTION

The observed intensity in a neutron-diffraction experiment is converted to a differential cross section $d\sigma/d\Omega$ by methods that are understood and used very widely. The numerical precision of the resulting $d\sigma/d\Omega$ varies from case to case, but the ways of assessing the precision are known. In contrast to this satisfactory situation, the step of converting $d\sigma/d\Omega$ to the structure factor $S(q)$ is not handled properly by most workers (theoretical or experimental) in this field. This appendix sets out the main steps in the conversion to $S(q)$, which, if followed, would allow most of the recently published data to be made more useful. An example of this improvement will be given.

For all elements except the lightest (essentially this means except H and D), the expansion of $d\sigma/d\Omega$ as a power series in $(\text{mass})^{-1}$ will allow the experimentalist to convert $d\sigma/d\Omega$ to $S(q)$. This series was written out by Placzek.[67] After inserting the detector efficiency in the coefficients, his series to $O(1/M)$ is

$$\frac{4\pi}{\sigma_s}\frac{d\sigma}{d\Omega} = S_0 - \frac{A\hbar S_1}{2E_0} - \frac{4my\dot{S}_1}{\hbar} - \frac{B\hbar^2\ddot{S}_2}{8E_0^2}$$

$$+ \frac{m(1-2y+4Ay)}{2E_0}\dot{S}_2 + \frac{8my^2}{\hbar^2}\ddot{S}_2 + O(M^{-2}) \qquad (A.1)$$

where $\sigma_s = 4\pi\Sigma_{i=j}b_{ij}^2$, dots denote $\partial/\partial Q^2$, $y = \sin^2\frac{1}{2}\theta$, m is the neutron mass, M is the molecular mass, and S_n is defined by

$$S_n(q) = \int_{-\infty}^{+\infty}d\omega\sum_{ij}\omega^n S_{ij}(q,\omega)C_{ij} \qquad (A.2)$$

where $C_{ij} = b_{ij}^2/\Sigma_{i=j}b_{ij}^2$ and $b_{ij}^2 = b_i^c b_j^c + b_i^n b_j^n \delta_{ij}$. In this equation b_i^c and b_i^n are the coherent and incoherent scattering lengths respectively of atom i, and the sums run over the atoms in a molecule. [This normalization via C_{ij} ensures that $S(q) \to 1$ as $q \to \infty$.] The scattering law $S_{ij}(q,\omega)$ may be defined in terms of the double differential scattering cross section

$$\frac{d^2\sigma}{d\Omega\,d\omega} = \frac{k}{k_0}f(\omega)\sum_{ij}b_{ij}^2 S_{ij}(q,\omega) \qquad (A.3)$$

with $\hbar\omega = E_0 - E$ and $\mathbf{q} = \mathbf{k}_0 - \mathbf{k}$. $f(\omega)$ is the detector efficiency and is defined by

$$f(\omega) = [1 - \exp(-u/k)]\varepsilon_0^{-1} \qquad (A.4)$$

Here $f(\omega) = 1$ when $\omega = 0$ or $\varepsilon_0 = 1 - \exp(-u/k_0)$, and u is calculated from the absorption cross section, density, and dimensions of the detector material. The detector coefficients A and B are defined in terms of ε_0 as

$$\begin{aligned} A &= 1 + (\varepsilon_0^{-1} - 1)\ln(1-\varepsilon_0) \\ B &= A + (\varepsilon_0^{-1} - 1)\ln^2(1-\varepsilon_0) \end{aligned} \qquad (A.5)$$

Experiments need to be done so that only terms $O(1/M)$ are needed, because higher terms involve details of the fluid dynamics. This requirement defines E_0 in Eq. (A.1).

The above discussion is widely accepted and understood, and easily applied to atomic fluids: the formulae derived from it for atomic fluids are accepted generally. In evaluating (A.1) the expressions for S_n are required.

Placzek[67] showed that the terms $O(1/M)$ (for classical translation and rotation—see ref. 1 for the nonclassical part of S_2) are

$$S_0 = 1 + \gamma [S(q) - 1] \qquad (A.6)$$

$$S_1 = \frac{\hbar q^2}{2M} X \quad \text{and} \quad S_2 = \frac{kTq^2}{M} X + O(M^{-2}) \qquad (A.7)$$

where γ is the ratio of coherent to total cross sections for the molecule, $S(q)$ is the required structure factor, and X is a quantity[1] that varies with molecular structure and is unity for atomic fluids. The relationship between the $O(1/M)$ terms of S_2 and S_1, given in (A.7), follows from the principle of detailed balance.[68] This simple requirement has not been met, for example, in the treatment given by Blum and Narten,[3] and is one example of the errors in the literature in the application of Eq. (A.1) to molecular liquids. The misunderstandings arise from two aspects of (A.1):

1. Equation (A.1) corrects $d\sigma/d\Omega$ to give $S(q)$ evaluated at q for elastic scattering: it is not necessary to correct q for inelastic scattering, and in this respect the Placzek correction is unlike the Breit–Dirac correction.
2. Equation (A.2) does not include the sampling factor $F(\omega)$, [Eq. (2.3) or Fig. 3 for $a \to \infty$]. For atomic fluids the sampling factor is normally not significant, but for molecular fluids it is very important.

We consider the conventional case here ($a \to \infty$ in Fig. 3), but similar conclusions apply to other values of a. Some papers are in error with regard to both items 1 and 2, while many theoretical papers are in error because Eq. (A.1) has not been modified correctly by the sampling factor (e.g. Blum and Narten[3] and Powles[2]). Egelstaff and Soper[1] include the sampling factor in (A.2) by multiplying the integrand by $F(\omega)$, which is a systematic modification required by Eq. (2.2b) and consistent with the derivation of the Placzek series. Because of the importance of this factor they divide experiments into two regimes:

Regime 1—molecular vibrational energies much greater than E_0 and kT.
Regime 2— E_0 much greater than molecular vibrational energies.

These regimes may be understood by reference to Fig. 3 for $a \to \infty$ and moderately heavy nuclei. In regime 1 the part of $S_{ij}(q, \omega)$ associated with the vibrational ground state occurs for $F(\omega) = 1$, while parts associated with excited states are negligible or occur for $F(\omega) = 0$. The function X in Eq. (A.7) is computed separately for each regime. In regime 2 virtually all of $S_{ij}(q, \omega)$ occurs for $F(\omega) = 1$, and a negligible part occurs for $F(\omega) = 0$. For light nuclei the moment of inertia is small and asymmetric broadening of $S_{ij}(q, \omega)$ by rotational terms occurs, except at low angles. This case will be discussed later.

An examination of their results shows that Blum and Narten[3] take the first moment from regime 2 and the second moment from regime 1. The correction used by Narten[61] for CCl_4 is of this kind and therefore incorrect, giving significant errors at high q. This error has led to the publication of a number of incorrect statements about the Placzek series. Powles[2] distinguishes between the two regimes and handles the self-term correctly, but then for his distinct term he takes the first moment from regime 1 and the second moment from regime 2 to obtain an incorrect result. The size of the numerical error this makes turns out to be larger than Blum and Narten's error in many cases. Later he amended his formula by omitting the distinct second moment to give a result correct in the limit $kT \ll E_0$ for regime 1. The mistake of taking moments from different regimes is equivalent to taking different $F(\omega)$'s or E_0's for S_1 and S_2. Also it is important to notice that the same $F(\omega)$ must be used to define S_0, and hence $S(q)$, as was used for S_1 and S_2. Thus the "$S(q)$" depends on whether the experiment is done in regime 1 or 2. Actually regime 2 gives the true $g(r)$ including the usual intramolecular term, whereas regime 1 gives a $g(r)$ that has a different intramolecular term, and so the bond length d is different in the two regimes. The definition of the bond lengths in regimes 1 and 2 has been given by Sullivan and Egelstaff[19] (their Appendix 1). If the value of E_0 is chosen to lie between the regimes, the meaning of "$S(q)$" depends on the number of vibrational levels excited, although the differences between these meanings are small.

Many experiments (e.g., Narten,[61] Gibson and Dore,[64] Clarke et al.[24]) have been done with incident energies that fall between the regimes. Egelstaff and Soper[1] recommend that the moments be computed for each term of the vibrational series[69] and at each energy the correct number of terms be taken to correspond to the cutoff in $F(\omega)$. Because of the complications in doing this and because the translational broadening of each term limits the precision, they recommend that in the interests of precision E_0 be chosen to confine experiments to either regime 1 or 2. However, Powles[2] works with energies spanning several vibrational levels and obtains an approximate correction formula that is a function of q. Since the position of the cutoff in $F(\omega)$ is a function of E_0 only (for a given molecule), this is not a logical procedure; it relies on the fact that the amplitude of the step in $d\sigma/d\Omega$ is q-dependent. Powles[70] uses this formula in an analysis of the high-q cross section of heavy water for $\lambda = 0.35$ and 0.7Å, or 0.7 and 0.17 eV respectively. The intramolecular bond lengths are derived from data taken over a region where the correction term is about 6 times larger than $|S(q)-1|_{max}$. Thus the derived bond length is that length which makes an approximate correction formula fit the data. This is an example of a molecule with a low moment of inertia, so that the energy transfer is shared between rotation and vibration (translation is usually less important than rotation). Thus although the energy of 0.7-Å neutrons (0.17 eV) is larger than the bending-mode energy in D_2O (0.14 eV), this mode is not excited significantly. However, for 0.35 Å (0.7 eV) the first excited states of D_2O should be considered. For low-moment-of-inertia molecules the cutoff in $F(\omega)$ is not easily related to the molecular vibrational energies unless the momentum transfer is small. But the separation between regimes 1 and 2 is important, and because rotational broadening is large and depends on the environment, the inelasticity correction has to be calculated through a good dynamical model. In this case $F(\omega)$ is included automatically with the model $S_{ij}(q, \omega)$ in Eq. (2.2b).

In the case of time-of-flight measurements the energy is varied continuously, and for much of the published work the energy runs between regimes 1 and 2 (e.g., Refs. 62, 64, 71). Present-style experiments could be restricted to either regime, but it would appear to be better to conduct time-of-flight experiments with continuous scanning of θ. In this way a diffraction pattern may be obtained for each time-of-flight channel, and then some of them combined after an appropriate selection and correction procedure.

The M^{-1} term is the largest term after the first in Eq. (A.1), and it is independent of the system dynamics (i.e., independent of gas, liquid, or solid states). It is important for this reason to choose E_0 so that the series can be truncated at the M^{-1} term, and this decision will depend upon some estimate of the error introduced by the truncation. Although terms $O(M^{-2})$ depend upon the detailed molecular dynamics, it is believed that this dependence is weak except at low θ [see Eq. (2.5)]. Therefore the usual procedure is to estimate the error from the difference between numerical integration of Eq. (2.2b) using a free-molecule model for $S_{ij}(q, \omega)$ and the Placzek-series result. It would be useful to use other models, but they must be selected with care, as they must have the correct behavior at high ω. Since free-molecule formulae can be written out from "exact" theory, they meet this requirement, but several liquid-state models (e.g., those using an empirical memory function) do not. If the difference between the $O(M^{-2})$ terms for the free-molecule model and other suitable models is found to be small, the numerical integration of a model will give a better estimate of the correction than the series. So far models that are sufficiently different from one another have not been used, and the above conclusion is a conjecture. (In principle the harmonic solid, evaluated properly in the high-frequency limit, would be a suitable case. But this calculation is rather involved and has not been carried out except for the self-term of an atomic solid.[72]) As an example of the present work the calculations of Sullivan and Egelstaff[18] for nitrogen are given in Fig. 24. Although the Sachs–Teller effective mass is not high ($\frac{3}{5} \times 28$)

the difference between the model (dashed line) and the series cut at M^{-1} (full line) is ~ 1% of the value of $S(q)$ at large q.

To illustrate the inappropriateness of some of the correction formulae now in use, we consider the data of Clarke et al.[24] on liquid nitrogen. They used an approximate correction formula involving five constants (taken from Ref. 2—many of the formulae there are approximations valid in the limits $T \to 0$ and $q \to \infty$). Two of the constants were "effective masses" for the self- and distinct terms and were treated as free constants, one was a scale factor, and the other two were adjusted but kept close to their known values. After fitting the formula to data taken at four different wavelengths (0.35, 0.70, 0.84, and 1.06 Å), they deduced values of the masses. These were (in terms of the neutron mass) 19, 14.6, 21, and 16.8 respectively for the self-mass and 12, 15.4, 27.4, and 28 respectively for the distinct mass. In contrast to this procedure, they could have concluded that 0.35 Å (or 0.67 eV) lies outside regime 1 ($E_0 < 0.28$ eV) and discarded it, and used the elementary formula[1] deduced from (A.1) for the remaining cases. This gives self-consistent results using the molecular mass and no adjustable constants. For example, their case $\lambda = 0.70$ Å is given in Fig. 25, after correction as in Ref. 1 with $M = 28$ for all terms. It can be seen that $D(q) = 0$ at high q as required, and the overall symmetry of $D(q)$ is satisfactory. In addition a scale factor of 1.0 was used, which contrasts with the factor given by Clarke et al. and shown by the arrow in the figure. These data have been used to find $g(r)$ for liquid nitrogen,[28] and this result was given in Fig. 10b.

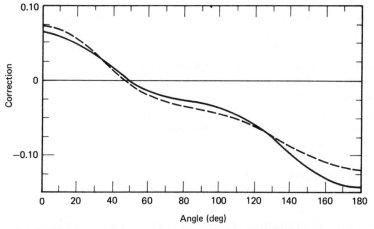

Fig. 24. Calculated inelasticity corrections for nitrogen. The full line is the Placzek series truncated at the M^{-1} term [Eq. (A.1)], and the dashed line is a free-molecule calculation[1]. (Reproduced from ref. 18 by permission of Taylor & Francis, Ltd.).

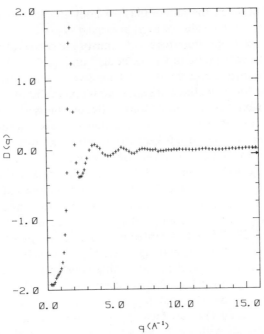

Fig. 25.   The intermolecular-interference part (or van Hove distinct part) of the structure factor of liquid nitrogen at $T = 77$ K and $\rho = 17.3 \times 10^{27}$ molecules/m³. The reduction from $d\sigma/d\Omega$ is described in Appendix A, and the arrow marks the level corresponding to the arbitrary scale factor used in the published analysis.[24]

Many papers have been criticized above because they did not apply Eq. (A.1), including the sampling factor $F(\omega)$, in a self-consistent way, or because they did not collect all the $1/M$ terms together properly. This means that most of the data on molecular fluids have not been corrected properly and therefore are less valuable than they should be. In some cases the conclusions reached by the authors should be revised, while in others the modifications to the data are fairly small. When the authors present a table of $d\sigma/d\Omega$ results, it is possible to make the inelasticity correction post facto, and this was done as an example in Fig. 25.

APPENDIX B: TRUNCATION EFFECTS IN FOURIER TRANSFORMATIONS

Since the pair correlation function $g(r)$ and the structure factor $S(q)$ are related by a Fourier transformation [Eq. (2.1a)], it is necessary to know either function over its whole range in order to find the other. Unfortunately, computer simulation data are truncated in r-space through cost and time restrictions, while experimental data are truncated in q-space through experimental limitations, so that neither can be compared with the other. The effects produced by truncation have been studied for many years, and a common practice is to multiply one function by a "modification factor" which introduces a cutoff via a known mathematical function. Then in the other space the result obtained is a convolution of the true transform with the transform of the mathematical function. Thus if the q-space results are multiplied by a function $M(q)$, the transform $g_T(r)$ is related to $g(r)$ by

$$g_T(r) = \int g(|\mathbf{r}-\mathbf{r}'|)\hat{M}(r')\,d\mathbf{r}' \tag{A.8}$$

where $\hat{M}(r)$ is the transform of $M(q)$. Some simple examples are listed in the Table III and the functions $r^2\hat{M}(r)$ are plotted in nondimensional units in Fig. 26.

When the convolution in Eq. (A.8) is carried through, $g_T(r)$ is broadened relative to $g(r)$ and some "Fourier oscillations" are evident, especially at small r. The broadening is $\sim \pi/q_m$, but it increases through cases $a \to b \to c$ in Table III. Conversely, the Fourier oscillations decrease through cases $a \to b \to c$, as is evident from the shape and size of $r^2\hat{M}(r)$ in Fig. 26. These points are demonstrated in Fig. 27, where $g_T(r)$ is compared with $g(r)$ for these three cases using a theoretical $g(r)$ for liquid chlorine at 297 K and 1 atm as an example. It is clear that simple truncation (case a) is most advantageous in this example, as it distorts the "interesting" section of $g(r)$ the least while

TABLE III
Examples of Modification Functions

Case	$M(q)$	$\hat{M}(r)$
a	$\begin{aligned}q \leqslant q_m: &\ 1 \\ q > q_m: &\ 0\end{aligned}$	$4\pi q_m^3 j_1(q_m r)/q_m r$
b	$q \leqslant q_m: \dfrac{3j_1(\pi q/q_m)}{\pi q/q_m}$	$r \leqslant \dfrac{\pi}{q_m}: \ \dfrac{6q_m^3}{\pi^2}[\mathrm{Si}(\pi + q_m r) + \mathrm{Si}(\pi - q_m r)]$
	$q > q_m: \ 0$	$r > \dfrac{\pi}{q_m}: \ \dfrac{6q_m^3}{\pi^2}[\mathrm{Si}(q_m r + \pi) - \mathrm{Si}(q_m r - \pi)]$
c	$\begin{aligned}q \leqslant q_m: &\ j_0(\pi q/q_m) \\ q > q_m: &\ 0\end{aligned}$	$\dfrac{4\pi q_m^3 j_0(q_m r)}{\pi^2 - (q_m r)^2}$

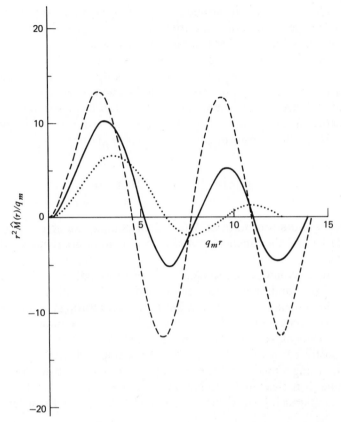

Fig. 26. The functions $r^2\hat{M}(r)$ of Table 3. Dashed line, case a; full line, case b; dotted line, case c. The first peak moves to higher $q_m r$ as we move from a to b to c, showing that the resolution is worsened; but the higher peaks decrease in amplitude, showing that the Fourier oscillations decrease.

distorting the "uninteresting" section at low r the most. The examples given in the text were treated this way; that is, we compared experimental and theoretical $g_T(r)$'s using case a of Table III. By interchanging r and q this discussion may be applied to q-space.

   It is implicit in this discussion of modification functions that, to be useful, they need to be precisely known mathematical functions that may be applied to both experimental and theoretical data. In this respect the practice of calling X-ray form factors "modification factors" is misleading, as they are not known precisely, and also the convolution equation involving $g(r)$ and the X-ray cross section is not an exact equation.

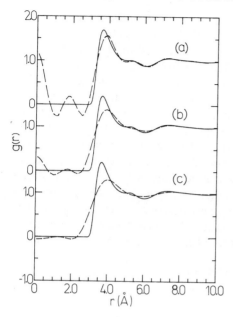

Fig. 27. An example of the use of the modification functions of Table III. The full lines are an ISA calculation of $g(r)$ for liquid chlorine at 297 K and $\rho = 11.9 \times 10^{27}$ molecules/m^3. These data were transformed to q-space and multiplied by $M(q)$ (Table III) using $q_m = 4.2 \text{ Å}^{-1}$. Then $g_T(r)$ was calculated and is shown by the dashed lines for cases a, b, and c of Table III. Note how the resolution in case a is worsened successively in b and c, while the size of the Fourier oscillations is reduced also.

It is instructive to compare the methods used with fluids with the interpretative methods used for Bragg scattering from crystals. The elastic scattering by a crystal involves a three-dimensional Fourier transformation and produces a "reciprocal lattice" in the laboratory. This is a real lattice into which the scattered radiation is split, and it can be photographed and seen easily because the distances between the lattice sites are macroscopic. In contrast the crystal lattice involves microscopic separations which are difficult to see by direct means. A full picture usually involves a discussion of both "real" and "reciprocal" space; for example, in one of the methods used by crystallographers part of the reciprocal lattice is measured and the data are compared with theoretical predictions. Subsequently the rest of the reciprocal lattice is deduced and the whole is Fourier-transformed to r-space in order to discuss the crystal lattice. The step that is difficult to take in work on the structure of fluids is the deduction of that part of reciprocal or q-space which is not measured. This depends on fitting the measured part with a good theoretical model, and for the krypton-gas experiments this has been done successfully.[11] In the other examples discussed here this was not possible, as the theory is not as satisfactory as for krypton. In fluids research we are at the stage of improving the methods by which theoretical predictions are compared with experimental data either in q-space or in r-space, and here the truncation problem is of great importance.

In order to provide reliable data in q-space for $q > 0.2$ Å^{-1}, a theoretical curve must extend to ~ 20 Å in r-space. While it is possible for the analytical theories (e.g. virial expansion and PY) to extend to 20 Å, it is unusual for computer simulations to do so. Most of the published simulation data are truncated tables of $g(r)$, but sometimes $\langle e^{i\mathbf{q}\cdot\mathbf{r}} \rangle$ is calculated. This gives a better q-space calculation but does not avoid the basic truncation problem due to the finite size of the box. An advance would be made if computer simulations were taken to ~ 20 Å or were extrapolated to large r by reliable theoretical methods. If this were done and the computer models were reasonably satisfactory, it is possible that the crystallographers method described above could be used for fluids. Meanwhile, for r-space comparisons, it is necessary to transform the calculation to q-space, truncate it in the same manner as for the experimental data, and transform back to r-space. This procedure allows a meaningful comparison between theory and experiment to be made, and was adopted for most of the comparisons in this article.

References

1. P. A. Egelstaff and A. K. Soper, *Molec. Phys.* **40**, 553, 569 (1980).

2. J. G. Powles, *Molec. Phys.* **36**, 1161 (1978); **37**, 623 (1979).

3. L. Blum and A. H. Narten, *J. Chem. Phys.* **64**, 2804 (1976).

4. J. A. Janik and A. Kowalska, *Thermal Neutron Scattering*, P. A. Egelstaff, Ed., Academic Press, New York, 1965, Chap. 10, p. 480.

5. H. Fredrikze, C. D. Andriesse, and E. Legrand, *Physica* **62**, 474 (1972).

6. J. F. Karnicky, H. H. Reamer, and C. J. Pings, *J. Chem. Phys.* **64**, 4592 (1976).

7. J. L. Yarnell, M. J. Katz, R. G. Wenzel, and S. H. Koenig, *Phys. Rev. A* **7**, 2130 (1973).

8. J. Ram, personal communication.

9. G. G. Chell and I. J. Zucker, *J. Phys. C* **1**, 35 (1968).

10. J. A. Barker, R. O. Watts, J. K. Lee, T. P. Schafer, and Y. T. Lee, *J. Chem. Phys.* **61**, 3081 (1974).

11. A. Teitsma and P. A. Egelstaff, *Phys. Rev. A* **21**, 367 (1980).

12. P. A. Egelstaff, A. Teitsma, and S. S. Wang, *Phys. Rev. A* **22**, 1702 (1980).

13. R. Aziz, *Molec. Phys.* **38**, 177 (1979); K-C. Ng, W. J. Meath, and A. R. Allnatt, *Molec. Phys.* **37**, 237 (1979).

14. S. K. Sinha, J. Ram, and Y. Singh, *J. Chem. Phys.* **66**, 5013 (1977).

15. P. A. Egelstaff, W. Gläser, D. Litchinsky, E. Schneider, and J. B. Suck, *Phys. Rev. A* **27**, No. 2 (1983).

16a. F. Mulder, G. van Dijk, and Ad van der Avoird, *Molec. Phys.* **39**, 407 (1980).

16b. R. M. Berns and Ad van der Avoird, *J. Chem. Phys.* **72**, 6107, (1982).

17. K. E. Gubbins, C. G. Gray, P. A. Egelstaff, and M. S. Ananth, *Molec. Phys.* **25**, 1353 (1973).

18. J. D. Sullivan and P. A. Egelstaff, *Molec. Phys.* **39**, 329 (1980).

19. J. D. Sullivan and P. A. Egelstaff, *Molec. Phys.* **44**, 287 (1981). Appendix 1 gives a relation between r_e and the bond length seen in a diffraction experiment.

20. J. G. Powles, J. C. Dore, and E. K. Osae, *Molec. Phys.* **40**, 193 (1980).

21. P. S. Y. Cheung and J. G. Powles, *Molec. Phys.* **32**, 1383 (1976).

22. J. C. Raich and N. S. Gillis, *J. Chem. Phys.* **66**, 846 (1977).

23. C. S. Murthy, K. Singer, M. L. Klein, and I. R. McDonald, *Molec. Phys.* **40**, 1517 (1980).

24. J. H. Clarke, J. C. Dore, and H. Egger, *Molec. Phys.* **39**, 533 (1980).

25. A. H. Narten, E. Johnson, and A. Habenschuss, *J. Chem. Phys.* **73**, 1248 (1980).

26. P. A. Egelstaff, *J. Phys. Chem. Liq.* **11**, 353 (1982).

27. J. M. Haile, personal communication.

28. P. A. Egelstaff, *Faraday Discuss. Chem. Soc. (London)* **66**, 7 (1978).

29. J. D. Sullivan and P. A. Egelstaff, *J. Chem. Phys.* **76**, 4631 (1982).

30. J. D. Sullivan and P. A. Egelstaff, to be published.

31. K. Singer, A. Taylor, and J. V. L. Singer, *Molec. Phys.* **33**, 1757 (1977).

32. G. Vineyard, *Phys. Rev.* **96**, 93 (1954).

33. I. A. Blech and B. L. Averbach, *Phys. Rev. A* **137**, 1113 (1965).

34. V. F. Sears, *Adv. Phys.* **24**, 1 (1975).

35. A. K. Soper and P. A. Egelstaff, *Nucl. Inst. & Meth.* **178**, 415 (1980).

36. A. K. Soper and P. A. Egelstaff, *Molec. Phys.* **42**, 399 (1981). The triple-sample difference method is explained in an appendix.

37. A. K. Soper and P. A. Egelstaff, *Molec. Phys.* **39**, 1201 (1980).

38. J. G. Powles, E. K. Osae, J. C. Dore, and P. Chieux, *Molec. Phys.* **43**, 1051 (1981).

39. J. G. Powles, J. C. Dore, E. K. Osae, J. H. Clarke, P. Chieux, and S. Cummings, *Molec. Phys.* **44**, 1131 (1981).

40. C. Votava and R. Ahlrichs, Proceedings of the 14th Jerusalem Symposium in Quantum Chemistry and Biochemistry, 1981.

41. C. Votava, A. Geiger and R. Ahlrichs, personal communication.

42. I. R. McDonald, S. F. O'Shea, D. G. Bounds, and M. L. Klein, *J. Chem. Phys.* **72**, 5710 (1980).

43. M. L. Klein and I. R. McDonald, *Molec. Phys.* **42**, 243 (1981).

44. E. Kálmán, G. Pálinkás, and P. Kovács, *Molec. Phys.* **34**, 505, 525 (1977).

45. A. H. Narten and H. Levy, *J. Chem. Phys.* **55**, 2263 (1971).

46. A. H. Narten, *J. Chem. Phys.* **56**, 5681 (1972).

47. G. C. Lie, E. Clementi, and M. Yoshimine, *J. Chem. Phys.* **64**, 2314 (1976).

48. O. Matsuoka, E. Clementi, and M. Yoshimine, *J. Chem. Phys.* **64**, 1351 (1976).

49. J. G. Powles, J. C. Dore, and D. I. Page, *Molec. Phys.* **24**, 1205 (1972).

50. A. K. Soper and R. N. Silver, *Phys. Rev. Lett.* **49**, 471 (1982).

51. A. Rahman, *J. Nucl. Energy A* **13**, 128 (1961).

52. M. Rovere, L. Blum, and A. H. Narten, *J. Chem. Phys.* **73**, 3729 (1980).

53. P. A. Egelstaff and M. J. Poole, *Experimental Neutron Thermalisation*, Pergamon Press, 1969.

54. R. W. Impey, M. L. Klein, and I. R. McDonald, *J. Chem. Phys.* **74**, 647 (1981).

55. J. R. Reimers, R. O. Watts, and M. L. Klein, *Chemical Physics* **64**, 95 (1982).

56. G. C. Lie and E. Clementi, *J. Chem. Phys.* **64**, 5308 (1976).

57. R. O. Watts, *Chem. Phys.* **26**, 367 (1977).

58. P. A. Egelstaff and J. H. Root, *Chem. Phys. Letts.* **91**, 96 (1982) and to be published. They point out that the center of electron density is at the oxygen atom for the model of Eq. (14), while for the IAM it is 0.11 Å away from the oxygen, compared to 0.09 Å for the real molecule.

59. P. A. Egelstaff, J. A. Polo, J. H. Root, and L. J. Hahn, *Phys. Rev. Lett.* **47** 1733 (1981).

60. J. A. Barker and D. Henderson, *Rev. Mod. Phys.* **48**, 587 (1976).

61. A. H. Narten, *J. Chem. Phys.* **65**, 573 (1976).

62. J. B. Van Tricht, *J. Chem. Phys.* **66**, 85 (1977).

63. J. H. Clarke, J. R. Granada, and J. C. Dore, *Molec. Phys.* **37**, 1263 (1979).

64. I. P. Gibson and J. C. Dore, *Molec. Phys.* **37**, 1281 (1979).

65. J. R. Granada, G. W. Stanton, J. H. Clarke, and J. C. Dore, *Molec. Phys.* **37**, 1297 (1979).

66. A. Habenschuss, E. Johnson, and A. H. Narten, *J. Chem. Phys.* **XX**, XXX (1981).

67. G. Placzek, *Phys. Rev.* **86**, 377 (1952).

68. P. A. Egelstaff, in *Inelastic Scattering of Neutrons in Solids and Liquids*, I.A.E.A. Vienna, 1961, p. 25.

69. A. C. Zemach and R. J. Glauber, *Phys. Rev.* **101**, 118 (1956).

70. J. G. Powles, *Molec. Phys.* **42**, 757 (1981).

71. G. Walford, J. H. Clarke, and J. C. Dore, *Molec. Phys.* **36**, 1581 (1978).

72. P. A. Egelstaff and P. Schofield, *Nucl. Sci. Eng.* **12**, 260 (1962).

CHEMICAL REACTION DYNAMICS
IN LIQUID SOLUTION

STEVEN A. ADELMAN*

*Department of Chemistry
Purdue University
West Lafayette, Indiana 47907*

CONTENTS

*John Simon Guggenheim Memorial Fellow 1982–1983.

I. INTRODUCTION

Gas-phase theoretical chemistry has made impressive advances in the past 20 years.[1,2] Condensed-phase theoretical chemistry, which we loosely define as the science devoted to the understanding and calculation of the structures, properties, spectra, and reaction dynamics of molecules in condensed phases, is by comparison in a primitive state of development. This is unfortunate, since most real-world chemistry occurs in condensed phases. The present situation is fortunately undergoing a transformation. The area of liquid-solution chemistry, in particular, is on the verge of a renaissance.

Solution chemistry is one of the oldest areas of physical chemistry. Thus its traditional subject matter (e.g. Henry's-law constants, activity coefficients, and freezing-point depressions) is by today's standards pedestrian.

The emerging *new* solution chemistry, however, is hardly a pedestrian subject. Rather it is perhaps the most challenging area of modern theoretical chemistry. To work effectively in this area one must possess a rather large theoretical–computational background based on state-of-the-art techniques from both gas-phase theory and statistical mechanics. In addition, some of the most fundamental problems in the field are either unsolved or only partially solved. Thus future progress hinges on the development of new concepts in the many-body problem as well as the integration and application of existing methodologies.

Modern solution chemistry, moreover, is concerned with issues that in our opinion are among the most important unresolved questions in chemistry today. Briefly stated, the main goal of modern solution chemistry is to obtain a molecular understanding of the nature of solvent effects on chemical equilibria and chemical-reaction dynamics. Such an understanding is certain to have far-reaching consequence for diverse areas of chemistry, biology, and related sciences.

Significant advances in solution chemistry are now possible because of progress in the gas phase and also because of several key calculational and conceptual advances in statistical mechanics.

The advances in the gas phase that we believe will be of greatest importance for progress in solution chemistry include the emergence of quantum chemistry as a quantitative tool for the calculation of molecular structures and spectra,[1] and also the conceptual[2,3] and calculational[2,4] development of gas-phase chemical-reaction theory.

Two major advances in statistical mechanics that occurred mainly in the 1960s are importantly influencing the development of solution chemistry:

1. The first is the development of the powerful Monte Carlo (MC) and molecular-dynamics (MD) methods for the computer simulation of the equilibrium and dynamical properties of liquids. These machine simulation methods were first developed for the rigid disk and rigid sphere fluids by Metropolis and co-workers (MC)[5] and by Alder and Wainwright (MD).[6] The MD method was later extended to more realistic Lennard–Jones models for simple liquids by Rahman,[7] Levesque and Verlet,[8] and others. The MC and MD methods are now being frequently applied to the complex liquids of importance is chemistry. Early and noteworthy work along these lines is the MD study of waterlike liquids by Rahman and Stillinger.[9]

2. The second major development is the formulation of the modern statistical-mechanical theory of irreversible processes. While there were many contributions that led to this advance, the key syntheses in our opinion are due to Kubo,[10] Mori,[11] and Zwanzig.[12] The central achievement of modern irreversibility theory, in our opinion, is the establishment by Mori[11] of rigorous reduced equations of motion[13] that resemble the equations of phenomenological transport theory.[14] These equations of motion, along with related results and theorems, constitute the subject matter of the conventional theory of generalized Brownian motion. The equations of generalized Brownian motion theory are of prime importance to solution chemistry since they provide a starting point for the theory of condensed phase chemical reaction dynamics. This theory, in turn, is basic to a molecular understanding of solvent effects on liquid-solution reaction dynamics.

The gas-phase and statistical-mechanical advances just discussed provide a *partial* rather than a full foundation for the molecular understanding of liquid-solution phenomena. This is because real progress on at least two of the most fundamental problems in the field will require the development of new concepts and new methodologies as well as innovative use of what is now available.

1. The first of the two problems is the problem of developing general concepts and calculational methods for the theoretical treatment of those liquids whose behavior is influenced in an essential way by

strong (for example, Coulombic or dipolar) intermolecular forces. Such liquids include most of the solvents of greatest interest in chemistry—for example, hydrogen-bonded solvents like water and methanol, and dipolar aprotic solvents like acetone and dimethyl sulfoxide. The development of generally valid theoretical concepts for strong-force liquids would be an advance of major importance. The reason is that such concepts would be basic to an improved understanding of a host of important and largely unsolved problems in chemistry. These include the water-structure problem,[15] the problem of hydrophobic interaction on nonpolar species,[16] the ion-solvation problem,[17] and so on. The theory of strong-force liquids, however, appears to be much more difficult than the theory of simple liquids.[18] While some progress has been made and a few general features have emerged,[19] no large-scale theoretical synthesis has yet appeared. The most promising approach to strong-force liquids at present thus appears to lie in the careful application of the numerical simulation methods to specific systems. This approach is currently being pursued, for example, by Jorgensen,[20] who is investigating chemical structures in organic liquids by the MC method. The results of such simulations, in addition to being of direct chemical interest, will provide a data base that may aid in the development and testing of theories.

2. Liquid-state chemical reactions are, of course, many-body dynamical events. Thus a second fundamental problem in solution chemistry requiring the development of new concepts and methodologies is the problem of condensed-phase chemical-reaction dynamics. In brief, one must somehow generalize gas-phase molecular-collision theory[2] to produce a new theory that is capable of dealing with the many-body problems characteristic of condensed-phase chemistry. The new theory should provide *concepts* that help one understand complex many-body events in simple terms. It should also provide *algorithms* that permit one to conveniently simulate condensed-phase chemical events and thus to compute reaction rates and other observables.

While the problem of strong-force liquids has, so far, proven recalcitrant, considerable recent progress has been made on the problem of condensed-phase chemical-reaction dynamics. This progress is the main topic of the present review.

In particular, a general conceptual and calculational framework for condensed-phase (both solid and liquid state) chemical-reaction dynamics has been developed.[21-23] This framework is based on the modern statistical-mechanical theory of irreversible processes.[10-11] We call this framework the molecular-time-scale generalized-Langevin-equation (MTGLE) theory.[21-23] The reason for this name will become apparent below.

The MTGLE theory has already been applied to a number of basic liquid-state chemical processes. For example, we shall discuss in Chapter 8 and MTGLE stochastic classical trajectory simulation of vibrational-energy relaxation of highly vibrationally excited diatomic molecules in simple solvents. This energy relaxation occurs following photolysis and subsequent recombination of the molecules.

The MTGLE theory at present is restricted to the approximation of classical mechanics. However, first steps in the formulation of a quantum generalization have been taken.[24] Generalization of the theory to quantum mechanics is a problem that is both challenging and interesting, and also very important. For example, one expects that quantum tunneling effects are important in such fundamental proton-transfer processes an protonic conduction in ice and water[25] and acid–base reactions.[26]

We wish to stress at the outset that many questions, at the level of development and application and also at the level of fundamental formulation,[‡] remain to be answered. This is expected, since the field of condensed-phase chemical reaction dynamics is still in its infancy.

Before presenting the detailed development of the theory, begin with a few general comments about the physical ideas and mathematical methods upon which it is based. (It will probably be useful to reread these comments after completeing the review article.)

1. The general problem of condensed-phase chemical-reaction dynamics may at first glance appear hopelessly complex. Physical simplifications, however, exist that render the general problem manageable. The simplifications arise from the fact that condensed-phase chemical reactions occur in spatially localized regions of the many-body system. This fact suggests the following general strategy for simplifying problems in condensed-phase reaction dynamics. Treat the influence of the local (i.e., molecular distance scale) condensed-phase environment on the reaction in detail. Model the influence of the bulk environment by that of a macroscopic energy reservoir. For the case of liquid-solution reactions this strategy may be restated as follows: Treat the influence of the first one or two solvation shells on the reagents in detail; model the influence of the bulk solvent.

2. The mathematical formulation of the above strategy is based on a mental partitioning of the full many-body system into primary and

---

[‡]For example, the approximations made in the formulation of the equation of motion for chemical reactions in liquids, Eq. (6.15), may be too severe for certain reactions, and improved equations of motion may be required.

secondary parts. The primary part or primary system is composed of those few molecules that are intimately involved in the chemical process. The secondary system is composed of the many ($\sim 10^{24}$) remaining molecules, which function as a heatbath for the primary system. A complete solution of a problem in condensed-phase chemical-reaction dynamics amounts to a simulation of the motion of the primary system.

3. If the primary system could be uncoupled from the heatbath, it would obey few-body dynamics and thus could, in principle, be treated by the methods of gas-phase theory. Therefore the new aspects of condensed-phase reaction theory (i.e., aspects absent in the gas phase) arise solely from the influence of the heatbath on the motion of the primary system. This influence is of a qualitative character and thus cannot be ignored. A completely rigorous treatment of the heatbath influence, however, is tantamount to an exact solution of the full many-body problem. Thus a central problem in the theory of condensed-phase reaction dynamics is to develop systematic, practical, and rigorously founded methods for realistically modeling many-body heatbaths by much simpler dynamical systems.

4. We have mentioned that the new aspects of condensed-phase reactions arise from coupling of the primary system and the heatbath. A basic consequence of this coupling is that primary-system motion has a dissipative component, that is, energy can flow *irreversibly* from the primary system to the heatbath. Condensed-phase chemical reactions are therefore irreversible processes. Thus the natural basis of condensed-phase chemical-reaction theory is the modern statistical-mechanical theory of irreversible processes[10-12] and, in particular, the theory of generalized Brownian motion.[10-11]

5. Generalized Brownian-motion theory, as originally formulated by Mori,[11] was designed to provide an exact foundation based on the fundamental laws of dynamics (i.e., $i\hbar\, \partial\psi/\partial t = H\psi$) for Onsager's[14] general phenomenological theory of linear irreversible processes. Onsager's theory, however, is in essence an explicit statement of the general structure of the laws and equations of phenomenological transport theory. These laws, in turn, are ultimately based on macroscopic distance and time-scale measurements. Thus it is only on long time and distance scales that the phenomenological transport equations, and hence the equations of the Onsager theory, reliably describe many-body dynamics. A consequence is that the conventional theory of generalized Brownian motion yields exact equations of motion that emphasize the long time scale, or more properly the adiabatic limit [see Section III.F.3], of the full many-body problem.

6. Condensed-phase chemical reactions are, however, subpicosecond-time-scale rather than macroscopic-time-scale events. For this reason the conventional theory of generalized Brownian motion provides a less than optimal foundation for the construction of condensed-phase reaction theory. What is instead required is a new rigorous formulation of irreversible dynamics that emphasizes the short-time-scale rather than the adiabatic limit of the full many-body problem. The MTGLE theory is, in essence, this now formulation. The MTGLE theory thus automatically incorporates the ideas of primary system and heatbath described above, since the short-time-scale limit of the full many-body response is dominated by the motion of the condensed-phase environment in the immediate vicinity of the reaction.

7. The central result of the MTGLE theory, so far as condensed-phase chemical dynamics is concerned, is a reduced equation of motion for the reagents. This takes the form of a fictitious *equivalent harmonic chain* equation of motion. Chain atom 0 simulates the dynamics of the reagents; chain atom 1, roughly speaking, simulates the dynamics of those condensed-phase molecules that are nearest neighbors to the reagents; chain atom 2, roughly speaking, simulates second-nearest-neighbor influence; and so on. The equivalent-chain formalism thus isolates the interaction between the reagents and the local condensed environment in an explicit and physically clear manner, and includes the effect of the bulk environment in a manner that permits one to construct practical and systematic heatbath modeling schemes.

8. The effects of the bulk environment on reagent motion may be modeled by approximating the dynamics of distant equivalent-chain atoms. Details are described below. Such approximations reduce the rigorous equivalent-chain equations to *effective few-body* truncated-chain models. These models provide a physically realistic and calculationally practical reduction of the many-body problem. The models play a role in condensed-phase reaction theory somewhat analogous to the role played by the basic equations of dynamics in gas-phase theory. The chain models, for example, may be directly solved by stochastic classical trajectory methods to obtain a computer simulation of reagent motions during a condensed-phase reaction. Such simulations are a natural extension of the gas-phase classical trajectory method.[4] Alternatively the MTGLE chain models may be made the basis for the development of theories of condensed-phase reaction dynamics that are analogous to gas-phase theories.[3]

9. For the case of chemical reaction dynamics in liquids, the truncated-chain models simulate in an essentially correct manner the correlated

motions of the reagents and the first few solvation shells (see, e.g., Fig. 7) which form their solvent cage. The force constants that characterize the truncated-chain models are determined by the structure of the cage. The cage structure and hence the force constants, however, change with the solvent thermodynamic state, solvent composition, and so on. One may, however, use intuition about the behavior of mechanical systems to understand changes in reagent dynamics as the force constants change. Thus one may correlate changes in the reagent dynamics with changes in the solvent in a meaningful manner.

Thus the MTGLE chain models provide a tool for probing the molecular nature of the solvent effect on liquid-solution chemical-reaction dynamics.

The plan of this review article is as follows.

Section II provides a glossary of those symbols most frequently used in the article.

Section III provides a simplified introduction to the physical concepts and mathematical methods that are fundamental to condensed-phase reaction theory. Included is a simplified introduction to some basic methods and concepts from the theory of irreversible processes.

Section IV presents a review of selected topics from statistical mechanics. This review is very concise and is thus no substitute for available textbooks.[27] The review is included for two reasons. First it provides the background required to understand the remainder of the review article. Second it is intended as a road map for beginners entering the field of solution chemistry. Novices entering the field of statistical mechanics are confronted with the task of mastering a bewildering range of topics and formalisms. Section IV is intended to direct the reader to those limited areas of statistical-mechanical theory that are likely to be of greatest importance for the development of solution chemistry. Only the bare bones of the most basic results in these areas are presented. Thus, as mentioned, Section IV is no substitute for more detailed treatments.[27]

Finally, Sections V–VIII concern the detailed development of the MTGLE approach to liquid-solution chemical-reaction dynamics. Section V presents a rigorous development of the basic theorems of the MTGLE representation of irreversible dynamics. This development is based on the methods of modern irreversibility theory, which are outlined in Section IV. In Section VI we present an approximate theoretical framework for the treatment of liquid-solution reactions based on the theory of Section V. Section VII deals with several topics concerning the calculation and modeling of reagent-configuration-dependent correlation functions, which are required to practically implement the theory of Section VI. Finally, in Section VIII we present a numerical simulation of a simple liquid-state chemical reaction based on the methods developed in this article.

II. GLOSSARY OF SYMBOLS

Here we present a glossary of those symbols used most frequently in the article. Some infrequently used symbols are not included.

$\mathbf{A}(t) = (A_1(t), \ldots, A_n(t)) =$ arbitrary set of n solvent dynamical variables.
$\alpha =$ Morse potential repulsion parameter.

$b = R_{eq} =$ bond length of a diatomic molecule.
$\beta_1(t) =$ friction kernel of an isolated solute atom.
$\hat{\beta}_1(z) = \int_0^\infty e^{-zt} \beta_1(t)\, dt = \mathcal{L}\beta_1(t)$.
$\beta_{p+1}(t) =$ friction kernel of pth equivalent-chain atom.
$\beta_1(t; r_0) = 3n \times 3n$ friction kernel matrix of reagent atoms.
$\beta_p(t; r_0) = 3n \times 3n$ friction kernel matrix of pth vector "atom" of equivalent chain for interacting reagents.
$\beta = (k_B T)^{-1}$.
$\beta_1 =$ friction coefficient of an isolated solute atom.
$\beta_{p+1} =$ friction coefficient of pth equivalent-chain atom.
$\beta_1(r_0) = 3n \times 3n$ friction coefficient matrix of interacting reagent atoms.
$\beta_{p+1}(r_0) = 3n \times 3n$ friction coefficient matrix of pth vector "atom" of equivalent chain for interacting reagents.

$\dot{\chi}(t) =$ normalized velocity autocorrelation function of an isolated solute atom.
$\hat{\chi}(z) = \int_0^\infty e^{-zt} \chi(t)\, dt = \mathcal{L}\chi(t)$.

$D =$ Morse potential parameter for the dissociation energy of a diatomic molecule.
$D_1 =$ diffusion coefficient for an isolated solute atom.
$\delta(t) =$ Dirac δ-function.

$\varepsilon =$ Lennard-Jones well depth.

$F(r_0) = F = (\mathbf{F}_1, \ldots, \mathbf{F}_n) =$ instantaneous total force acting on reagent atoms.
$F_g(r_0) = F_g = (\mathbf{F}_{g1}, \ldots, \mathbf{F}_{gn}) =$ gas phase force acting on reagent atoms.
$\mathcal{F}(r_0) = \mathcal{F} = (\mathcal{F}_1, \ldots, \mathcal{F}_n) =$ instantaneous solvent force acting on the reagent atoms.
$\mathcal{F}_0(t; r_0) = \mathcal{F}_0(t) = (\mathcal{F}_{01}(t), \ldots, \mathcal{F}_{0n}(t)) =$ instantaneous solvent force acting on the clamped reagents at time t.
$\tilde{\mathcal{F}}_0(t; r_0) = \tilde{\mathcal{F}}_0(t) = \mathcal{F}_0(t) - \langle \mathcal{F} \rangle_{r_0} = (\tilde{\mathcal{F}}_{01}(t), \ldots, \tilde{\mathcal{F}}_{0n}(t)) =$ fluctuating solvent force acting on the clamped reagents at time t.

$\langle F \rangle_{r_0} = F_g + \langle \mathcal{F} \rangle_{r_0} = -\dfrac{\partial W^{(n)}(r_0)}{\partial r_0} = (\langle \mathbf{F}_1 \rangle_{r_0}, \dots, \langle \mathbf{F}_n \rangle_{r_0}) = $ total mean force acting on the reagents.

$\langle \mathcal{F} \rangle_{r_0} = -\dfrac{\partial w^{(n)}(r_0)}{\partial r_0} = (\langle \mathcal{F}_1 \rangle_{r_0}, \dots, \langle \mathcal{F}_n \rangle_{r_0}) = $ liquid-state contribution to the mean force acting on the reagents.

$f[p, q; t] = $ time-dependent phase-space probability distribution function of the solvent.

$f_{CA}[p, q] = $ canonical-ensemble distribution function for the pure solvent.

$f_{CA}[p, q; m\dot{r}_0, r_0] = $ canonical-ensemble distribution function of the solution.

$f_{CA}^{(0)}[p, q] = $ canonical-ensemble distribution function of the solvent in the presence of the clamped reagents.

$\mathbf{f}_1(t) = $ white-noise Gaussian random force acting on an isolated solute atom.

$\mathbf{f}_{p+1}(t) = $ white-noise Gaussian random force acting on pth equivalent-chain atom.

$f_1(t; r_0) = (\mathbf{f}_{11}(t; r_0), \dots, \mathbf{f}_{1n}(t; r_0)) = 3n$-dimensional vector white-noise Gaussian random force acting on the reagent atoms.

$f_p(t; r_0) = (\mathbf{f}_{p1}(t; r_0), \dots, \mathbf{f}_{pn}(t; r_0)) = 3n$-dimensional vector white-noise Gaussian random force acting on $(p-1)$th "atom" of equivalent chain for interacting reagents.

$g_{ij}(r) = \exp[-\beta W_{ij}(r)] = $ radial distribution function for solute species i and j.

$H[p, q] = H = $ classical Hamiltonian of pure solvent.

$H[p, q; m\dot{r}_0, r_0] = H = $ classical Hamiltonian of the solution.

$H_0[p, q] = $ Hamiltonian of the solvent in the presence of the clamped reagents.

$i = \sqrt{-1}$.

$i\hat{\omega} = $ Mori frequency matrix.

$k_B = $ Boltzmann's constant.

$L = $ Liouville operator of a dynamical system.

$L_0 = $ Liouville operator of the solvent in the presence of the clamped reagents.

$L^{(p)} = $ Liouville operator of the pth clamped equivalent chain.

$L^{(1)}(t) = $ perturbation correction to the Liouville operator of the solvent due to an external force.

$\mathcal{L} = $ Laplace transform.

$\mathcal{L}^{-1} = $ inverse Laplace transform.

$M = $ mass of a solvent atom.

$m = $ mass of an isolated solute atom.

m_i = mass of the ith reagent atom.

m = n-dimensional diagonal mass matrix of the reagents.

N = number of equivalent-chain heatbath atoms = $1, 2, \ldots$.

N_s = number of solvent atoms.

n = number of reagent atoms.

Ω_0 = adiabatic frequency of an isolated primary-system atom.

Ω_p = adiabatic frequency of pth equivalent-chain atom.

$\Omega_0(r_0)$ = $3n \times 3n$ adiabatic frequency matrix of the interacting reagents.

$\Omega_p(r_0)$ = $3n \times 3n$ adiabatic frequency matrix of the pth vector "atom" of the equivalent chain for interacting reagents.

ω^2 = dynamical matrix of the equivalent chain for an isolated solute atom.

$[\omega^{(p)}]^2$ = dynamical matrix of the pth clamped equivalent chain for an isolated solute atom.

ω_{c1}^2 = coupling constant linking an isolated solute atom to its first solvation shell.

ω_{cp+1}^2 = coupling constant linking pth and $(p+1)$th equivalent chain atoms.

ω_{e0} = Einstein frequency of an isolated solute atom.

ω_{ep}^2 = Einstein frequency of the pth equivalent chain atom.

$\omega_{c1}^2(r_0)$ = $3n \times 3n$ coupling-constant matrix linking reagent atoms to the first shell of their common cage.

$\omega_{cp+1}^2(r_0)$ = $3n \times 3n$ coupling constant matrix linking pth and $(p+1)$th vector "atoms" of equivalent chain for interacting reagents.

$\omega_{e0}(r_0)$ = $3n \times 3n$ Einstein frequency matrix of interacting reagents.

$\omega_{ep}(r_0)$ = $3n \times 3n$ Einstein frequency matrix of pth vector "atom" of equivalent chain for interacting reagents.

$P[\ ]$ = PDF = probability distribution function.

$p(t) = \left(\mathbf{p}_1(t), \ldots, \mathbf{p}_{N_s}(t)\right)$ = momenta of the solvent molecules.

$p_0 = p(t=0)$ = initial momenta of the solvent molecules.

$\mathbf{p}_\lambda(t) = M\dot{\mathbf{q}}_\lambda(t)$ = momentum of solvent molecule λ at time t.

$\mathbf{p}_{\lambda 0} = \mathbf{p}_\lambda(t=0) = M\dot{\mathbf{q}}_{\lambda 0}$ = initial momentum of solvent molecule λ.

$q(t) = \left(\mathbf{q}_1(t), \ldots, \mathbf{q}_{N_s}(t)\right)$ = coordinates of the solvent molecules.

$q_0 = q(t=0)$ = initial coordinates of the solvent molecules.

$\mathbf{q}_\lambda(t)$ = coordinate of solvent molecule λ at time t.

$\mathbf{q}_{\lambda 0} = \mathbf{q}_\lambda(t=0)$ = initial coordinate of solvent molecule λ.

$R = |\mathbf{r}_{02} - \mathbf{r}_{01}|$ = internuclear separation of a pair of reagent atoms.

$\mathbf{R}_p(t)$ = abstract coordinate of the pth MTGLE heatbath.

$\mathbf{R}_p(t; r_0) = \left(\mathbf{R}_{p1}(t; r_0), \ldots, \mathbf{R}_{pn}(t; r_0)\right)$ = $3n$-dimensional vector coordinate of pth heatbath for interacting reagents.

$\mathbf{r}_0(t)$ = coordinate of an isolated solute atom.

$\mathbf{r}_p(t)$ = coordinate of pth equivalent-chain atom.

$r_0(t) = (\mathbf{r}_{01}(t), \ldots, \mathbf{r}_{0n}(t))$ = coordinates of the reagent atoms.

$r_p(t) = (\mathbf{r}_{p1}(t), \ldots, \mathbf{r}_{pn}(t))$ = $3n$-dimensional coordinate of pth vector heat-bath "atom" in equivalent chain for interacting reagents.

$\rho = N_s/V$ = solvent number density.

$\rho(r) = \rho g(r)$ = local solvent density near an isolated solute atom.

$\rho(\omega)$ = spectral density of $\dot{\chi}(t)$.

$\rho^{(n+1)}(\mathbf{q}: r_0)$ = local solvent density near clamped reagents.

$\rho_q(t)$ = coordinate of the qth atom in the equivalent-chain representation for $\dot{\chi}(t)$.

$\rho_q^{(p)}(t)$ = coordinate of the qth atom $(q \geqslant p)$ in the equivalent-chain representation for $\dot{\theta}_p(t)$.

$\rho_q(t) = \rho_q(t; r_0) = (\rho_{q1}(t), \ldots, \rho_{qn}(t))$ = $3n$-dimensional vector coordinate in equivalent-chain representation for $\theta_1(t; r_0)$.

σ = Lennard–Jones molecular diameter.

$\sigma_p(\omega)$ = spectral density of $\dot{\theta}_p(t)$.

T = Kelvin temperature.

^{T} = matrix transpose.

t = time.

τ = time variable.

$\dot{\theta}_p(t)$ = response function of the pth heatbath of an isolated solute atom.

$\tilde{\theta}_p(z) = \int_0^\infty e^{-zt}\theta_p(t)\,dt = \mathcal{L}\theta_p(t)$.

$\theta_p(t; r_0)$ = $3n \times 3n$ response-function matrix of pth heatbath of the MTGLE hierarchy for interacting reagents.

$U_{UU}(r_0)$ = gas-phase potential-energy function of the reagents.

$U_{VU}(q, r_0)$ = solvent–reagent gas-phase potential-energy function.

$U_{VV}(q)$ = solvent gas-phase potential-energy function.

$u_{uv}[|\mathbf{q} - \mathbf{r}_{0j}|]$ = gas-phase pair potential between reagent atom j and a solvent molecule.

V = volume of liquid solution.

$\mathbf{v}_0(t) = \dot{\mathbf{r}}_0(t)$ = velocity of an isolated solute atom.

$W^{(n+1)}(\mathbf{q}, r_0) = (n+1)$-atom solute–reagent potential of mean force.

$W_j^{(2)}(|\mathbf{q} - \mathbf{r}_{0j}|)$ = pair potential of mean force between reagent atom j and a solvent molecule.

$W^{(n)}(r_0)$ = n-atom reagent potential of mean force.

$w^{(n)}(r_0)$ = n-point reagent cavity function.

$w_{ij}^{(2)}(|\mathbf{r}_{0i} - \mathbf{r}_{0j}|)$ = pair cavity function between reagent atoms i and j.

$Z(\beta)$ = partition function of solution.

$Z_0(\beta)$ = partition function of solvent in the presence of the clamped rea-
        gents.

$\zeta_\lambda(t)$ = λth normal-mode coordinate of equivalent chain.

III.   BASIC CONCEPTS

This section presents a simplified introductory survey of the physical concepts and mathematical methods of molecular-time-scale irreversibility theory and of the application of the theory to problems in condensed-phase
chemical-reaction dynamics. More detailed and rigorous discussions will be
presented in later sections.

We begin by defining a model many-body system, which will permit us to
illustrate basic concepts in a simple manner.

A.   The Harmonic Chain, a Model Many-Body System

Our model many-body system is an $(N + 1)$-atom $(N \to \infty)$ nearest-
neighbor harmonic chain at Kelvin temperature T. Analysis of the classical
statistical dynamics of the chain will expose fundamental aspects of condensed-phase dynamics, which are of far-reaching generality. This generality
follows because molecular motion in an arbitrary condensed-phase environment may be rigorously recast as equivalent harmonic-chain dynamics (Section V).

Denote the coordinates and momenta of the chain atoms by

$$
r = \begin{pmatrix} r_0 \\ r_1 \\ \vdots \\ r_N \end{pmatrix}, \qquad p = m\dot{r} = m \begin{pmatrix} \dot{r}_0 \\ \dot{r}_1 \\ \vdots \\ \dot{r}_N \end{pmatrix} \tag{3.1a}
$$

where each chain atom is assumed to have mass m.

The primary system is taken as chain atom 0, coordinate r_0. The remaining chain atoms with coordinates

$$
r_Q = \begin{pmatrix} r_1 \\ r_2 \\ \vdots \\ r_N \end{pmatrix} \tag{3.1b}
$$

constitute the heatbath.

The force-constant matrix of the chain is $m\omega^2$, where ω^2 is the tridiagonal matrix

$$\omega^2 = \begin{pmatrix} \omega_{e0}^2 & -\omega_{c1}^2 & 0 & 0 & \cdots \\ -\omega_{c1}^2 & \omega_{e1}^2 & -\omega_{c2}^2 & 0 & \cdots \\ 0 & -\omega_{c2}^2 & \omega_{e2}^2 & -\omega_{c3}^2 & \cdots \\ 0 & 0 & -\omega_{c3}^2 & \omega_{e3}^2 & \cdots \\ \vdots & \vdots & \vdots & \vdots & \ddots \end{pmatrix} \tag{3.2a}$$

We shall also require the force-constant matrix of the chain conditional on the assumption that atoms $0, 1, 2, \ldots, p-1$ are clamped at equilibrium. This matrix, which we denote by $m[\omega^{(p)}]^2$, is obtained by striking out the first p rows and columns of the matrix $m\omega^2$:

$$[\omega^{(p)}]^2 = \begin{pmatrix} \omega_{ep}^2 & -\omega_{cp+1}^2 & 0 & 0 & \cdots \\ -\omega_{cp+1}^2 & \omega_{ep+1}^2 & -\omega_{cp+2}^2 & 0 & \cdots \\ 0 & -\omega_{cp+2}^2 & \omega_{ep+2}^2 & -\omega_{cp+3}^2 & \cdots \\ 0 & 0 & -\omega_{cp+3}^2 & \omega_{ep+3}^2 & \cdots \\ \vdots & \vdots & \vdots & \vdots & \ddots \end{pmatrix} \tag{3.2b}$$

The canonical ensemble distribution function for the chain is $[\beta = (k_B T)^{-1}]$

$$f_{CA}[p, r] = Z^{-1}(\beta)\exp[-\beta H(p, r)] \tag{3.3}$$

where $Z(\beta) = \int dp\, dr \exp[-\beta H(p, r)]$ is the classical partition function of the chain, and where the Hamiltonian of the chain is given by

$$H[p, r] = \frac{1}{2m} p^T p + \tfrac{1}{2} m r^T \omega^2 r. \tag{3.4}$$

An analogous canonical distribution function for the clamped chain $f_{CA}^{(p)}$ may be defined by setting $r_0 = \cdots = r_{p-1} = p_0 = \cdots p_{p-1} = 0$ in $H[p, r]$.

The canonical distribution functions for both full and clamped chains are multivariate Gaussian probability distribution functions (PDF). Thus the chain phase-space coordinates p, r are Gaussian random variables. Properties of Gaussian distribution functions and random variables will often be used below. These properties are reviewed elsewhere.[28]

Here and below we denote canonical ensemble averages by $\langle \cdots \rangle$. There is one exception. We use the special symbol $\langle \cdots \rangle_{r_0, r_1, \ldots, r_{p-1}}$ to denote a canonical average for the clamped chain, i.e., an average over $f_{CA}^{(p)}$.

Finally, using properties of Gaussian integrals, one may verify the following averages for chain-atom phase-space coordinates $(q, q' = 0, 1, \ldots, N)$:

$$\langle r_q \rangle = \langle p_q \rangle = 0 \tag{3.5a}$$

$$\frac{1}{2m} \langle p_q p_{q'} \rangle = \tfrac{1}{2} k_B T \delta_{qq'} \tag{3.5b}$$

$$\langle r_q p_{q'} \rangle = 0 \tag{3.5c}$$

$$\frac{1}{2m} \langle r_q r_{q'} \rangle = \tfrac{1}{2} k_B T [\omega^{-2}]_{qq'} \tag{3.5d}$$

The analogous averages for the clamped chain are $(q, q' = p, p+1, \ldots, N)$

$$\langle r_q \rangle_{r_0, r_1, \ldots, r_{p-1}} = \langle p_q \rangle_{r_0, r_1, \ldots, r_{p-1}} = 0 \tag{3.6a}$$

$$\frac{1}{2m} \langle p_q p_{q'} \rangle_{r_0, r_1, \ldots, r_{p-1}} = \tfrac{1}{2} k_B T \delta_{qq'} \tag{3.6b}$$

$$\langle r_q p_{q'} \rangle_{r_0, r_1, \ldots, r_{p-1}} = 0 \tag{3.6c}$$

$$\tfrac{1}{2} m \langle r_q r_{q'} \rangle_{r_0, r_1, \ldots, r_{p-1}} = \tfrac{1}{2} k_B T \left[(\omega^{(p)})^{-2} \right]_{qq'} \tag{3.6d}$$

B. The Dissipative–Stochastic Character of Primary-System Motion

The new features of condensed-phase dynamics (i.e., features absent in the gas phase) arise from primary-system–heatbath coupling. A fundamental point is that the forces exerted on the primary system due to this coupling may be decomposed into *systematic* and *random* components.

A consequence is that primary-system motion may be regarded as a superposition of *dissipative* and *stochastic* components. The dissipative motion is induced by the systematic force; the stochastic motion, by the random force. The dissipative character of primary-system motion manifests itself as irreversible energy flow from the primary system to the heatbath. We next explain what we mean by stochastic motion.

Consider, as an example, an ensemble of liquid solutions composed of a single solute atom with coordinate $r_0(t)$ dissolved in a solvent whose N_s molecules are described by the coordinates $q(t) \equiv (q_1(t), \ldots, q_{N_s}(t))$.

The ensemble is described by the solvent thermodynamic state T, the solvent number density ρ, and the solute initial conditions $r_0, \dot{r}_0$. That is, each member of the ensemble has the same value for T, ρ, $\mathbf{r}_0$, $\dot{\mathbf{r}}_0$. The initial conditions of the solvent molecules $\dot{q}, q$ are, however, not fixed, but rather are distributed thermally. Thus the solute-atom trajectory

$$\mathbf{r}_0(t) = \mathbf{r}_0[\dot{\mathbf{r}}_0, \mathbf{r}_0, \dot{q}, q; t] \tag{3.7}$$

is different for different members of the ensemble. A unique trajectory is, of course, generated from $\dot{\mathbf{r}}_0, \mathbf{r}_0$ in the gas phase. This is the sense in which primary-system dynamics may be regarded as stochastic.

A consequence is that solution of a problem in condensed-phase chemical dynamics always involves calculation of a probability distribution function (PDF) for groups of trajectories; solution to a problem in gas phase dynamics may involve calculation of a single trajectory.

An exception is at $T = 0$ K. For this case $\dot{q}$ and q are identical for all members of the ensemble,[‡] the stochastic character disappears and primary system dynamics is purely dissipative.

We close this section by noting that the systematic and random forces have an approximate physical interpretation for solute motion in liquids. This interpretation is important for later work.

The systematic force is approximately the force exerted on the solute by the time-dependent mean-field (i.e., average) solvent density induced by solute motion. The random force is approximately the force exerted on a *clamped* solute due to thermal temporal fluctuations of the local solvent density about its equilibrium value (Fig. 1).

This approximate physical interpretation is exact for perfectly harmonic many-body systems such as the chain model.

C. The Velocity Autocorrelation Function

Time correlation functions provide a convenient mathematical tool for condensing information contained in large numbers of many-body trajectories into physically meaningful form. We next present a simplified discussion of time correlation functions designed to make their physical meaning clear.

Our discussion will focus on the velocity autocorrelation function (VAF). The VAF is chosen to illustrate general features of time correlation functions because it is basic to the theoretical description of molecular motion in condensed phases.

[‡]This is only true for the hypothetical case of a $T = 0$ K system that obeys classical statistical mechanics. For such a system the heatbath initial conditions are the same for all members of the ensemble and thus the random force vanishes. For quantum systems [see Ref. 24] the heatbath initial conditions are distributed rather than fixed at $T = 0$ K, due to quantum zero-point motion.

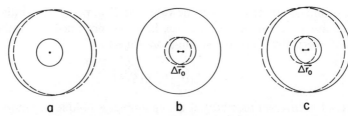

Fig. 1. Approximate density-response interpretation of the systematic and random solvent forces exerted on an isolated solute atom. (a) Solute atom of mass m (small circle) is initially clamped at point $\mathbf{r}_0$. Fluctuations (dashed circle) of the local solvent density about its equilibrium value $\rho g_1(r) = \rho^{(2)}(\mathbf{q}; \mathbf{r}_0)$ (large solid circle) give rise to a random force $\tilde{\mathcal{F}}_0(t)$ acting on the solute atom. The random force $\tilde{\mathcal{F}}_0(t)$ is approximately equal to $\omega_{cl}^2 \mathbf{R}_1(t)$. ($b$) Solute atom initially located at point $\mathbf{r}_0$ (small dashed circle) is slightly displaced to point $\mathbf{r}_0 + \Delta\mathbf{r}_0$ (small solid circle). The initial mean-field solvent density $\rho^{(2)}(\mathbf{q}; \mathbf{r}_0)$ (large solid circle) exerts an instantaneous systematic restoring force $\mathcal{F}_I$ on the solute atom. This force is rigorously equal to $-m\omega_{e0}^2\Delta\mathbf{r}_0$. ($c$) The initial mean-field density $\rho^{(2)}(\mathbf{q}; \mathbf{r}_0)$ (large dashed circle) relaxes in response to the solute-atom displacement to produce a time-dependent mean-field density (large solid circle). This density produces a systematic relaxing restoring force on the solute atom, $\mathcal{F}_D(t)$. The relaxing restoring force $\mathcal{F}_D(t)$ is approximately equal to $[m\omega_{cl}^4 \int_0^t \theta_1(\tau)\,d\tau]\,\Delta\mathbf{r}_0$.

Consider, as an example, the motion of a strongly solvated ion in a polar liquid [Fig. 2b and d]. The dissipative component of the ion motion is of damped oscillatory character. This dissipative motion may be regarded as the projection onto the ion coordinate of the coupled vibratory motion of the ion and the local mean-field solvent density produced by the solvation shell(s) that surround and cage the ion. The dissipative motion thus reflects the basic physics of coupled ion–cage motion. The basic ion motion is, however, masked in individual ion trajectories by the stochastic component of ion motion, which is governed by thermal fluctuations in solvent cage density.

To obtain a useful dynamical measure of ion–solvent interaction, one must somehow average over the thermal cage fluctuations and thus remove the stochastic component. This leads to the concept of the VAF.

The normalized VAF $\dot{\chi}(t)$ is the dissipative component of the primary-system velocity $\dot{r}_0(t)$ given an initial unit velocity, $\dot{r}_0 = 1$, and given that the primary system is initially in equilibrium with the mean-field density of its heatbath, that is, $r_0 = 0$. Thus to specify the VAF we must precisely define the dissipative component of $\dot{r}_0(t)$. The rigorous definition is based on Mori's[11] elegant method of formal projection operation.[12] This method is reviewed in Section IV.B.8.

The dissipative component may be *qualitatively* defined by averaging $\dot{r}_0(t)$ over heatbath fluctuations, that is, over the canonical distribution of heatbath initial conditions $\dot{q}, q$. We shall denote such an average by $\langle \cdots \rangle_{\dot{q}q}$. The

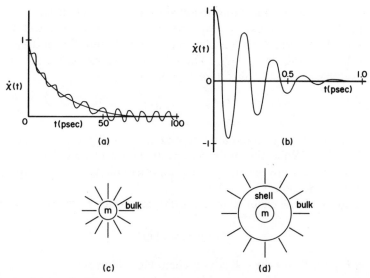

Fig. 2. Solute velocity autocorrelation functions and the local solvent environment. (a) Normalized velocity autocorrelation function (VAF) $\dot{\chi}(t) = \dot{\chi}_B(t) = e^{-\beta_1 t}$ for a Brownian (i.e., semimacroscopic) particle immersed in a liquid. A typical Brownian velocity that displays thermal fluctuations arising from the random force $\mathbf{f}_1(t)$ is also plotted. (b) Normalized VAF $\dot{\chi}(t)$ for an ion dissolved in a polar solvent. This displays oscillations characteristic of strong caging. (c) Brownian particle of mass m. The liquid is modeled by a bulk continuum. This model, which is consistent with the picture of a semimacroscopic Brownian particle immersed in a molecular liquid, gives rise to the exponentially decaying normalized VAF of Fig. 1a. (d) Ion in a polar solvent. Solvent is modeled by a single solvation shell surrounded by a bulk continuum. This model, which roughly corresponds to a two-atom truncated MTGLE chain, can give rise to the damped oscillatory VAF of Fig. 1b, which is due to ion–solvation-shell correlated vibrations.

normalized VAF is then qualitatively defined as the average velocity given the initial conditions $r_0 = 0$, $\dot{r}_0 = 1$:

$$\dot{\chi}(t) \simeq \langle \dot{r}_0[\dot{r}_0 = 1, r_0 = 0, \dot{q}, q; t]\rangle_{\dot{q}q} \qquad (3.8)$$

Note that $\dot{\chi}(0) = 1$ according to Eq. (3.8), which is why $\dot{\chi}(t)$ is described as normalized. Also, since the stochastic component of $\dot{r}_0(t)$ is suppressed by the averaging, $\dot{\chi}(t)$ is expected to be a decaying function: $\dot{\chi}(t) < 1$ for $t > 0$ and $\lim_{t \to \infty} \dot{\chi}(t) = 0$.

The qualitative definition of $\dot{\chi}(t)$, Eq. (3.8), has been introduced because it makes the physical interpretation of the VAF transparent. This definition is, moreover, exact for a number of simple dynamical systems. These include the Langevin model discussed in Section IIIC.1. Equation (3.8) is also exact for harmonic systems, including the chain model of Section IIIA. For the chain model $\langle \cdots \rangle_{\dot{q}q}$ must, however, be interpreted as a canonical average conditional that chain atom 0 is clamped at equilibrium. Recall that we denote such an average by $\langle \cdots \rangle_{r_0}$. The following expression for the normalized VAF chain atom 0 is therefore rigorous:

$$\dot{\chi}(t) = \langle \dot{r}_0 [\dot{r}_0 = 1, r_0 = 0; \dot{r}_Q, r_Q; t] \rangle_{r_0} \tag{3.9}$$

We next derive three alternative representations of the normalized VAF from Eq. (3.9). While the present derivations are restricted to the chain model, the results are generally valid representations of time correlation functions (see Sections IV and V).

We begin with the chain equation of motion, which in matrix notation is

$$\ddot{r}(t) = -\omega^2 r(t) \tag{3.10}$$

Solving Eq. (3.10) for the velocity of chain atom 0 gives

$$\dot{r}_0(t) = \{[\cos \omega t]_{00} \dot{r}_0 - [\omega \sin \omega t]_{00} r_0 \}$$
$$+ \left[\sum_{p=1}^{N} ([\cos \omega t]_{0p} \dot{r}_p - [\omega \sin \omega t]_{0p} r_p) \right] \tag{3.11}$$

The term in braces is the dissipative component of $\dot{r}_0(t)$; the term in square brackets, linear in the heatbath initial conditions, is the stochastic component.

The first representation is derived by noting that since $\langle \dot{r}_Q \rangle_{r_0} \equiv \langle r_Q \rangle_{r_0} = 0$ [Eqs. (3.1b) and (3.6a)], Eqs. (3.9) and (3.11) yield

$$\dot{\chi}(t) = [\cos \omega t]_{00} \tag{3.12}$$

The second representation is based on the observation that $\chi(t)$ is a particular dissipative trajectory of chain atom 0 and that $\dot{\chi}(t)$ is therefore the velocity which corresponds to that trajectory. Explicitly

$$\dot{\chi}(t) = \dot{r}_0(t) \tag{3.13a}$$

given the initial conditions

$$\dot{r}_0 = 1, \qquad r_0 = 0$$
$$\dot{r}_p = r_p = 0, \qquad p \geq 1 \tag{3.13b}$$

Equations (3.13) follow from Eqs. (3.11) and (3.12).

The third representation is an expression for $\dot{\chi}(t)$ as a full rather than a clamped canonical average. This representation is derived by constructing the quantity $\langle \dot{r}_0(t + \tau)\dot{r}_0(\tau)\rangle$ using Eqs. (3.5), (3.11), and (3.12). A brief calculation yields

$$\dot{\chi}(t) = \frac{m}{k_B T}\langle \dot{r}_0(t + \tau)\dot{r}_0(\tau)\rangle$$

$$= \frac{m}{k_B T}\langle \dot{r}_0(t)\dot{r}_0\rangle \tag{3.14}$$

Note that $\langle \dot{r}_0(t + \tau)\dot{r}_0(\tau)\rangle$ is independent of the time origin τ. This is an example of the stationarity property of time correlation functions. This property, which holds if the underlying Hamiltonian is independent of time, is the basis of the theoretical developments of Section V.

Equation (3.14) is the conventional form for the classical VAF[‡] for a particle moving in one dimension. For a solute, with mass m and coordinate $r_0(t)$, translating in an isotropic liquid, Eq. (3.14) generalizes to

$$\dot{\chi}(t) = \frac{m}{3k_B T}\langle \dot{\mathbf{r}}_0(t + \tau)\cdot\dot{\mathbf{r}}_0(\tau)\rangle$$

$$= \frac{m}{3k_B T}\langle \dot{\mathbf{r}}_0(t)\cdot\dot{\mathbf{r}}_0\rangle \tag{3.15}$$

Note that $\dot{\chi}(0) = 1$, since the equilibrium kinetic energy $\frac{1}{2}m\langle \dot{r}_0^2\rangle = \frac{3}{2}k_B T$ for a classical particle moving in three dimensions.

D. The Langevin Model

We next present an elementary discussion of the Langevin equation and phenomenological Brownian-motion theory.[29] This discussion illustrates the general points raised in Sections III.B and III.C.

The Langevin equation is the simplest and oldest[30] primary-system equation of motion based on a decomposition of the heatbath force into systematic and random components. The Langevin equation is not, however, a rigorous result. Rather it is a phenomenological model based on experimental evidence[31] and constructed via physical intuition. This model, however, has proven to be seminal in many areas of mathematics, science, and engineering.

We shall analyze the Langevin dynamics of a rigid spherical "Brownian" particle [mass m, coordinate $\mathbf{r}_0(t)$] translating in an isotropic liquid. For this simplest case, the Langevin equation is

$$m\ddot{\mathbf{r}}_0(t) = -m\beta_1\dot{\mathbf{r}}_0(t) + m\mathbf{f}_1(t) \tag{3.16}$$

[‡]The VAF takes a different form for quantum systems, as discussed in Ref. 24.

The systematic and random forces acting on the Brownian particle are, respectively, $-m\beta_1 \dot{r}_0(t)$ and $mf_1(t)$. The systematic force is of the simple frictional form expected for a macroscopic sphere. Thus β_1 is the friction coefficient (per unit mass) of the Brownian particle. The random force simulates the effects of uncertainty in the solvent initial conditions $q, \dot{q}$. Equivalently, $f_1(t)$ simulates the influence of thermal fluctuations of the local solvent density (Fig. 1).

The Langevin model is most appropriate for semimacroscopic particles, for example, polymer molecules.[32] Such particles are large enough to be governed by the law of macroscopic friction, but are small enough to be sensibly influenced by the molecular distance and time-scale fluctuations described by $f_1(t)$.

We next solve the Langevin equation. Recall that this implies calculation of the PDFs for Brownian particle velocities, trajectories, and so on. The formal solution of Eq. (3.16) is[‡]

$$\dot{r}_0(t) = e^{-\beta_1 t}\dot{r}_0 + \int_0^t e^{-\beta_1(t-\tau)}f_1(\tau)\,d\tau \tag{3.17}$$

The quantity $e^{-\beta_1 t}$ is the dissipative component of $\dot{r}_0(t)$; the integral term is the stochastic component.

To proceed further we require the statistical properties of $f_1(t)$. These are chosen to simulate the influence of solvent fluctuations within the bulk soltbvent model (Fig. 2c) appropriate for semimacroscopic solutes.

The first two statistical properties follow from Onsager's[14] basic hypothesis concerning regression of fluctuations for semimacroscopic primary systems. This hypothesis states that both average motion and time correlation functions satisfy the macroscopic transport equations that would hold if heatbath fluctuations could be ignored.

The macroscopic transport equation for the present case is $\ddot{r}_0(t) = -\beta_1 \dot{r}_0(t)$, that is, Eq. (3.16) with the fluctuation term $f_1(t)$ suppressed. Thus according to the Onsager hypothesis, the average velocity $\langle \dot{r}_0(t) \rangle_{\dot{q}q}$ and the normalized VAF $\chi_B(t)$ for the Brownian particle respectively satisfy $\langle \ddot{r}_0(t) \rangle_{\dot{q}q} = -\beta_1 \langle \dot{r}_0(t) \rangle_{\dot{q}q}$ and $\dot{\chi}_B(t) = -\beta_1 \chi_B(t)$. Solving these equations yields

$$\langle \dot{r}_0(t) \rangle_{\dot{q}q} = e^{-\beta_1 t}\dot{r}_0 \tag{3.18a}$$

[‡]Equation (3.17) may be derived in the following manner. Defining $v_0(t) = \dot{r}_0(t)$, Eq. (3.16) may be rewritten as $\dot{v}_0(t) = -\beta_1 v_0(t) + f_1(t)$. The Laplace space form of this equation is $[z + \beta_1]\hat{v}_0(z) = v_0 + \hat{f}_1(z)$. Solving this equation for $\hat{v}_0(z)$ and inverting the Laplace transform yields Eq. (3.17).

and

$$\dot{\chi}_B(t) = \frac{m}{3k_BT}\langle \dot{\mathbf{r}}_0(t)\cdot\dot{\mathbf{r}}_0\rangle = e^{-\beta_1 t} \tag{3.18b}$$

Note that Eqs. (3.18) show that β_1^{-1} is the velocity relaxation time of the Brownian particle.

The first two statistical properties of $\mathbf{f}_1(t)$ follow from Eqs. (3.17) and (3.18). These are

$$\langle \mathbf{f}_1(t)\rangle_{\dot{q}q} = 0 \tag{3.19a}$$

$$\langle \mathbf{f}_1(t)\cdot\dot{\mathbf{r}}_0\rangle = 0 \tag{3.19b}$$

Equation (3.19b), in the language of modern irreversibility theory[11] (Section IV), states that the random force is *orthogonal* to the initial velocity $\dot{\mathbf{r}}_0$.

The remaining two statistical properties are

$$\langle \mathbf{f}_1(\tau)\cdot\mathbf{f}_1(\tau')\rangle_{\dot{q}q} = 2A\delta(\tau - \tau') \tag{3.19c}$$

where $\delta(\tau - \tau')$ is the Dirac δ-function and A is a constant to be determined below, and

$$P[\mathbf{f}_1(t)] \sim \exp\left[-\frac{3}{2}\frac{f_1^2(t)}{\langle f_1^2\rangle_{\dot{q}q}}\right] \tag{3.19d}$$

where $P[\mathbf{f}_1(t)]$ is the PDF for $\mathbf{f}_1(t)$.

The "white noise" approximation, Eq. (3.19c), is based on the physical idea that the local solvent density fluctuations responsible for $\mathbf{f}_1(t)$ relax on a molecular time scale [~ 1 psec], which is negligible compared to Brownian time scales [$\sim \beta_1^{-1} \sim 100$–$1000$ psec].

The Gaussian noise approximation, Eq. (3.19d), is also based on the physical picture of a semimacroscopic solute moving in molecular solvent. This picture suggests that $\mathbf{f}_1(t)$ is a superposition of many small contributions arising from solvent-molecule collisions with the Brownian particle. Thus by the central limit theorem[29] $\mathbf{f}_1(t)$ is a Gaussian random variable. This random variable is also stationary [e.g., $\langle f^2(t)\rangle_{\dot{q}q} = \langle f^2\rangle_{\dot{q}q}$] and has zero mean [Eq. (3.19a)].

We next construct the PDF for Brownian velocities. Defining the velocity fluctuation about the average velocity $\langle \dot{\mathbf{r}}_0(t)\rangle_{\dot{q}q} = e^{-\beta_1 t}\dot{\mathbf{r}}_0$ by

$$\dot{\mathbf{y}}_0(t) = \dot{\mathbf{r}}_0(t) - e^{-\beta_1 t}\dot{\mathbf{r}}_0 \tag{3.20}$$

we rewrite Eq. (3.17) as

$$\dot{\mathbf{y}}_0(t) = \int_0^t e^{-\beta_1(t-\tau)} \mathbf{f}_1(\tau)\, d\tau \qquad (3.21)$$

Note that $\langle \dot{\mathbf{y}}_0(t) \rangle_{\dot{q}q} = 0$. Moreover, since $\dot{\mathbf{y}}_0(t)$ is a linear superposition of Gaussian random variables $\mathbf{f}_1(\tau)$, $0 \leqslant \tau \leqslant t$, it is also Gaussian random variable.[28] The PDF for $\dot{\mathbf{y}}_0(t)$ or equivalently $\dot{\mathbf{r}}_0(t)$ is therefore a normalized Gaussian determined by the second moment $\langle \dot{y}_0^2(t) \rangle_{\dot{q}q}$, namely,

$$P[\dot{\mathbf{r}}_0(t)] = \left\{ \frac{3}{2\pi \langle \dot{y}_0^2(t) \rangle_{\dot{q}q}} \right\}^{3/2} \exp\left[-\frac{3}{2} \frac{\dot{y}_0^2(t)}{\langle \dot{y}_0^2(t) \rangle_{\dot{q}q}} \right] \qquad (3.22)$$

The second moment $\langle \dot{y}_0^2(t) \rangle_{\dot{q}q}$ may be determined using Eqs. (3.19c) and (3.21). A brief calculation yields[‡]

$$\langle \dot{y}_0^2(t) \rangle_{\dot{q}q} = A\beta_1^{-1}[1 - e^{-2\beta_1 t}] \qquad (3.23)$$

We next show that

$$A = \frac{3k_B T}{m} \beta_1 \qquad (3.24)$$

The proof is based on the fact that the mean kinetic energy $\frac{1}{2}m\langle \dot{r}_0^2(t) \rangle_{\dot{q}q}$ must relax to $\frac{3}{2}k_B T$, the equilibrium result, for $t \gg \beta_1$. Also [Eq. (3.20)] $\dot{\mathbf{y}}_0(t) \to \dot{\mathbf{r}}_0(t)$ for $t \gg \beta_1$. Thus $\lim_{t \to \infty} \langle \dot{y}_0^2(t) \rangle_{\dot{q}q} = \lim_{t \to \infty} \langle \dot{r}_0^2(t) \rangle_{\dot{q}q} = 3k_B T/m$. However [Eq. (3.23)], it is also true that $\lim_{t \to \infty} \langle \dot{y}_0^2(t) \rangle_{\dot{q}q} = A\beta_1^{-1}$. Thus $A\beta_1^{-1} = 3k_B T/m$, which is equivalent to Eq. (3.24).

Using Eq. (3.24), Eq. (3.19c) may be rewritten as

$$\langle \mathbf{f}_1(\tau) \cdot \mathbf{f}_1(\tau') \rangle_{\dot{q}q} = \frac{6k_B T}{m} \beta_1 \delta(\tau - \tau') \qquad (3.25)$$

[‡]The derivation of Eq. (3.23) from Eq. (3.21) proceeds as follows. Squaring Eq. (3.21) and averaging over heatbath initial conditions yields

$$\langle \dot{y}_0^2(t) \rangle_{\dot{q}q} = \int_0^t e^{-\beta_1(t-\tau)}\, d\tau \int_0^t e^{-\beta_1(t-\tau')}\, d\tau' \langle \mathbf{f}_1(\tau) \cdot \mathbf{f}_1(\tau') \rangle_{\dot{q}q}$$

Using Eq. (3.19c), this equation simplifies to

$$\langle \dot{y}_0^2(t) \rangle_{\dot{q}q} = 2A \int_0^t e^{-2\beta_1(t-\tau)}\, d\tau = A\beta_1^{-1}[1 - e^{-2\beta_1 t}]$$

which is Eq. (3.23).

Equation (3.25) is an example of the second fluctuation–dissipation theorem, a fundamental theorem in irreversibility theory. This theorem plays a central role in condensed-phase reaction theory. The key points concerning it are:

1. The random $[\mathbf{f}_1(t)]$ and systematic $[\beta_1]$ components of the heatbath force are interrelated rather than independent; a larger frictional damping is compensated for by a stronger random force, so that thermal equilibrium is maintained. The physical basis of this interrelation is that both systematic and random forces arise from the same underlying molecular processes, namely, collisions between heatbath and primary-system molecules.

2. Because the condition $\lim_{t \to \infty} \frac{1}{2} m \langle \dot{r}_0^2(t) \rangle = \frac{3}{2} k_B T$ is built into the second fluctuation–dissipation theorem, this theorem guarantees that an initial *nonequilibrium* primary-system phase-space distribution will relax (as $t \to \infty$) to the canonical-ensemble distribution function.

3. The random force is $\propto T^{1/2}$. Thus, as mentioned, the stochastic component of primary-system motion vanishes at $T = 0$ K (see footnote in Section III.B).

The PDF $P[\dot{\mathbf{r}}_0(t)]$ may now be determined from Eqs. (3.22)–(3.24) as

$$P[\dot{\mathbf{r}}_0(t)] = \left\{ \frac{m}{2\pi k_B T [1 - e^{-2\beta_1 t}]} \right\}^{3/2} \exp\left[-\frac{1}{2} \frac{m}{k_B T} \frac{[\dot{\mathbf{r}}_0(t) - e^{-\beta_1 t} \dot{\mathbf{r}}_0]^2}{1 - e^{-2\beta_1 t}} \right]$$

(3.26)

$P[\dot{\mathbf{r}}_0(t)]$ is plotted vs. t in Fig. 3.

The main points concerning Eq. (3.26) and Fig. 3 are as follows: (i) $\lim_{t \to 0} P[\dot{\mathbf{r}}_0(t)] = \delta[\dot{\mathbf{r}}_0(t) - \dot{\mathbf{r}}_0]$. (ii) As t increases, the initial δ-function broadens due to the stochastic effect, and the peak position, (i.e., $\langle \dot{\mathbf{r}}_0(t) \rangle_{\dot{q}q}$ $= e^{-\beta_1 t} \dot{\mathbf{r}}_0$) decays to zero due to the dissipative effect. (iii) As $t \to \infty$, $P[\dot{\mathbf{r}}_0(t)]$ properly relaxes to the Maxwell–Boltzmann velocity distribution. (iv) At $T = 0$ K, the stochastic broadening disappears and the PDF reduces to $\delta[\dot{\mathbf{r}}_0(t) - e^{-\beta_1 t} \dot{\mathbf{r}}_0]$. Equivalently, the velocity is purely dissipative and is fixed at $e^{-\beta_1 t} \dot{\mathbf{r}}_0$.

A stochastic calculation analogous to that just presented by $P[\dot{\mathbf{r}}_0(t)]$ may be carried out to determine the PDF $P[\mathbf{r}_0(t)]$ for Brownian trajectories. This calculation begins with the time integral of Eq. (3.17). The general result for $P[\mathbf{r}_0(t)]$ involves both $\mathbf{r}_0$ and $\dot{\mathbf{r}}_0$. For $t \gg \beta_1^{-1}$, the dependence on initial velocity is of negligible importance and $P[\mathbf{r}_0(t)]$ reduces to the following

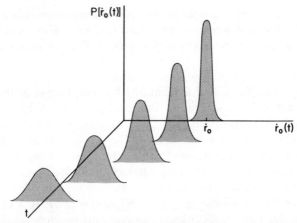

Fig. 3. The probability distribution function for Brownian velocities [Eq. (3.26)], $P[\dot{r}_0(t)]$ vs. t. At $t = 0$, the PDF is sharply peaked at $\dot{r}_0$. As t increases, the initial sharply peaked function broadens due to the stochastic effect, and the position of its peak $\langle \dot{r}_0(t) \rangle_{\dot{q}q} = e^{-\beta_1 t}\dot{r}_0$ decays to zero due to the dissipative effect. At $t = \infty$, $P[\dot{r}_0(t)]$ properly relaxes to the Maxwell–Boltzmann distribution characteristic of the solvent temperature T.

normalized Gaussian distribution function:

$$\lim_{t \to \infty} P[\mathbf{r}_0(t)] = \left\{ \frac{3}{2\pi \langle [\mathbf{r}_0(t) - \mathbf{r}_0]^2 \rangle_{\dot{q}q}} \right\}^{3/2} \exp\left[-\frac{3}{2} \frac{[\mathbf{r}_0(t) - \mathbf{r}_0]^2}{\langle [\mathbf{r}_0(t) - \mathbf{r}_0]^2 \rangle_{\dot{q}q}} \right]$$

(3.27)

where

$$\lim_{t \to \infty} \langle [\mathbf{r}_0(t) - \mathbf{r}_0]^2 \rangle_{\dot{q}q} = 6D_1 t \tag{3.28}$$

and where

$$D_1 = \frac{k_B T}{m\beta_1} \tag{3.29}$$

Equation (3.28) is a statement that on a time scale that is coarse-grained relative to the velocity relaxation time β_1^{-1}, the Brownian particle executes diffusive motion (root-mean-square displacement $\propto t^{1/2}$) with diffusion coefficient D_1. (Equivalently one may show that $P[\mathbf{r}_0(t)]$ satisfies the diffusion equation.)

Equation (3.29) is the Einstein formula that relates the diffusion and friction coefficients of the Brownian particle. Although the present derivation of the Einstein relation is based on the Langevin model, the relation holds generally for solute motion in a liquid.

An alternative expression for D_1 in terms of the normalized VAF follows from Eqs. (3.18b) and (3.29). This expression is $[\dot{\chi}(t) = \dot{\chi}_B(t)]$

$$D_1 = \frac{k_B T}{m} \lim_{z \to 0} \int_0^\infty e^{-zt} \dot{\chi}(t)\, dt$$

$$= \frac{k_B T}{m} \lim_{z \to 0} [z\hat{\chi}(z)]. \qquad (3.30)$$

[Note that here and below we denote the Laplace transform of an arbitrary function $f(t)$ by $\hat{f}(z)$. Thus, for example, $\hat{\chi}(z) = \int_0^\infty e^{-zt} \chi(t)\, dt \equiv \mathcal{L}\chi(t)$.] Equation (3.30) is again a generally valid relation rather than a restricted result of the Langevin model.

Finally we note that the Langevin theory may be generalized to treat Brownian motion in external force fields. We shall require below the Langevin equation for a harmonically bound particle, which is

$$\ddot{r}_0(t) = -\Omega_0^2 r_0(t) - \beta_1 \dot{r}_0(t) + f_1(t). \qquad (3.31)$$

This concludes our introductory discussion of the Langevin theory.

E. Short-Time-Scale Many-Body Dynamics

The Langevin equation describes solute motion on time scales long relative to solvent motional times ~ 1 psec. The making and breaking of chemical bonds, however, occurs on subpicosecond time scales. Thus we expect the Langevin model to be inadequate as a basis for the theory of condensed-phase reaction dynamics.

We next discuss the basic physics of ultrashort-time-scale many-body dynamics using the harmonic chain model to illustrate the main points. This basic physics has far-reaching implications for condensed-phase reaction dynamics.

The nature of the many-body problem on short time scales is illustrated in "pure" form by the VAF of chain atom 0, which is $\dot{\chi}(t) = [\cos \omega t]_{00}$ [Eq. (3.12)]. This, we recall, is the particular dissipative velocity ($\dot{r}_Q = r_Q = 0$) that derives from the initial conditions $\dot{r}_0 = 1$, $r_0 = 0$ [Eq. (3.13)]. Immediately after preparation of this initial state, chain atom 0 begins to execute oscillatory motion with frequency ω_{e0} where $\omega_{e0}^2 = [\omega^2]_{00}$ [see Eq. (3.2a)]. This oscillatory motion is eventually damped due to flow of the initial kinetic energy of atom 0 ($\frac{1}{2} m \dot{r}_0^2$) into chain atoms $1, 2, \ldots$.

The damping does not, however, begin until atom 1 "gives" in response to the motion of atom 0. Before this response occurs, atom 1 acts like a rigid wall. Thus initially atom 0 moves like an independent or Einstein harmonic oscillator with frequency ω_{e0}. We therefore call ω_{e0} the *Einstein frequency* of chain atom 0.

The frequency ω_{ep} analogously governs the initial isolated oscillator motion of chain atom p conditional on the assumption that chain atoms $0, 1, \ldots, p - 1$ are clamped at equilibrium. Thus we call ω_{ep} the Einstein frequency of chain atom p.

The initial reduction of primary-system motion to effective gas-phase motion of the isolated oscillator type, which we call the Einstein limit, is a fundamental and general feature of the many-body problem on molecular time scales. The cage oscillations of a solute atom in a liquid (Fig. 2b) are one manifestation of this feature.

Because of the time scale of condensed-phase chemical events, any realistic theory of condensed-phase reaction dynamics must properly describe the Einstein limit. The Einstein oscillations, for example, are intimately related to the dynamical cage effects, which profoundly influence photodissociation dynamics in liquid solvents (for a computational illustration see Section VIII).

The Einstein limit is *not* properly described by the Langevin model. This is because the systematic force $-\beta_1 \dot{r}_0(t)$ acts on even the shortest time scale. Thus the effective gas-phase limit (no dissipation) is not reproduced.

Energy flow from the primary system to the heatbath results in eventual breakdown of the Einstein limit and damping of the Einstein oscillations. For the harmonic chain model, this breakdown is easily visualized as the flow of initial chain energy into the bulk chain. The energy flow may be broken down into stages. The initial stage, flow from chain atom 0 to chain atom 1, is governed by the Einstein frequency ω_{e0} and the *coupling constant* ω_{c1}^2 linking chain atoms 0 and 1. The next stage, flow from chain atom 1 to chain atom 2, is governed by the Einstein frequency ω_{e1} of chain atom 1 and the coupling constant ω_{c2}^2 linking atoms 1 and 2, and so on.

Thus systematic primary-system motion may be described from the short-time-scale (MTGLE) viewpoint as Einstein oscillations that are damped by stepwise energy flow through the heatbath. The stepwise flow, for the case of solute dynamics in a liquid, may be qualitatively interpreted as energy transfer from the solute to its first solvation shell, from the first shell to the second shell, and so on (see Fig. 4).

A key point is that only the initial stages of the flow, involving energy transfer within the local condensed-phase environment, importantly influence the primary-system dynamics. The detailed pattern of energy flow through distant regions of the heatbath is largely irrelevant.

NESTED-SHELL PICTURE

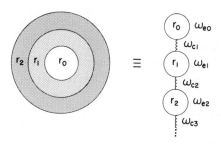

Fig. 4. Nested-solvation-shell interpretion of the MTGLE equivalent chain. A solute atom in a liquid with coordinate r_0 (left) is surrounded by its first, second,... solvation shells. The dynamics of the solute atom is rigorously equivalent to that of the MTGLE equivalent chain (right). Chain atom coordinates $r_1, r_2,...$ are qualitatively collective coordinates that describe the dynamics of the first, second,... solvation shells.

Equivalently the primary-system motion depends sensitively on the structure and dynamics of the local many-body environment, but is insensitive to the detailed nature of the bulk environment. The bulk environment, however, cannot be ignored, since it provides an energy reservoir for the primary system.

These considerations suggest that one may replace the huge number of degrees of freedom that represent the distant regions of the heatbath by a small number of effective degrees of freedom that model its energy-reservoir function. This type of heatbath modeling, which we develop in Section III.G, is the key to the simplification of problems in condensed-phase reaction theory.

F. The MTGLE Representation[21-24]

We next develop a rigorous representation of primary-system dynamics. A central result of this representation is an exact primary-system equation of motion [Eq. (3.37) below], which, like the phenomenological Langevin equation, is based on a decomposition of the heatbath force into systematic and random components. For this reason, the equation of motion is usually referred to as a *generalized* Langevin equation.

We present in this section an elementary development of the generalized Langevin representation based on the harmonic chain model. The representation, however, is in no way tied to the chain model, but rather is of great generality (Section V).

1. Basic Results

The present derivation is based on a projection-operator decomposition of the chain equation of motion, Eq. (3.10). We thus define $(N+1) \times (N+1)$

component projection matrices P and Q by

$$P = \begin{array}{cc} & \begin{array}{cc} P & Q \end{array} \\ \begin{array}{c} P \\ Q \end{array} & \left(\begin{array}{c|c} 1 & 0 \\ \hline 0 & 0 \end{array} \right) \end{array} \tag{3.32a}$$

and

$$Q = \begin{array}{cc} & \begin{array}{cc} P & Q \end{array} \\ \begin{array}{c} P \\ Q \end{array} & \left(\begin{array}{c|c} 0 & 0 \\ \hline 0 & 1 \end{array} \right) . \end{array} \tag{3.32b}$$

The matrix P projects onto chain atom 0 [1 is a 1×1 unit matrix in Eq. (3.32a)], while its orthogonal complement Q projects onto the heatbath atoms $1, 2, \ldots$ [1 is an $N \times N$ unit matrix in Eq. (3.32b)]. The matrices P, Q have the standard projection properties $P^2 = P$, $Q^2 = Q$, $PQ = 0$, and $P + Q = 1$. Using these projection properties, one may decompose Eq. (3.10) as follows:

$$\ddot{x}_P(t) = -\omega_{PP}^2 x_P(t) - \omega_{PQ}^2 x_Q(t) \tag{3.33a}$$

and

$$\ddot{x}_Q(t) = -\omega_{QQ}^2 x_Q(t) - \omega_{QP}^2 x_P(t) \tag{3.33b}$$

where $x_P = Pr$, $x_Q = Qr$, $\omega_{PP}^2 = P\omega^2 P$, and so on.

The breakup of the total heatbath force $-\omega_{PQ}^2 x_Q(t)$ into systematic and random components may be accomplished by solving Eq. (3.33b) for $x_Q(t)$ and then constructing $-\omega_{PQ}^2 x_Q(t)$. This yields the following result:

$$-\omega_{PQ}^2 x_Q(t) = -\int_0^t \omega_{PQ}^2 \cos \omega_{QQ}(t-\tau) \, \omega_{QP}^2 x_P(\tau) \, d\tau - \omega_{PQ}^2 x_Q^{(0)}(t) \tag{3.34}$$

where

$$\omega_{PQ}^2 x_Q^{(0)}(t) = \omega_{PQ}^2 \left[\cos(\omega_{QQ}t) \, x_Q + \omega_{QQ}^{-1} \sin(\omega_{QQ}t) \, \dot{x}_Q \right]. \tag{3.35}$$

Substituting Eq. (3.34) in Eq. (3.33a) yields the generalized Langevin equation

$$\ddot{x}_P(t) = -\omega_{PP}^2 x_P(t) - \int_0^t \omega_{PQ}^2 \cos \omega_{QQ}(t-\tau) \, \omega_{QP}^2 x_P(\tau) \, d\tau - \omega_{PQ}^2 x_Q^{(0)}(t). \tag{3.36}$$

The random component of the heatbath force is $\omega_{PQ}^2 x_Q^{(0)}(t)$. This is the force that the heatbath would exert if the primary system were clamped at equilibrium $[x_P(t) = 0$ in Eq. (3.36)].

The integral term in Eq. (3.36) is the delayed systematic component of the heatbath force. It is the component induced by response of the heatbath to primary-system motion. Thus the quantity $\omega_{PQ}^2 \cos \omega_{QQ} t\, \omega_{QP}^2$ may be interpreted as a response function for the heatbath.

Notice that Eq. (3.36), in contrast to Eq. (3.33a), is a closed equation of motion for $x_P(t)$. This is because $\omega_{PQ}^2 x_Q^{(0)}(t)$ is independent of the primary-system initial conditions $[x_Q^{(0)}(t)$ is the trajectory of the heatbath atoms conditional on the assumption that the primary system is clamped at equilibrium] and therefore functions as an external force acting on the primary system.

So far our development is valid for arbitrary harmonic systems: we have not yet used the tridiagonal property of the chain-model dynamical matrix ω^2 [Eq. (3.2a)]. Using Eq. (3.2a), Eq. (3.36) simplifies to

$$\ddot{r}_0(t) = -\omega_{e0}^2 r_0(t) + \omega_{c1}^4 \int_0^t \theta_1(t - \tau) r_0(\tau)\, d\tau + \omega_{c1}^2 R_1(t). \quad (3.37)$$

We next discuss the terms appearing in Eq. (3.37). The random force term, $\omega_{c1}^2 R_1(t)$, may now be interpreted as the force on chain atom 0 due to chain atom 1 conditional on the assumption that atom 0 is clamped at equilibrium $[r_0(t) = 0]$. Thus $R_1(t)$ is the trajectory of atom 1 in the clamped chain. The corresponding velocity $\dot{R}_1(t)$ is given by

$$\dot{R}_1(t) = [\cos \omega^{(1)} t]_{11} \dot{r}_1 - [\omega^{(1)} \sin \omega^{(1)} t]_{11} r_1$$
$$+ \sum_{p=2}^{N} \left([\cos \omega^{(1)} t]_{1p} \dot{r}_p - [\omega^{(1)} \sin \omega^{(1)} t]_{1p} r_p\right) \quad (3.38)$$

where $[\omega^{(1)}]^2$ is the dynamical matrix of the clamped chain given in Eq. (3.2b).

The heatbath response function $\dot{\theta}_1(t)$ appearing in Eq. (3.37) is given explicitly by

$$\dot{\theta}_1(t) = [\cos \omega^{(1)} t]_{11}. \quad (3.39)$$

Comparing Eqs. (3.38) and (3.39) shows that

$$\dot{\theta}_1(t) = \dot{R}_1(t) \quad (3.40a)$$

given the initial conditions

$$\dot{r}_1 = 1, \qquad r_1 = 0$$
$$\dot{r}_p = r_p = 0, \qquad p \geqslant 2. \tag{3.40b}$$

Thus $\theta_1(t)$ is a particular trajectory of atom 1 in the clamped chain.

Finally, using Eqs. (3.6), (3.38), and (3.39), one may show that $\dot{\theta}_1(t)$ is the normalized VAF corresponding to $R_1(t)$, that is,

$$\dot{\theta}_1(t) = \frac{m}{k_B T} \langle \dot{R}_1(t + \tau) \dot{R}_1(\tau) \rangle_{r_0}$$

$$= \frac{m}{k_B T} \langle \dot{R}_1(t) \dot{R}_1 \rangle_{r_0} \tag{3.41}$$

Comparison of Eqs. (3.11)–(3.14) with Eqs. (3.38)–(3.41) shows that $R_1(t)$ and $\dot{\theta}_1(t)$ play a role in the clamped chain analogous to that of $r_0(t)$ and $\dot{\chi}(t)$ in the full chain.

We shall refer to Eq. (3.37) as a molecular-time-scale generalized Langevin equation (MTGLE), because it emphasizes the short-time-scale aspects of condensed-phase dynamics.

The three force terms in the MTGLE may be interpreted as: (i) a systematic Einstein force $-\omega_{e0}^2 r_0(t)$, which governs the initial isolated-oscillator time dependence [Section IV.C] of the dissipative component of $\dot{r}_0(t)$; (ii) a systematic heatbath force $\omega_{c1}^4 \int_0^t \theta_1(t - \tau) r_0(\tau) \, d\tau$, which induces irreversible energy flow from the primary system to the heatbath, (iii) a random heatbath force $\omega_{c1}^2 R_1(t)$, which balances dissipation.

A key additional point is that the quantities governing the heatbath influence, $\theta_1(t)$ and $R_1(t)$, evolve in time according to *fictitious* (clamped chain) as opposed to physical (full chain) dynamics. This type of result is characteristic of the equations of modern irreversibility theory. The Mori random force terms (Section IV), for example, depend on quantities that evolve according to fictitious *projected* dynamics.

Equation (3.41), which is a statement that $\dot{\theta}_1(t)$ is the normalized VAF of $R_1(t)$, may alternatively be interpreted as a condition of balance between systematic $[\theta_1(t)]$ and random $[R_1(t)]$ heatbath forces. This condition guarantees that the primary system, if disturbed, will relax to thermal equilibrium. Equation (3.41) is therefore analogous to the phenomenological second fluctuation–dissipation theorem (3.25). We shall therefore also refer to Eq. (3.41) as a second fluctuation–dissipation theorem. The analogous relation for $\dot{\chi}(t)$, Eq. (3.14), is the *first* fluctuation–dissipation theorem.

The random force $\omega_{c1}^2 R_1(t)$ satisfies a second statistical constraint in addition to Eq. (3.41). This is the orthogonality relation

$$\langle R_1(t)\dot{r}_0 \rangle = 0 \tag{3.42}$$

which follows from Eqs. (3.5) and (3.38). Equation (3.42) is the rigorous analogue of the phenomenological relation (3.19b).

2. MTGLE *Hierarchy*

We next develop the hierarchy structure of the MTGLE theory. We use a simple argument based on the chain model. However, the hierarchy structure, like the basic results (3.37)–(3.42), is of general validity.

We develop the hierarchy structure by extending the projection process just discussed for the full chain to an infinite sequence of clamped chains. This sequence consists of chains governed by dynamical matrices $[\omega^{(1)}]^2$ (atom 0 clamped), $[\omega^{(2)}]^2$ (atoms 0 and 1 clamped), and so on.

We denote the coordinate of chain atom p in the pth clamped chain by $R_p(t)$. The successive projection processes yield a coupled sequence of MTGLEs for $R_1(t), R_2(t),\ldots$ analogous to Eq. (3.37). The sequence is

$$\ddot{R}_1(t) = -\omega_{e1}^2 R_1(t) + \omega_{c2}^4 \int_0^t \theta_2(t-\tau) R_1(\tau)\, d\tau + \omega_{c2}^2 R_2(t)$$

$$\vdots$$

$$\ddot{R}_{p-1}(t) = -\omega_{e\,p-1}^2 R_{p-1}(t) + \omega_{cp}^4 \int_0^t \theta_p(t-\tau) R_{p-1}(\tau)\, d\tau + \omega_{cp}^2 R_p(t)$$

$$\vdots \tag{3.43}$$

The response function $\dot{\theta}_p(t)$ is the normalized VAF of $R_p(t)$. Thus $\dot{\theta}_p(t)$ satisfies relationships analogous to those satisfied by $\dot{\chi}(t)$ and $\dot{\theta}_1(t)$. Namely [cf. Eqs. (3.39)–(3.41)],

$$\dot{\theta}_p(t) = [\cos \omega^{(p)} t]_{pp} \tag{3.44}$$

$$\dot{\theta}_p(t) = \dot{R}_p(t) \tag{3.45a}$$

given the following initial conditions for the chain atoms:

$$\dot{r}_p = 1, \qquad r_p = 0$$

$$\dot{r}_q = r_q = 0, \qquad q \geqslant p+1 \tag{3.45b}$$

and

$$\dot{\theta}_p(t) = \frac{m}{k_B T} \langle \dot{R}_p(t+\tau)\dot{R}_p(\tau) \rangle_{r_0, r_1, \ldots, r_{p-1}} = \frac{m}{k_B T} \langle \dot{R}_p(t)\dot{R}_p \rangle_{r_0, r_1, \ldots, r_{p-1}}$$

(3.46)

Equation (3.46) is the $(p+1)$th fluctuation–dissipation theorem, which guarantees thermal balance in the $(p-1)$th clamped chain. The pth heat-bath coordinate $R_p(t)$ satisfies one rigorous statistical constraint in addition to Eq. (3.46). This is the orthogonality relation

$$\langle R_p(t)\dot{R}_{p-1} \rangle_{r_0, r_1, \ldots, r_{p-2}} = 0,$$

(3.47)

which is the analogue of Eq. (3.42).

3.   The Adiabatic Limit and the Langevin Model

We next sketch the connection between the rigorous generalized Langevin theory and the Langevin model. This connection is made by integrating the kernel term in Eq. (3.37) by parts to yield

$$\ddot{r}_0(t) = -\Omega_0^2 r_0(t) - \int_0^t \beta_1(t-\tau)\dot{r}_0(\tau)\,d\tau + f_1(t).$$

(3.48)

The new quantities in Eq. (3.55) are the *friction kernel*

$$\beta_1(t) = \omega_{c1}^4 \int_t^\infty \theta_1(\tau)\,d\tau$$

(3.49)

the *adiabatic frequency*

$$\Omega_0^2 = \omega_{e0}^2 - \beta_1(t=0)$$

(3.50)

and the Mori[11] form for the random force,

$$f_1(t) = \omega_{c1}^2 R_1(t) - \beta_1(t)r_0$$

(3.51)

The Mori random force satisfies the following second fluctuation–dissipation and orthogonality theorems:[‡]

$$\langle f_1(\tau)f_1(\tau') \rangle = \frac{k_B T}{m}\beta_1(\tau-\tau')$$

(3.52a)

$$\langle f_1(t)\dot{r}_0 \rangle = \langle f_1(t)r_0 \rangle = 0$$

(3.52b)

Note that the Mori fluctuation–dissipation theorem (3.52a), in contrast to Eq. (3.41), involves a full rather than a clamped canonical average.

[‡]These may be derived from Eqs. (3.11) and (3.12) of Ref. 11a by taking $\mathbf{A} = (r_0, \dot{r}_0)$.

The Langevin model may now be recovered by assuming that $f_1(t)$ has the Gaussian white-noise statistical properties described in Section III. In particular, one makes the *Markovian approximation*

$$\beta_1(t) \simeq 2\beta_1 \delta(t) \tag{3.53}$$

where the *friction coefficient*,

$$\beta_1 = \lim_{z \to 0} \int_0^\infty \beta_1(t) e^{-zt} dt = \hat{\beta}_1(z=0) \tag{3.54}$$

is chosen so that $\int_0^\infty \beta_1(t)\,dt = \int_0^\infty 2\beta_1 \delta(t)$.

Given Eq. (3.53), Eq. (3.48) reduces to the Langevin equation (3.31) for a one-dimensional Brownian oscillator. Equation (3.52a) reduces to the associated phenomenological second fluctuation–dissipation theorem.

We next compare the two forms for the generalized Langevin equation, Eqs. (3.37) (MTGLE) and (3.48) (Mori). Compare, for example, the systematic restoring forces $-m\Omega_0^2 r_0(t)$ and $-m\omega_{e0}^2 r_0(t)$. The former quantity is the systematic restoring force exerted on the primary system given that it is displaced *infinitely slowly* [$\dot{r}_0(t) = 0$ in Eq. (3.48)]. This is why we call Ω_0 the adiabatic frequency. The latter quantity is the *instantaneous* systematic restoring force [$t = 0$ in Eq. (3.37)].

Equation (3.37) and (3.48) thus represent mathematically equivalent but physically distinct decompositions of the heatbath force into systematic and random components. Equation (3.37) emphasizes those features of primary-system dynamics that are important on time scales short compared to the relaxation time of the local heatbath density. These short-time-scale features are important for molecular motion during a chemical reaction. Equation (3.48), in contrast, emphasizes the adiabatic limit (heatbath density can nearly perfectly follow primary-system motion) of the many-body problem. The adiabatic limit describes the dynamics of macroscopic or semimacroscopic primary systems. Thus Mori-type generalized Langevin equation [see Eq. (4.134) below] permit one to connect rigorous dynamics with phenomenological transport equations[14] like the Langevin equation.

Finally we note that the adiabatic frequency must vanish for isolated solute motion in a liquid; that is,

$$\Omega_0 = 0 \qquad \text{(liquids)} \tag{3.55}$$

This is because solute motion in a liquid is diffusive rather than harmonically bound as $t \to \infty$.

Equation (3.55) may be proven as follows. Equations (3.14), (3.48), and (3.52b) yield the following equation of motion for $\dot{\chi}(t)$:

$$\ddot{\chi}(t) = -\Omega_0^2\chi(t) - \int_0^t \beta_1(t-\tau)\dot{\chi}(\tau)\,d\tau \qquad (3.56a)$$

The Laplace-space form of Eq. (3.56a) is

$$\hat{\chi}(z) = \left[z^2 + \Omega_0^2 + z\hat{\beta}_1(z)\right]^{-1} \qquad (3.56b)$$

Combining Eqs. (3.30) and (3.56b) yields the following result for the solute diffusion coefficient D_1:

$$D_1 = \frac{k_B T}{m} \lim_{z \to 0} \left\{ z\left[z^2 + \Omega_0^2 + z\hat{\beta}(z)\right]^{-1} \right\} \qquad (3.57)$$

Equation (3.57) shows that $D_1 = 0$ unless $\Omega_0 = 0$. This proves Eq. (3.55). Moreover, for $\Omega_0 = 0$, Eq. (3.57) is the Einstein relation (3.29).

4. The Equivalent Harmonic Chain

The dynamics of an arbitrary primary system moving in an arbitrary heatbath may be rigorously recast as the dynamics of the terminal atom 0 of an $(N+1)$-atom nearest-neighbor equivalent harmonic chain. This equivalent-chain representation is a basic result of the MTGLE theory.

Chain atoms $1, 2, \ldots, N$ constitute an equivalent heatbath for the primary system, which rigorously simulates the influence of the true many-body heatbath. The number of heatbath atoms, N, can equal $1, 2, \ldots$. Thus the equivalent-chain representation is actually an infinite class of representations.

The chain representation may be readily developed from the MTGLE hierarchy. First consider the case $N = 1$. *Define* the coordinate of chain atom 1 by

$$r_1(t) = \omega_{c1}^2 \int_0^t \theta_1(t-\tau)r_0(\tau)\,d\tau + R_1(t) \qquad (3.58a)$$

The MTGLE, Eq. (3.37), may be rewritten using Eq. (3.58a) as

$$\ddot{r}_0(t) = -\omega_{e0}^2 r_0(t) + \omega_{c1}^2 r_1(t). \qquad (3.59a)$$

Using Eq. (3.43) one may show (see Section V) that $r_1(t)$ satisfies the following equation of motion:

$$\ddot{r}_1(t) = \omega_{c1}^2 r_0(t) - \omega_{e1}^2 r_1(t) + \omega_{c2}^4 \int_0^t \theta_2(t-\tau)r_1(\tau)\,d\tau + \omega_{c2}^2 R_2(t). \qquad (3.59b)$$

Equations (3.59) constitute the two-atom equivalent-chain representation.

The general $(N+1)$-atom representation may be similarily derived. Define the coordinate of the general qth chain atom, $r_q(t)$, by [cf. Eq. (3.58a)]

$$r_q(t) = \omega_{cq}^2 \int_0^t \theta_q(t-\tau) r_{q-1}(\tau) \, d\tau + R_q(t). \qquad (3.58b)$$

Equations (3.58) provide a recursive definition of the chain coordinates $r_1(t), r_2(t),\ldots$ in terms of the physical coordinate $r_0(t)$. Equations (3.37), (3.43), and (3.58) then yield (for details see Section V) the following $(N+1)$-atom chain representation:

$$\ddot{r}_0(t) = -\omega_{e0}^2 r_0(t) + \omega_{c1}^2 r_1(t)$$

$$\ddot{r}_1(t) = \omega_{c1}^2 r_0(t) - \omega_{e1}^2 r_1(t) + \omega_{c2}^2 r_2(t)$$

$$\vdots \qquad\qquad\qquad (3.60)$$

$$\ddot{r}_N(t) = \omega_{cN}^2 r_N(t) - \omega_{eN}^2 r_N(t) + \omega_{cN+1}^2$$
$$\times \int_0^t \theta_{N+1}(t-\tau) r_N(\tau) \, d\tau + \omega_{cN+1}^2 R_{N+1}(t)$$

The equivalent chain [Eq. (3.60] has a number of rigorous formal properties that are identical to properties of physical chains discussed earlier. These include the following: (i) The VAF $\dot{\chi}(t)$ is the particular dissipative velocity of chain atom 0, $\dot{r}_0(t)$, which evolves from the initial conditions $\dot{r}_0 = 1$, $r_0 = 0$; that is,

$$\dot{\chi}(t) = \dot{r}_0(t) \qquad\qquad (3.61a)$$

given the conditions

$$\dot{r}_0 = 1, \qquad\qquad r_0 = 0$$
$$\dot{r}_q = r_q = 0, \qquad q = 1, 2, \ldots, p$$

and the additional condition

$$R_{N+1}(t) = 0. \qquad\qquad (3.61b)$$

(ii) The qth heatbath variable $R_q(t)$ $(1 \leqslant q \leqslant N)$ is the trajectory of chain atom q conditional on the assumption that chain atoms $0, 1, \ldots, q-1$ are clamped at equilibrium. (iii) The heatbath response function $\theta_q(t)$ is, correspondingly, the particular dissipative velocity $\dot{R}_q(t)$ that evolves from the initial state $\dot{R}_q = 1$, $R_q = 0$. (iv) The binary statistical properties of the full chain are identical in form to those given in Eqs. (3.5b, c, d); the corresponding statistical properties of the clamped chain are identical to those given in Eqs. (3.6b, c, d).

Finally we note that by passing to the limit of an infinite chain, $N \to \infty$ in Eq. (3.60), one sees that *all* information about heatbath influence on primary system dynamics is carried by the MTGLE parameters $\{\omega_{ep}^2, \omega_{cp+1}^2\}$, $p = 0, 1, \ldots, \infty$.

We have not yet shown how these parameters may be determined for arbitrary (i.e., nonchain) many-body systems. The parameters, in fact, may be constructed for general systems from the primary-system normalized VAF $\dot{\chi}(t)$. This is reasonable given Eqs. (3.60) and (3.61). An algorithm for constructing the parameters is given in Appendix A.

G. Heatbath Modeling

The key to simplification of problems in condensed-phase reaction dynamics (Section III.) is to treat the influence of the local condensed-phase environment on the reaction in detail, while modeling the influence of the distant environment by that of a bulk energy reservoir.

This type of heatbath modeling may be systematically implemented within the equivalent-chain formalism. The key point is that *qualitatively* chain atom 1 simulates the influence of nearest-neighbor heatbath molecules, chain atom 2 simulates next-nearest-neighbor influence, and so on.

For solute motion in a liquid, for example, the fictitious chain atom coordinates $r_1(t), r_2(t), \ldots, r_p(t)$ may be qualitatively interpreted as collective variables describing the dynamics of the first, second, $\ldots, p$th solvation shells (see Fig. 4).[‡]

The heatbath variables $R_p(t)$ have a corresponding rough interpretation based on the fact that $R_p(t)$ is the trajectory of chain atom p in the pth clamped equivalent chain. Specifically $R_p(t)$ is interpretable as a collective variable describing temporal density fluctuations in the pth shell given that the solute is clamped in the liquid and that the density of the first, second, $\ldots, (p-1)$th shells are fixed at their equilibrium values.

The collective variable $R_p(t)$ rigorously evolves according to generalized Langevin dynamics [see Eq. (3.43)]. For sufficiently large p, however, little error in the *solute* dynamics is incurred if fluctuations in the pth shell are modeled by Brownian-oscillator dynamics. Thus we adopt the following

---

[‡]The interpretation of the coordinates $r_1(t), r_2(t), \ldots$ as collective variables describing solvation-shell dynamics is only approximate for solute motion in liquids. It is exact, however, for solutes linearly coupled to perfectly harmonic "solvents" with only nearest-neighbor interactions.

model equation of motion [cf. Eq. (3.31)] for $R_p(t)$:

$$\ddot{R}_p(t) = -\Omega_p^2 R_p(t) - \beta_{p+1}\dot{R}_p(t) + f_{p+1}(t). \qquad (3.62)$$

The random force $f_{p+1}(t)$ is assumed to have standard (Section III.) Langevin-model statistical properties, for example

$$\langle f_{p+1}(\tau) f_{p+1}(\tau') \rangle = \frac{2k_B T}{m} \beta_{p+1}\delta(\tau - \tau'). \qquad (3.63)$$

Utilization of Eq. (3.62) for liquids roughly corresponds to modeling the solvent external to the pth shell by a bulk thermal reservoir.

The parameters Ω_p (adiabatic frequency of the pth shell) and β_{p+1} (friction coefficient of the pth shell) are constructible from the primary-system VAF via an algorithm sketched in Appendix A.

Equation (3.62) is a model for the time development of chain atom p in the pth clamped chain. The analogous model for the time development of chain atom p in the full chain is $[R_p(t) \to r_p(t)]$

$$\ddot{r}_p(t) = \omega_{cp}^2 r_{p-1}(t) - \Omega_p^2 r_p(t) - \beta_{p+1}\dot{r}_p(t) + f_{p+1}(t). \qquad (3.64)$$

Given Eq. (3.64), the rigorous equivalent-chain equations (3.60) are approximated by the following set of model equations:

$$\ddot{r}_0(t) = -\omega_{e0}^2 r_0(t) + \omega_{c1}^2 r_1(t)$$
$$\ddot{r}_1(t) = \omega_{c1}^2 r_0(t) - \omega_{e1}^2 r_1(t) + \omega_{c2}^2 r_0(t) \qquad (3.65)$$
$$\vdots$$
$$\ddot{r}_p(t) = \omega_{cp}^2 r_{p-1}(t) - \Omega_p^2 r_p(t) - \beta_{p+1}\dot{r}_p(t) + f_{p+1}(t).$$

These equations require several comments.

1. The equations provide an effective few-body $[(p+1)$-atom] model for the original many-body problem of primary-system motion. The value of this model is that it permits one to reduce problems in condensed-phase reaction dynamics to manageable classical stochastic trajectory calculations (see Sections III.I and VIII below).

2. Even the simplest useful model ($p = 1$, a two-atom chain) simulates the chemically important short-time-scale aspects of the many-body problem in a qualitatively correct manner. The Einstein limit, for example, is precisely reproduced. For solute motion in a liquid, the

$p = 1$ approximant roughly corresponds to the single-solvation-shell solvent model depicted in Fig. 2d. This model can correctly describe the correlated solute–cage motions that are responsible for the oscillatory character of the VAF $\dot{\chi}(t)$ of a solvated ion. Thus ion dynamics in a polar solvent may be usefully represented by a two-atom MTGLE chain model. (This has been shown for Na$^+$ motion in the solvent pyridine by Edgell and Balk.[33]) Moreover, the two-atom chain model accounts for the dynamical cage effects, missing in the phenomenological Langevin equation, which profoundly influence reagent motion during liquid-solution reactions. These considerations suggest that low-order truncations of the equivalent chain, which permit one to simulate condensed-phase reactions via relatively small-scale stochastic classical trajectory calculations, can provide a chemically useful reduction of the many-body problem. Moreover, a model may be systematically improved by increasing p. For example, the two-atom chain model permits one to reduce the problem of molecular-iodine recombination dynamics in a liquid solvent[34] to an effective *four-body* classical trajectory problem arising from the collision of a pair of two-atom chain models [see Section III.I].

3. The truncated chain models *exactly* simulate the adiabatic, as well as the short-time-scale, limit of primary-system motion if one chooses the truncation parameters, Ω_p^2 and β_{p+1}, according to the prescription of Appendix A. Choosing the truncation parameters via this prescription guarantees, for the case of solute motion in a liquid, that the chain models diffuse, $\Omega_0 = 0$, with correct diffusion coefficient $D_1 = (k_B T/m) \lim_{z \to 0} \hat{\chi}(z)$ on a coarse-grained time scale while simulating the coupled atom–solvation-shell oscillatory motion on a subpicosecond time scale.

We next discuss convergence in p of the MTGLE chain models. The most direct way to examine convergence is to perform simulations of a chemical process [e.g. I_2 recombination, successively using chain models with two, three,... atoms, and then to check the convergence of computed dynamical attributes such as gas–solid sticking probabilities and liquid-state rate constants. Convergence studies of this type have been discussed elsewhere[22(b)]; the chain approximants are usually found to converge rapidly.

An alternative and simpler approach is to compare the normalized VAF $\dot{\chi}_p(t)$ for a $(p+1)$-atom truncated-chain model with the corresponding exact normalized VAF $\dot{\chi}(t)$. The chain-model VAF $\dot{\chi}_p(t)$ may be computed as a particular velocity of chain atom 0 in the *truncated* chain [Eq. (3.65)], just as $\dot{\chi}(t)$ may be computed as the corresponding velocity in the equivalent chain [Eq. (3.60)]. Explicitly [cf. Eq. (3.61)],

$$\dot{\chi}_p(t) = \dot{r}_0(t) \tag{3.66a}$$

given the conditions

$$\dot{r}_0 = 1, \qquad r_0 = 0$$
$$\dot{r}_q = r_q = 0, \qquad q = 1, 2, \ldots, p \tag{3.66b}$$
$$f_{p+1}(t) = 0.$$

Comparisons of $\dot{\chi}_p(t)$ with the corresponding exact atomic VAF $\dot{\chi}(t)$ are made for two model condensed-phase systems in Fig. 5a and b. The MTGLE parameters $\{\omega_{ep}^2, \omega_{cp+1}^2\}$ and the truncation parameters $\{\Omega_p^2, \beta_{p+1}\}$ used to construct the chain models are listed elsewhere.[35]

The key points concerning Fig. 5 are: (i) A two-atom chain model is able to provide a reasonable approximant for atomic dynamics in an argonlike (Lennard–Jones type) liquid. (ii) Few atom-chain models are able to reproduce the complex oscillatory VAF's that describe atomic motion in solids.

H. Atom–Chain Collisions: A Prototype for Problems in Condensed-Phase Chemical Dynamics

Our discussion, so far, has focused on: (i) the basic physics of ultrashort-time-scale condensed-phase dynamics; (ii) how this basic physics may be formulated mathematically via the equivalent-harmonic-chain representation; (iii) how this basic physics may be exploited via the chain modeling methods to reduce the problem of atomic motion in a condensed-phase environment to an effective few-body problem. We next begin to connect these general concepts in condensed-phase dynamics with the more specific problem of condensed-phase chemical-*reaction* dynamics.

Our development of general concepts in condensed-phase dynamics has been facilitated by an analysis of a prototype many-body system, the $(N + 1)$-atom ($N \to \infty$) harmonic-chain model. We next analyze another model problem, the classical collinear inelastic scattering of a gas atom with mass M and coordinate $R(t)$ off the $(N + 1)$-atom harmonic chain (see Fig. 6). This problem, studied in an important early paper by Zwanzig[36] as a model for atomic collisions at the gas–solid interface, is a prototype problem for the development of concepts in condensed-phase reaction dynamics, just as the isolated chain model is a prototype problem for development of general concepts in condensed-phase dynamics. The atom–chain collision problem will permit us to transparently illustrate fundamental points of great

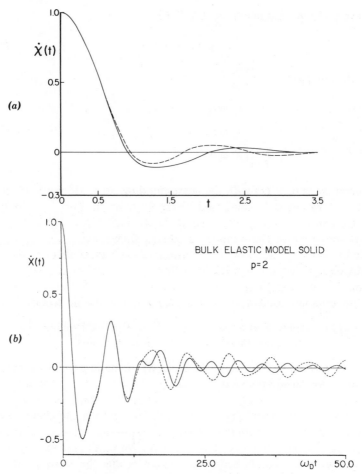

Fig. 5. Equivalent-chain modeling of normalized velocity autocorrelation functions $\dot{\chi}(t)$. (a) The "exact" normalized VAF $\dot{\chi}(t)$ (solid curve) for an argon atom in Levesque–Verlet[8] liquid argon in the thermodynamic state $T^* = 0.76$, $\rho^* = 0.85$ (Lennard–Jones units) is compared with its two-atom ($p = 1$) truncated-chain approximant (dashed curve). (b) The exact normalized VAF $\dot{\chi}(t)$ (short-dashed curve) for an atom moving in a bulk elastic model solid[22(a)] is compared with its three-atom ($p = 2$) truncated-chain approximant (solid curve).

generality concerning the physics of condensed-phase energy transfer and reaction processes. The generality is, of course, a consequence of the existence of the equivalent-harmonic-chain representation.

We assume, for simplicity, that the gas atom is *directly* coupled only to the terminal chain atom 0; that is, we assume the atom–chain potential-energy function is of the form $U(r_0, R)$. The primary system is then the gas

ATOM–CHAIN COLLISIONS

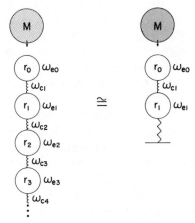

Fig. 6.   Atom–chain inelastic collisions: a prototype problem for condensed-phase reaction theory. An atom of mass M impinges on a collinear semiinfinite harmonic chain. The many-body collision problem (left) is approximately equivalent to an effective three-body problem (right), which corresponds to Eqs. (3.68) for the case $p = 1$.

atom and chain atom 0. The heatbath is the remainder of the chain. Let the incident gas energy be E_i, and let the Kelvin temperature of the chain be T. The problem is then to calculate the PDF $P[E_f, E_i; T]$ for the final gas-atom energy E_f.

The exact many-body classical equations of motion for the atom–chain inelastic-collision problem are

$$\ddot{R}(t) = - M^{-1} \frac{\partial U(r_0, R)}{\partial R} \tag{3.67a}$$

and

$$\ddot{r}_0(t) = - m^{-1} \frac{\partial U[r_0, R]}{\partial r_0} - \omega_{e0}^2 r_0(t) + \omega_{c1}^2 r_1(t)$$

$$\ddot{r}_1(t) = - \omega_{e1}^2 r_1(t) + \omega_{c1}^2 r_0(t) + \omega_{c2}^2 r_2(t) \tag{3.67b}$$

$$\vdots$$

$$\ddot{r}_N(t) = - \omega_{eN}^2 r_N(t) + \omega_{cN}^2 r_{N-1}(t)$$

We first consider the atom–chain collision problem for the case of $T = 0$ K scattering (see footnote in Section III.B). This case illustrates the physics of the many-body problem in its most transparent form, since stochastic "contamination" of the basic dynamics is suppressed at $T = 0$ K.

The importance of short-time-scale chain dynamics, in particular the magnitudes of the force constants $m\omega_{e0}^2$ and $m\omega_{c1}^2$, for the energy transfer process is evident for the $T = 0$ K trajectories. The Einstein frequency ω_{e0} governs the efficiency of *initial* energy transfer from the gas atom to the chain. Increasing ω_{e0} (i.e., increasing the initial chain "stiffness"), for example, decreases the initial energy-transfer efficiency. This efficiency completely determines the collision dynamics in the limit of very high incident gas energy E_i. For this limit the collision is completed before the transferred energy can flow out of chain atom 0 and penetrate the heatbath. Thus in the limit of large E_i, an Einstein-oscillator model for the chain with frequency ω_{e0} is sufficient to describe the energy-transfer process, and the many-body collision problem simplifies to a gas-phase (two-body) problem.

At lower values of E_i, transferred energy penetrates the heatbath before the collision is completed. The penetration efficiency increases with increasing coupling constant ω_{c1}^2. Once the transferred energy has penetrated the heatbath, it is dissipated by stepwise flow through chain atoms $1, 2, \ldots$.

Penetration and subsequent dissipation of energy deposited into the chain increases the atom–chain energy transfer efficiency over that predicted by the Einstein model for the chain. Moreover, energy dissipation permits gas-atom adsorption for sufficiently low E_i. Adsorption is a qualitative effect of the many-body nature of the chain that is absent for the Einstein model.

The dissipative $T = 0$ K energy-transfer dynamics just described is modified by the stochastic effect at finite temperature. One consequence of the stochastic effect is that energy is transferred from the chain to the gas atom in certain trajectories.

We have just discussed the basic physics of atom–chain collisions. (Because of the generality of the equivalent-chain representation, our description also applies to the basic physics of H_2 collisions with the surface of a single crystal of metallic copper, adsorption of CO by the iron atom in the protein hemoglobin, etc.) We next discuss reduction of the many-body problem to an effective few-body problem. The basic idea, first discussed in Section III.E, is that the details of energy flow through distant chain atoms is unimportant for the collision dynamics. Thus one may model the chain as in Eq. (3.65). Given this modeling of the chain, Eqs. (3.67) are approximated by the following set of few-body [$(p + 2)$-atom] effective equations of motion:

$$\ddot{R}(t) = - M^{-1} \frac{\partial U(r_0, R)}{\partial R} \tag{3.68a}$$

and

$$\ddot{r}_0(t) = -m^{-1}\frac{\partial U[r_0, R]}{\partial r_0} - \omega_{e0}^2 r_0(t) + \omega_{c1}^2 r_1(t)$$

$$\ddot{r}_1(t) = -\omega_{e1}^2 r_1(t) + \omega_{c1}^2 r_0(t) + \omega_{c2}^2 r_2(t)$$

$$\vdots$$

$$\ddot{r}_p(t) = -\Omega_p^2 r_p(t) - \beta_{p+1}\dot{r}_p(t) + \omega_{cp}^2 r_{p-1}(t) + f_{p+1}(t)$$

(3.68b)

Eqs. (3.68) may be solved[22(b)] by Monte Carlo classical stochastic trajectory methods for a given gas-atom incident energy E_i and chain Kelvin temperature T.

The initial conditions for chain atoms $0, 1, \ldots, p$ may be sampled from the effective Boltzmann distribution [cf. Eq. (3.3)]

$$f_{CA}^{eff}[p_0, \ldots p_p; r_0, \ldots r_p] = [Z^{eff}(\beta)]^{-1}\exp[-\beta H^{eff}]$$ (3.69)

where $Z^{eff}(\beta) = \int dp_0 \ldots dp_p\, dr_0 \ldots dr_p \exp[-\beta H^{eff}]$, and where H^{eff} is given by

$$H^{eff} = \sum_{\lambda=0}^{p} \frac{p_\lambda^2}{2m} + \tfrac{1}{2}m \sum_{\lambda,\mu=0}^{p} r_\lambda[\omega_{eff}^2]_{\lambda\mu} r_\mu.$$ (3.70)

The dynamical matrix of the truncated chain is given by [cf. Eq. (3.2a)]

$$\omega_{eff}^2 = \begin{bmatrix}
\omega_{e0}^2 & -\omega_{c1}^2 & 0 & 0 & \cdots & 0 & 0 \\
-\omega_{c1}^2 & \omega_{e1}^2 & -\omega_{c2}^2 & 0 & \cdots & 0 & 0 \\
0 & -\omega_{c2}^2 & \omega_{e2}^2 & -\omega_{c3}^2 & \cdots & 0 & 0 \\
0 & 0 & -\omega_{c3}^2 & \omega_{e3}^2 & \cdots & 0 & 0 \\
\vdots & \vdots & \vdots & \vdots & & \vdots & \vdots \\
0 & 0 & 0 & 0 & \cdots & -\omega_{cp}^2 & \Omega_p^2
\end{bmatrix}.$$ (3.71)

The random force $f_{p+1}(t)$ may be Monte Carlo samples using techniques[37] based on the Gaussian-white-noise statistical properties

$$\langle f_{p+1}(t+\tau)f_{p+1}(\tau)\rangle = \frac{2k_B T}{m}\beta_1\delta(t)$$ (3.72a)

$$\langle f_{p+1}(t)\rangle = 0$$ (3.72b)

$$P[f_{p+1}(t)] \sim \exp\left[-\frac{1}{2}\frac{f_{p+1}^2(t)}{\langle f_{p+1}^2\rangle}\right]$$ (3.72c)

Using the trajectory method just outlined, the required PDF describing the energy-transfer process, $P[E_f, E_i, T]$, may be constructed. This Monte Carlo trajectory construction of $P[E_f, E_i; T]$ is a more sophisticated analogue of the Langevin-model stochastic calculation outlined in Section III.D.

The key point is that the tractable effective few-body collision problem described by Eqs. (3.68)–(3.72) simulates the essential features of the intractable many-body atom–chain problem.

Finally, the reason that we have discussed atom–chain collisions in such detail is that they provide a paradigm for general methods of solution of many-body problems in condensed-phase reaction theory.

I. Model Treatment of Atomic Recombination Dynamics in Liquid Solution

The truncated-chain Equations (3.65) provide an approximate representation of the dynamics of an *isolated* solute atom that, roughly speaking, explicitly accounts for the influence of the first p solvation shells on the motion of the atom. If one assume that these *isolated*-solute-atom chain equations may be used to treat the dynamics of *interacting* solute atoms, then oversimplified but instructive model treatments of liquid-solution reactions may be developed. These model treatments ignore effects arising from modification of the isolated-atom solvation shells due to shell overlap and common-cage formation (see Fig. 7). The incorporation of common-cage effects into the MTGLE formalism will be discussed in Sections III.J and VI.

Consider, for example, the liquid-state recombination dynamics of a pair of identical solute atoms of mass m with coordinates $r_{01}(t)$ and $r_{02}(t)$. Denote the interatomic separation by $R(t) = |r_{02}(t) - r_{01}(t)|$, and the interatomic potential by $V[R(t)]$. We shall model the dynamics of a solute atom via a two-atom truncated chain [Eq. (3.65) with $p = 1$]. Recall that this roughly corresponds to the single-solvation-shell model of Fig. 2d. Denote the coordinate of the solvation shell surrounding atom 1 by $r_{11}(t)$, and the coordinate of the solvation shell surrounding atom 2 by $r_{12}(t)$. Then within our model the equations of motion for the recombination dynamics are

$$\ddot{r}_{01}(t) = -m^{-1}\frac{\partial V[R(t)]}{\partial r_{01}(t)} - \omega_{e0}^2 r_{01}(t) + \omega_{c1}^2 r_{11}(t)$$

$$\ddot{r}_{11}(t) = \omega_{c1}^2 r_{01}(t) - \Omega_1^2 r_{11}(t) - \beta_2 \dot{r}_{11}(t) + f_{21}(t)$$

(3.73a)

and

$$\ddot{r}_{02}(t) = -m^{-1}\frac{\partial V[R(t)]}{\partial r_{02}(t)} - \omega_{e0}^2 r_{02}(t) + \omega_{c1}^2 r_{12}(t)$$

$$\ddot{r}_{12}(t) = \omega_{c1}^2 r_{02}(t) - \Omega_1^2 r_{12}(t) - \beta_2 \dot{r}_{12}(t) + f_{22}(t)$$

(3.73b)

COMMON CAGE EFFECT

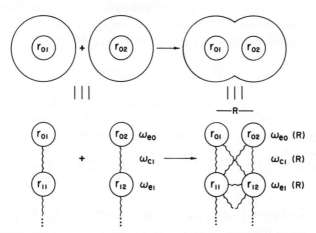

Fig. 7. Equivalent-chain representation of the common-cage effect. Isolated solute atoms (top left) approach to an internuclear distance R to form a commonly caged encounter complex. The dynamics of the isolated solute atoms are rigorously equivalent to noninteracting MTGLE chains (bottom left) characterized by frequencies $\omega_{e0}, \omega_{c1}, \omega_{e1}, \ldots$. A commonly caged solute pair is equivalent within the clamping approximation [see Sections III.J and VI] to matrix R-dependent MTGLE chains (bottom right) characterized by matrix R-dependent force constants $\omega_{e0}^2(R), \omega_{c1}^2(R)$.

The chain truncation parameters Ω_1^2 and β_2 are determined by specializing the results of Appendix A, Eqs. (A.8) and (A.14), to the case of a liquid (i.e., $\Omega_0 = 0$):

$$\Omega_1^2 = \frac{\omega_{c1}^4}{\omega_{e0}^2} \tag{3.74a}$$

and

$$\beta_2 = \frac{k_B T}{m D_1} \left(\frac{\omega_{c1}}{\omega_{e0}} \right)^4 \tag{3.74b}$$

Equations (3.74) show that the model equations of motion, Eqs. (3.73), are completely specified by the basic MTGLE parameters ω_{e0}^2 and ω_{c1}^4, by the solute diffusion coefficient D_1, and by the interatomic potential $V(R)$. The Langevin model for solute-recombination dynamics is, in contrast, specified for a given $V(R)$ solely by the friction coefficient β_1, or equivalently by the diffusion coefficient $D_1 = k_B T / m \beta_1$.

In order to illustrate that short-time-scale cage effects, absent in the Langevin model, influence solute recombination dynamics, Monte Carlo stochastic classical trajectory calculations based on Eqs. (3.73) were performed for several values of ω_{e0}^2 and ω_{c1}^4 with D_1 *fixed* at

$$D_1 = 5 \times 10^{-5} \text{ cm}^2/\text{sec} \qquad (3.75)$$

This value of D_1 provides a reasonable estimate of typical iodine-atom diffusion rates in hydrocarbon solvents at room temperature. The basic MTGLE parameters ω_{e0}^2 and ω_{c1}^2 were then varied about the reasonable reference values of $10^{26}/\text{sec}^2$. The atomic mass was chosen as $m = 2.11 \times 10^{-22}$ g, the mass of an iodine atom. The potential $V(R)$ was chosen to be of the Morse form

$$V(R) = D\{\exp[-2\alpha(R-b)] - 2\exp[-\alpha(R-b)]\} \qquad (3.76)$$

with parameters ($\alpha = 1.867$ Å^{-1}, $b = 2.67$ Å, and $D = 60k_BT$ with $T = 300$ K) chosen to simulate the interaction of two iodine atoms in the gas phase. (Use of a gas-phase potential, rather than a liquid-state potential of mean force, is consistent with neglect to the common-cage effect.) The initial conditions of the four chain atoms and the values of the random forces $f_{21}(t)$ and $f_{22}(t)$ were chosen by Monte Carlo methods [see Section III.H] to simulate a liquid solution at $T = 300$ K.

The recombining solute atoms were started at $R = 7$ Å. The radicals were assumed to have reacted at first contact, $R = 3.7$ Å. Typical reactive stochastic trajectories are plotted in Fig. 8. Notice that these trajectories display, as expected, molecular-time-scale cage oscillations superimposed on motion that is of diffusive character on a coarse-grained time scale.

The probability of finding an unreacted pair at time t, $P(t)$, is plotted for three of the simulation cases in Fig. 9. These results illustrate the general finding that for fixed D_1 and fixed $V(R)$:

1. Increasing ω_{e0}^2 at constant ω_{c1}^2 decreases the reaction rate.
2. Increasing ω_{c1}^2 for fixed ω_{e0}^2 increases the reaction rate.

The interpretation of these results is as follows. If ω_{e0}^2 increases, then the restoring force confining a solute atom to its cage increases. Thus the pair must approach more closely before the attractive potential $V(R)$ can overcome the cage forces. Thus an increase in ω_{e0}^2 leads to a decrease in the reaction rate.

If the atom–cage coupling constant is increased, then the cage is more strongly "pulled" by solute-atom displacements. This weakens the effective cage restoring force and hence speeds up the reaction rate.

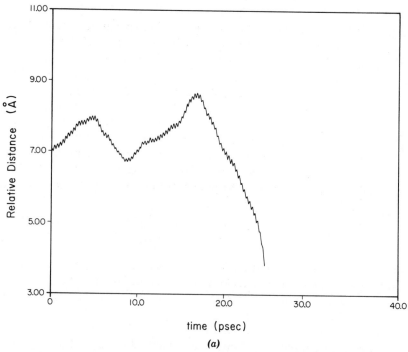

Fig. 8. Typical free-radical trajectories for model simulations of I_2 recombination dynamics. Relative I_2 separation (Å) is plotted vs. time (psec). Basic MTGLE parameters are (a) $\omega_{e0} = 2 \times 10^{13}$ sec^{-1}, $\omega_{c1} = 10^{13}$ sec^{-1}; (b) $\omega_{e0} = 10^{13}$ sec^{-1}, $\omega_{c1} = 10^{13}$ sec^{-1}. For both cases the truncation parameters Ω_1^2 and β_2 are determined from Eqs. (3.74) with $D_1 = 5 \times 10^{-5}$ cm^2/sec. Morse-potential parameters are given in the text after Eq. (3.76). Note that the trajectories show short-time-scale solute cage motions superimposed on a more coarse-grained diffusive motion.

The above discussion illustrates an important general point. The chain formalism helps one to comprehend, as well as compute, the many-body influence on reagent dynamics. The chain parameters, in particular, are determined by the structure and dynamics of the solvent cage. The basic parameters ω_{e0}^2 and ω_{c1}^4, for example, are determined by the solvent cage *equilibrium* structure via Eqs. (7.4)–(7.5) below. In fact one may build models for the full solvent dynamical response that depend only on the equilibrium cage structure [see Eq. (7.16) below]. Thus the chain formalism permits one to utilize physical intuition about the behavior of mechanical systems to qualitatively correlate changes in reagent dynamics with changes in solvation-shell structure. This is a valuable result, since a basic goal of liquid-state reaction theory is to predict how chemical rates and detailed dy-

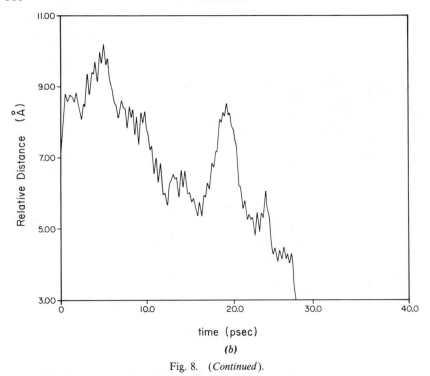

time (psec)

(b)

Fig. 8. (*Continued*).

namics vary with changes in solvation-shell structure as solvent thermodynamic state, solvent composition, and so on vary. This feature of the chain
formalism will be illustrated for a realistic treatment of atomic recombination dynamics in Section VIII.

Summarizing, our model treatment of atomic recombination dynamics in
liquids suggests that reaction rates depend not only on diffusive motion on a
coarse-grained time scale, as predicted by the Langevin model, but also on
correlated solute-shell molecular-time-scale motions, which depend on the
MTGLE parameters ω_{e0}^2 and ω_{c1}^4.

A much more realistic MTGLE trajectory study of atomic-recombination
dynamics, which accounts for common-cage effects, will be presented in
Section VIII. This study shows that the molecular-time-scale cage effects,
absent in Langevin dynamics, play an essential role in determining the dynamics of atomic recombination and subsequent vibrational-energy relaxation.

PROBABILITY P(t) OF AN UNREACTED PAIR

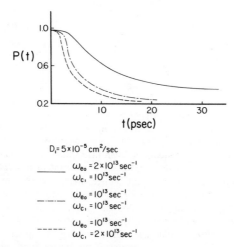

$D_i = 5 \times 10^{-5}$ cm^2/sec

——— $\omega_{e_0} = 2 \times 10^{13}$ sec^{-1}
$\omega_{c_1} = 10^{13}$ sec^{-1}

—·—·— $\omega_{e_0} = 10^{13}$ sec^{-1}
$\omega_{c_1} = 10^{13}$ sec^{-1}

— — — $\omega_{e_0} = 10^{13}$ sec^{-1}
$\omega_{c_1} = 2 \times 10^{13}$ sec^{-1}

Fig. 9. Probability $P(t)$ of an unreacted pair vs. time t (psec) for model simulations of I_2 recombination dynamics. Note that decreasing the cage restoring forces (by either decreasing ω_{e0} or increasing ω_{c1}) increases the reaction rate.

J.   The Clamping Approximation and the Common-Cage Effect

We have just discussed an oversimplified treatment of liquid-state recombination dynamics. This treatment ignores the effects of solvation-shell overlap and common-cage formation, important at small interatomic separations R.

We close Section III with an abbreviated description of how the effects of R-dependent caging may be introduced into the MTGLE formalism. A full quantitative treatment of reagent-configuration-dependent caging, of which R-dependent caging is a special case, will be given in Section VI. This treatment is based on linear-response theory, which will be reviewed in Section IV.

The theory developed in Section VI is based on the following physical facts:

1. The systematic and the random solvent forces exerted on an isolated solute atom have simple approximate physical interpretations based on the concept of solvent density response. This point was mentioned at the close of Section III.B.
2. The spherical local solvent density that cages an isolated solute atom is modified by the approach of a reagent to an "ellipsoidal" R-depen-

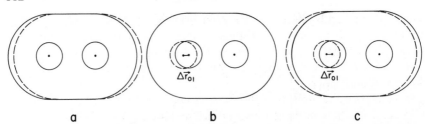

Fig. 10. Density-response model for the systematic and random solvent forces exerted on interacting solute atoms. (*a*) Solute atoms of mass m (small circles) are initially clamped at configuration point $r_0 = (\mathbf{r}_{01}, \mathbf{r}_{02})$. The equilibrium mean-field solvent density $\rho^{(3)}(\mathbf{q}; r_0)$ (solid "ellipse") exerts a mean force $\langle \mathscr{F} \rangle_{r_0} = (\langle \mathscr{F}_1 \rangle_{r_0} \langle \mathscr{F}_2 \rangle_{r_0})$ on the solute atoms. Fluctuations (dashed "ellipse") of the local solvent density about its mean-field value $\rho^{(3)}(\mathbf{q}; r_0)$ give rise to a random force $\widetilde{\mathscr{F}}_0(t) = \mathscr{F}_0(t) - \langle \mathscr{F} \rangle_{r_0} \equiv \omega_{c1}^2(r_0) R_1(t; r_0)$ acting on the interacting solute atoms. (*b*) Solute atom 1, initially located at point $\mathbf{r}_{01}$ (small dashed circle), is slightly displaced to point $\mathbf{r}_{01} + \Delta \mathbf{r}_{01}$ (small solid circle). The initial mean-field solvent density exerts an instantaneous systematic restoring force $\mathscr{F}_{f1} \equiv -m[\omega_{e_0}^2(r_0)]_{11} \cdot \Delta \mathbf{r}_{01}$ on atom 1. The initial force on atom 2 is due completely to the gas-phase potential [for pairwise additive forces.] Thus $[\omega_{e_0}^2(r_0)]_{12} = 0$. (*c*) The initial mean field density $\rho^{(3)}(\mathbf{q}; r_0)$ (dashed "ellipse") relaxes in response to the displacement of solute atom 1 to produce a time-dependent mean-field density (solid "ellipse"). The density produces a systematic relaxing restoring force $\mathscr{F}_D(t)$ on the interacting solute atoms. The relaxing restoring force exerted on atom 1 is $\mathscr{F}_{D1} \equiv m\int_0^t d\tau [\omega_{c1}^2(r_0)\theta_1(\tau; r_0)[\omega_{01}^2(r_0)]^\top]_{11} \cdot \Delta \mathbf{r}_{01}$. The relaxing restoring force exerted on atom 2 is $\mathscr{F}_{D2} \equiv m\int_0^t d\tau [\omega_{c1}^2(r_0)\theta_1(\tau; r_0)[\omega_{c1}^2(r_0)]^\top]_{12} \cdot \Delta \mathbf{r}_{01}$. The above identifications define the MTGLE quantities for interacting reagents, $\omega_{c1}^2(r_0)R_1(t; r_0)$, $\omega_{e_0}^2(r_0)$, $\omega_{c1}^4(r_0)$, and $\theta_1(t; r_0)$, within the clamping approximation.

dent density characteristic of common-cage formation (cf. Figs. 1 and 10).

3. The density-response interpretation may be extended to the solvent forces arising from the "ellipsoidal" density. This extended interpretation permits one to define *R-dependent* systematic and random solvent forces.

4. *Matrix R-dependent MTGLE parameters* $\{\omega_{ep}^2(R), \omega_{c\,p+1}^2(R)\}$, $p = 0, 1, \ldots$, may be constructed from the systematic part of the *R*-dependent solvent force. The construction is a matrix generalization of the procedure employed to construct the parameters for an isolated solute atom for $\dot{\chi}(t)$ (see Appendix A). Given this construction, the matrix equivalent-chain equations [(3.82) below] may be immediately written down. These equations provide a basic framework for liquid-state chemical-reaction dynamics.

The density-response interpretation is most simply illustrated for an isolated solute atom with mass m and coordinate $\mathbf{r}_0(t)$. We assume the atom is initially *clamped* at point $\mathbf{r}_0$. An equilibrium local solvent density

$\rho[|\mathbf{r}-\mathbf{r}_0|]$ surrounds and cages the atom [see Fig. 1]. This equilibrium density is spherically symmetric and thus by symmetry exerts zero net force on the atom. Thermal fluctuations in the local solvent density about its equilibrium value, however, exert a random force with zero mean, $\tilde{\mathscr{F}}_0(t)$, on the clamped atom.

Next assume the atom is slightly displaced to point $\mathbf{r}_0 + \Delta\mathbf{r}_0$. The mean-field density is now no longer in equilibrium with the atom. It thus exerts a systematic restoring force $\langle\mathscr{F}\rangle_t$ on the solute atom. This force depends on time, since the mean-field density relaxes in response to the solute displacement. (At $t = \infty$ the density is again in equilibrium with the atom, now centered at the new location $\mathbf{r}_0 + \Delta\mathbf{r}_0$. Thus $\lim_{t\to\infty}\langle\mathscr{F}\rangle_t = 0$.)

The force $\langle\mathscr{F}\rangle_t$ is determined in Section VI by a linear-response calculation in which $\Delta\mathbf{r}_0$ is the smallness parameter. This calculation yields

$$\langle\mathscr{F}\rangle_t = \frac{1}{3}\left\langle\frac{\partial}{\partial\mathbf{r}_0}\cdot\mathscr{F}\right\rangle_{\mathbf{r}_0}\Delta\mathbf{r}_0 + \frac{1}{3}\beta\int_0^t\left\langle\tilde{\mathscr{F}}_0(t-\tau)\cdot\dot{\tilde{\mathscr{F}}}_0\right\rangle_{\mathbf{r}_0}d\tau\Delta\mathbf{r}_0, \quad (3.77)$$

where $\langle\ \rangle_{\mathbf{r}_0}$ denotes a canonical average over solvent coordinates conditional on the assumption that the solute is *clamped* at point $\mathbf{r}_0$.

Notice that (i) the instantaneous restoring force on the solute due to the initial density is $\frac{1}{3}\langle(\partial/\partial\mathbf{r}_0)\cdot\mathscr{F}\rangle_{\mathbf{r}_0}\Delta\mathbf{r}_0$; (ii) the delayed restoring force on the solute due to density relaxation is $\frac{1}{3}\beta\int_0^t\langle\tilde{\mathscr{F}}_0(t-\tau)\cdot\dot{\tilde{\mathscr{F}}}_0\rangle_{\mathbf{r}_0}d\tau\Delta\mathbf{r}_0$.

Superimposed on the systematic restoring force $\langle\mathscr{F}\rangle_t$ is the random force $\tilde{\mathscr{F}}_0(t)$ due to temporal density fluctuations. Thus the total force acting on the displaced solute is

$$\mathbf{F}(t) = \frac{1}{3}\left\langle\frac{\partial}{\partial\mathbf{r}_0}\cdot\mathscr{F}\right\rangle_{\mathbf{r}_0}\Delta\mathbf{r}_0 + \frac{1}{3}\beta\int_0^t\langle\tilde{\mathscr{F}}_0(t-\tau)\cdot\dot{\tilde{\mathscr{F}}}_0\rangle_{\mathbf{r}_0}d\tau\Delta\mathbf{r}_0 + \tilde{\mathscr{F}}_0(t). \quad (3.78)$$

We next *assume* that the generalized Langevin equation (3.37), which rigorously describes the solvent response to spontaneous solute motion, also describes its response to forced motion. The displacement $\mathbf{r}_0 \to \mathbf{r}_0 + \Delta\mathbf{r}_0$ is a special case of forced motion. Thus given this assumption $\mathbf{F}(t)$ may also be written as [let $\mathbf{r}_0(t) \to \Delta\mathbf{r}_0$ in Eq. (3.37)]

$$m^{-1}\mathbf{F}(t) = -\omega_{e0}^2\Delta\mathbf{r}_0 + \omega_{c1}^4\int_0^t\theta_1(t-\tau)\,d\tau\Delta\mathbf{r}_0 + \omega_{c1}^2\mathbf{R}_1(t). \quad (3.79)$$

Comparing Eqs. (3.78) and (3.79) yields the following *approximate* expressions for the MTGLE quantities [$\mathscr{F} \equiv \tilde{\mathscr{F}}_0(t=0)$]:

$$\omega_{e0}^2 = -(3m)^{-1} \left\langle \frac{\partial}{\partial \mathbf{r}_0} \cdot \mathscr{F} \right\rangle_{\mathbf{r}_0} \tag{3.80a}$$

$$\omega_{c1}^4 \simeq (3mk_BT)^{-1} \langle \dot{\tilde{\mathscr{F}}}_0^2 \rangle_{\mathbf{r}_0} \tag{3.80b}$$

$$\omega_{c1}^4 \dot{\theta}_1(t) \simeq (3mk_BT)^{-1} \langle \dot{\tilde{\mathscr{F}}}_0(t+\tau) \cdot \dot{\tilde{\mathscr{F}}}_0(\tau) \rangle_{\mathbf{r}_0} \tag{3.80c}$$

$$\omega_{c1}^2 \mathbf{R}_1(t) \simeq m^{-1} \tilde{\mathscr{F}}_0(t) \qquad \text{(clamping approximation)} \tag{3.80d}$$

[Note that the expression (3.80a) for ω_{e0}^2 is actually exact. This is shown in Section VI.]

Equation (3.80) are approximate results, since the solvent's response to forced solute motion differs from its response to spontaneous motion.

We will refer to Eqs. (3.80) as the *clamping approximation* to the MTGLE quantities. The reason for this designation is that all terms in Eq. (3.80) may be computed from the space–time solvent-density fluctuations about the solute atom conditional on the assumption that it is clamped at point $\mathbf{r}_0$. Such computations may, in practice, be carried out by molecular-dynamics simulations of the solvent motion about a solute fixed in the liquid. Examples of such simulations, including numerical tests of the clamping approximation, are given in Section VII.

Notice that Eqs. (3.80c, d) show that the second fluctuation–dissipation theorem is preserved within the clamping approximation; that is,

$$\dot{\theta}_1(t) = \frac{m}{3k_BT} \langle \dot{\mathbf{R}}_1(t+\tau) \cdot \dot{\mathbf{R}}_1(\tau) \rangle_{\mathbf{r}_0} \qquad \text{(clamping approximation)}$$

$$\tag{3.81}$$

The clamping approximation requires several additional comments:

1. The approximation becomes exact for two limiting cases: infinitely massive solutes and solutes linearly coupled to perfectly harmonic solvents.

2. The approximation was suggested by the rigorous mathematical operation of chain-atom clamping. For example, $\omega_{c1}^2 \mathbf{R}_1(t)$ is rigorously the force on chain atom 0 exerted by atom 1, given that atom 0 is clamped in the equivalent chain. The clamping approximation to $\omega_{c1}^2 \mathbf{R}_1(t)$, $\tilde{\mathscr{F}}_0(t)$, is correspondingly the force exerted on the solute atom by the solvent, given that the solute atom is physically clamped at point $\mathbf{r}_0$.

3. The clamping approximation is a generalization of the linear-response method long used in hydrodynamic calculations of friction kernels.[38]

The clamping approximation, moreover, provides approximate physical interpretations of the rigorous MTGLE quantities for the case of solute motion in liquids. Briefly: (i) $-m\omega_{e0}^2\Delta\mathbf{r}_0$ is *rigorously* (since the clamping approximation for ω_{e0}^2 is exact) the average *instantaneous* restoring force exerted by the solvent cage on a slightly displaced solute (see Fig. 1). (ii) $m\omega_{c1}^2\int_0^t\theta_1(t-\dot{\tau})\,d\tau\Delta\mathbf{r}_0$ is approximately the modification of this average restoring force due to cage relaxation. (iii) $\omega_{c1}^2\mathbf{R}_1(t)$ is approximately the force on a clamped solute due to thermal cage fluctuations.

These interpretations, and the clamping approximation that underlies them, may be readily extended to the case of interacting reagents.

We next briefly discuss some of the key results of this extension, which is worked out in detail in Section VI. We employ the matrix notation defined in that section. We specialize the general results of Section VI to the case of identical solute atoms with mass m, coordinates $\mathbf{r}_{01}(t)$ and $\mathbf{r}_{02}(t)$, and internuclear separation $R(t)=|\mathbf{r}_{02}(t)-\mathbf{r}_{01}(t)|$. For this case *matrix R-dependent* MTGLE quantities such as $\omega_{e0}^2(R)$ may be defined within the clamping approximation. The definition of these quantities [see Eq. (6.3) below] are matrix generalizations of Eqs. (3.80). For the present case, $\omega_{e0}^2(R)$ and the rest are 2×2 matrices whose elements are three-dimensional Cartesian tensors.

These quantities have physical interpretations based on the density-response concept, which are simple generalizations of the interpretations of the isolated solute atom quantities:

1. $-m\omega_{e0}^2(R)\Delta r_0$ is the average instantaneous solvent restoring force on the reagents immediately after a small displacement $\Delta r_0 = (\Delta\mathbf{r}_{01},\Delta\mathbf{r}_{02})$. Thus the diagonal element $[\omega_{e0}^2(R)]_{11}$ determines the solvent contribution to the instantaneous restoring force on atom 1 immediately after it is displaced by $\Delta\mathbf{r}_{01}$. The instantaneous force on atom 2 after the displacement $\Delta\mathbf{r}_{01}$ is due solely to the gas-phase potential. Thus the off-diagonal element $[\omega_{e0}^2(R)]_{12}=0$.

2. The diagonal coupling-constant element $[\omega_{c1}^2(R)]_{11}$ may be roughly interpreted as measuring the strength of interaction of atom 1 with its own cage. The off-diagonal element $[\omega_{c1}^2(R)]_{12}$ may similarily be roughly interpreted as measuring the strength of interaction of atom 1 with the cage of atom 2. Thus $\lim_{R\to\infty}[\omega_{c1}^2(R)]_{12}=0$.

3. The diagonal response function element $[\omega_{c1}^2(R)\cdot\theta_1(t;R)\cdot[\omega_{c1}^2(R)^{\mathrm{T}}]]_{11}$ governs the force on atom 1 due to common-cage relaxation in response to the motion of atom 1. The corresponding off-diagonal element governs the relaxing force on atom 1 due to motion of atom 2. This off-diagonal response function vanishes as $R\to\infty$.

4. The random-force elements $[\omega_{c1}^2(R)R_1(t;R)]_{1,2}$ describe, respectively, the forces on atoms 1 and 2 due to thermal fluctuations of the

common-cage density about its equilibrium value, given that the reagents are clamped at internuclear separation R.

A matrix equivalent-chain equation of motion for interacting reagents may be constructed from the R-dependent MTGLE quantities. The derivation is completely analogous to the corresponding derivation for an isolated solute atom presented in Section V. The result in matrix notation is

$$\ddot{r}_0(t) = m^{-1}\langle F \rangle_R - \omega_{e0}^2[R(t)]r_0(t) + \omega_{c1}^2[R(t)]r_1(t)$$

$$\ddot{r}_1(t) = -\omega_{e1}^2[R(t)]r_1(t) + \left[\omega_{c1}^2[R(t)]\right]^{\mathrm{T}}r_0(t) + \omega_{c2}^2[R(t)]r_2(t)$$

$$\vdots \qquad (3.82)$$

Note that $\langle F \rangle_R$ is the *mean force* on the reagents, defined by

$$\langle F \rangle_R = -\frac{\partial U_{UU}(R)}{\partial r_0} + \langle \mathscr{F} \rangle_R \qquad (3.83)$$

where $U_{UU}(R)$ is the gas-phase solute–solute potential-energy function and where $\langle \mathscr{F} \rangle_R$ is the average force exerted by the solvent on the reagents when they are clamped at internuclear separation R.

The mean force $\langle F \rangle_R$ may alternatively be expressed in terms of the potential of mean force $W(R)$ via

$$\langle F \rangle_R = -\frac{\partial W(R)}{\partial r_0} \qquad (3.84)$$

The potential of the mean force, as discussed in Section IV.A, is directly related to the solute–solute radial distribution function.

Finally we note that the chain coordinates $r_p = (\mathbf{r}_{p0}, \mathbf{r}_{p1})$ may be roughly interpreted as collective coordinates describing density fluctuations in the pth shell of the common cage.

IV.   STATISTICAL MECHANICS

This section presents a concise summary of some basic methods and results from statistical mechanics that are important for the quantitative treatment of liquid-solution reaction dynamics.

To keep the notation simple, we often explicitly consider a pure classical monotomic liquid characterized by thermodynamic state T, ρ. Much of what is presented below, however, is generalizable to arbitrary many-body systems.

Assume the liquid is composed of N_s atoms, each of mass M, interacting via gas-phase pair potentials $u(r)$. Collectively denote the Cartesian phase-space coordinates of the liquid atoms by $q = (\mathbf{q}_1, \mathbf{q}_2, \ldots, \mathbf{q}_{N_s})$ and $p = (\mathbf{p}_1, \mathbf{p}_2, \ldots, \mathbf{p}_{N_s})$. The classical Hamiltonian of the liquid is

$$H[p,q] = \sum_{\lambda=1}^{N_s} \frac{p_\lambda^2}{2M} + \sum_{\lambda>\mu}^{N_s} u[q_{\lambda\mu}] \tag{4.1}$$

where $q_{\lambda\mu} = |\mathbf{q}_\lambda - \mathbf{q}_\mu|$.

A. Some Results from the Equilibrium Theory of Liquids

We first discuss at an introductory level[39] several topics concerning the theory of the equilibrium structural properties of liquids. We begin with a discussion of the pair distribution and correlation functions of a liquid.

1. Pair Correlation Functions

The *pair distribution function* $g(r)$ for a monatomic liquid has the following physical meaning. Assume a liquid atom is fixed at the origin. Denote the mean local density of liquid atoms located a distance r away from the fixed particle by $\rho(r)$. Then $g(r) = \rho^{-1}\rho(r)$, where, as before, ρ is the bulk number density. Since $\lim_{r\to\infty}\rho(r) = \rho$, $g(r)$ satisfies the asymptotic normalization condition $\lim_{r\to\infty}g(r) = 1$.

As $\rho \to 0$ (ideal-gas limit) the local density is determined by gas-phase Boltzmann statistics; that is, $\rho(r) \sim \rho\exp[-\beta u(r)]$. Equivalently

$$\lim_{\rho\to 0} g(r) = \exp[-\beta u(r)] \tag{4.2}$$

At elevated densities Eq. (4.2) rigorously generalizes to

$$g(r) = \exp[-\beta W(r)] \tag{4.3}$$

where $W(r)$ is an effective potential that accounts for the liquid-state (i.e. static-cage; see Fig. 10) modification of the gas-phase interaction between a pair of liquid atoms. $W(r)$ may be shown[40] to be identical to the potential of mean force [see Eq. (3.84)].

The liquid-state contribution to $W(r)$, denoted by $w(r)$, is usually referred to as the *cavity function*. It is defined by

$$W(r) = u(r) + w(r) \tag{4.4}$$

Note that $\lim_{\rho\to 0}w(r) = 0$, since Eqs. (4.2) and (4.3) must agree as $\rho \to 0$.

We shall also discuss the *pair correlation function* $h(r)$ defined by

$$h(r) = g(r) - 1 = \exp[-\beta W(r)] - 1 \qquad (4.5)$$

We next give the statistical-mechanical definition of $g(r)$. We begin with the canonical-ensemble phase-space distribution function of the liquid,

$$f_{CA}[p, q] = Z^{-1}(\beta)\exp[-\beta H(p, q)] \qquad (4.6)$$

where $Z(\beta) = \int dp\, dq \exp[-\beta H(p, q)]$ is the partition function of the liquid. The phase-space distribution function is the probability density that the liquid is located at the phase point p, q. The pair distribution function $g(r)$ $= g[|\mathbf{q}_1 - \mathbf{q}_2|]$ is derivable from $f_{CA}[p, q]$, since it is proportional to the probability density $P[\mathbf{q}_1, \mathbf{q}_2]$ that atoms 1 and 2 are located, respectively, at points $\mathbf{q}_1$ and $\mathbf{q}_2$. This probability density is defined by

$$P[\mathbf{q}_1, \mathbf{q}_2] = \int f_{CA}[p, q]\, dp\, d\mathbf{q}_3 \cdots d\mathbf{q}_N$$

Additionally, as $|\mathbf{q}_1 - \mathbf{q}_2| \to \infty$, $P[\mathbf{q}_1, \mathbf{q}_2] \to V^{-2}$, where V is the sample volume. In contrast, as $|\mathbf{q}_1 - \mathbf{q}_2| \to \infty$, $g[|\mathbf{q}_1 - \mathbf{q}_2|] \to 1$. These considerations imply that the proportionality factor between g and P is V^2; that is,

$$g(r) = g(|\mathbf{q}_1 - \mathbf{q}_2|) = V^2 P[\mathbf{q}_1, \mathbf{q}_2]$$

$$= V^2 \int f_{CA}[p, q]\, dp\, d\mathbf{q}_3 \cdots d\mathbf{q}_N \qquad (4.7)$$

Calculation of $g(r)$ from Eq. (4.7) is only practical at low densities. At liquid-state densities $g(r)$ may be determined from either brute-force molecular dynamics (MD)[6-8] or Monte Carlo (MC)[5] simulation of the full N-body problem, or from an approximate theory.

Most of the currently popular approximate theories are based on the *Ornstein–Zernike integral equation*,[40] which we now discuss.

2.   The Ornstein–Zernike Equation

The Ornstein–Zernike (OZ) integral equation is

$$h(|\mathbf{q}_\lambda - \mathbf{q}_\mu|) = c(|\mathbf{q}_\lambda - \mathbf{q}_\mu|) + \rho \int c(|\mathbf{q}_\lambda - \mathbf{q}|) h(|\mathbf{q} - \mathbf{q}_\mu|)\, d\mathbf{q} \qquad (4.8a)$$

where $c(|\mathbf{q}_\lambda - \mathbf{q}_\mu|) \equiv c(r)$ is the *direct correlation function*. Equation (4.8a) may be interpreted as the relation that *defines* $c(r)$ in terms of $h(r)$.

The direct correlation function $c(r)$ has a simple qualitative interpretation. This interpretation is apparent from the Neumann expansion of Eq. (4.8a), which is

$$h(|\mathbf{q}_\lambda - \mathbf{q}_\mu|) = c(|\mathbf{q}_\lambda - \mathbf{q}_\mu|) + \rho \int c(|\mathbf{q}_\lambda - \mathbf{q}|) c(|\mathbf{q} - \mathbf{q}_\mu|) \, d\mathbf{q}$$

$$+ \rho^2 \int c(|\mathbf{q}_\lambda - \mathbf{q}|) c(|\mathbf{q} - \mathbf{q}'|) c(|\mathbf{q}' - \mathbf{q}_\mu|) \, d\mathbf{q} \, d\mathbf{q}' + \cdots$$

$$(4.8b)$$

According to Eq. (4.8b) the pair correlations between atoms λ and μ may be decomposed into a *direct* contribution $c(|\mathbf{q}_\lambda - \mathbf{q}_\mu|)$, a contribution mediated by a third atom located at point $\mathbf{q}$ (this contribution is multiplied by the number of atoms at $\mathbf{q}$, $\rho \, d\mathbf{q}$, and then integrated over all points $\mathbf{q}$), a contribution mediated by a third atom at $\mathbf{q}$ and then a fourth atom at $\mathbf{q}'$, and so on.

The key point is that detailed many-body effects in $h(r)$ may be built up from relatively simple (e.g., low-density) forms for $c(r)$. This follows because of the iterative structure of Eqs. (4.8). This idea is the physical basis of the modern integral-equation theories.

Before exploring this idea, we first generalize the OZ theory to the multicomponent fluids of relevance to solution chemistry.

Consider an n-component fluid. Denote the pair potential between an atom of species i and a second atom of species j by $u_{ij}(r)$. Collectively denote the number densities of the individual components by the diagonal matrix

$$\rho = \begin{bmatrix} \rho_0 & 0 & 0 & \cdots & 0 \\ 0 & \rho_1 & 0 & \cdots & 0 \\ 0 & 0 & \rho_2 & \cdots & 0 \\ \vdots & \vdots & \vdots & & \vdots \\ 0 & 0 & 0 & \cdots & \rho_n \end{bmatrix} \qquad (4.9)$$

For each pair of components i and j, there is a pair $[h_{ij}(r)]$ and a direct $[c_{ij}(r)]$ correlation function. [The interpretation of $h_{ij}(r)$ is that $\rho_j g_{ij}(r)$ is the number density of j-atoms located a distance r from a fixed i-atom, and $h_{ij}(r) = g_{ij}(r) - 1$.] Thus the pure-fluid functions $h(r)$ and $c(r)$ generalize to $n \times n$ matrices of the form $[\mathbf{f}(r) = \mathbf{h}(r) \text{ or } \mathbf{c}(r)]$

$$
\mathbf{f}(r) = \begin{bmatrix} f_{11}(r) & f_{12}(r) & \cdots & f_{1n}(r) \\ f_{21}(r) & f_{22}(r) & \cdots & f_{2n}(r) \\ \vdots & \vdots & & \vdots \\ f_{n1}(r) & f_{n2}(r) & \cdots & f_{nn}(r) \end{bmatrix} \tag{4.10}
$$

One may define a mean-force potential $W_{ij}(r)$ and a cavity function $w_{ij}(r)$ for components i and j via [cf. Eqs. (4.4) and (4.5)]

$$
W_{ij}(r) = u_{ij}(r) + w_{ij}(r) \tag{4.11}
$$

and

$$
h_{ij}(r) = g_{ij}(r) - 1 = \exp\left[-\beta W_{ij}(r) \right] - 1 \tag{4.12}
$$

The OZ equation (4.8a) generalizes to the following $n \times n$ matrix integral equation:

$$
\mathbf{h}(r) = \mathbf{c}(r) + \int \mathbf{c}[|\mathbf{r} - \mathbf{r}'|] \rho \mathbf{h}[|\mathbf{r}' - \mathbf{r}|] \, d\mathbf{r}' \tag{4.13}
$$

which may be interpreted as the defining relation for $\mathbf{c}(r)$.

3. Simple Closures of the OZ Equation

We next illustrate the idea that simple, physically motivated approximations for $\mathbf{c}(r)$ can yield results for $\mathbf{h}(r)$ that incorporate detailed N-body effects. We discuss two examples: a low-ionic-strength electrolyte solution and a high-density hard-sphere fluid mixture. The former example is a prototype for liquids in which strong long-range attractive forces play an essential role. The latter example is a prototype for liquids (like argon) for which packing effects due to short-range excluded-volume forces are basic.

For all liquids of chemical interest, both strong attractive and packing forces influence the fluid behavior in an essential way. The development of a satisfactory general theory for such liquids remains a fundamental challenge.

We first discuss the electrolyte solution. We assume an n-component mixture of ions immersed in a continuum dielectric model solvent[41] with dielectric constant ε. Given this model, the ion–ion effective pair potentials are of the form[‡]

$$u_{ij}(r) = \frac{Z_i Z_j e^2}{\varepsilon r} \qquad (4.14)$$

where $Z_i e$ is the charge on an ion of type i. (Note that electroneutrality requires $\sum_{i=1}^{n} Z_i = 0$.)

We next present an elementary statistical-mechanical (as opposed to electrostatic) derivation of the Debye–Hückel theory result for $h_{ij}(r)$. This result is

$$h_{ij}(r) = \frac{-Z_i Z_j \beta e^2}{\varepsilon r} e^{-\kappa r}. \qquad (4.15)$$

Here κ is Debye's inverse screening length defined by

$$\kappa^2 = \frac{4\pi \beta e^2}{\varepsilon} \rho_{\text{eff}} \qquad (4.16)$$

where

$$\rho_{\text{eff}} = \sum_{i=1}^{n} Z_i^2 \rho_i \qquad (4.17)$$

We now show how a simple form for $c(r)$ that accounts only for the gas-phase ion–ion interaction can yield Eq. (4.15), which incorporates an N-body effect, Debye screening.

The simplest plausible model for $c_{ij}(r)$ is its low-density or gas-phase value; that is, we approximate $c_{ij}(r)$ by

$$c_{ij}(r) \simeq \lim_{\rho \to 0} c_{ij}(r) = \lim_{\rho \to 0} h_{ij}(r) = \exp\left[-\beta u_{ij}(r)\right] - 1. \qquad (4.18)$$

We may further simplify Eq. (4.18) by linearizing the exponential to yield

$$c_{ij}(r) \simeq -\beta u_{ij}(r)$$
$$= -\beta \frac{Z_i Z_j}{\varepsilon r} e^2 \qquad (4.19)$$

The justification for Eq. (4.19) is that at low concentration the ions are typically so well separated that $-\beta u_{ij}(r) \ll 1$.

[‡]One may define an effective pair potential for which the form of Eq. (4.14) holds rigorously as $r \to \infty$ for arbitrary ionic concentrations. The dielectric constant ε, however, depends on ionic concentration and composition. See Ref. 19.

Note that Eq. (4.19) is of the form

$$c_{ij}(r) = Z_i Z_j c(r) \tag{4.20a}$$

where $c(r) = -\beta e^2/\varepsilon r$ is independent of ionic type. Equation (4.20a) and the linear character of the entire problem suggest that $h_{ij}(r)$ has the matrix structure

$$h_{ij}(r) = Z_i Z_j h(r) \tag{4.20b}$$

Equation (4.20b) may be verified by inserting Eqs. (4.20) into the component form of the matrix OZ equation, which is

$$h_{ij}(r) = c_{ij}(r) + \sum_{k=1}^{n} \int d\mathbf{r}' c_{ik}[|\mathbf{r} - \mathbf{r}'|] \rho_k h_{kj}[|\mathbf{r}' - \mathbf{r}|]$$

This yields the following effective one-component OZ equation for $h(r)$:

$$h(r) = c(r) + \rho_{\text{eff}} \int d\mathbf{r}' c[|\mathbf{r} - \mathbf{r}'|] h[|\mathbf{r}' - \mathbf{r}|] \tag{4.21}$$

where ρ_{eff} is given in Eq. (4.17).

Equation (4.21) may be solved by Fourier transforms. The Fourier transform of Eq. (4.21) is $\hat{h}(k) = \hat{c}(k) + \rho_{\text{eff}} \hat{c}(k)\hat{h}(k)$, or equivalently

$$\hat{h}(k) = \frac{\hat{c}(k)}{1 - \rho_{\text{eff}} \hat{c}(k)} \tag{4.22}$$

The Fourier transform of $c(r)$ is determined from Eqs. (4.19) and (4.20a) to be

$$\hat{c}(k) = \frac{-4\pi\beta e^2}{\varepsilon k^2} \tag{4.23}$$

Combining Eqs. (4.22) and (4.23) yields

$$\hat{h}(k) = \frac{-4\pi\beta e^2}{\varepsilon} [k^2 + \kappa^2]^{-1} \tag{4.24}$$

where κ^2 is given by Eq. (4.16).

Equation (4.15) is now readily obtained by taking the Fourier transform of Eq. (4.24) and then using Eq. (4.20b). This completes our derivation of the Debye–Hückel model from the OZ equation.

We next turn to the hard-sphere fluid mixture. The interaction potentials for such a solution are

$$u_{ij}(r) = \begin{cases} 0 & \text{for} \quad r > \sigma_{ij} \\ \infty & \text{for} \quad r < \sigma_{ij} \end{cases} \tag{4.25}$$

where $\sigma_{ij} = \frac{1}{2}(\sigma_i + \sigma_j)$ and σ_i is the hard-sphere diameter of fluid component i. The pair correlation functions of the hard-sphere mixture are of the form

$$h_{ij}(r) = \begin{cases} -1 & \text{for} \quad r < \sigma_{ij} \\ ? & \text{for} \quad r > \sigma_{ij} \end{cases} \tag{4.26a}$$

[The constraint $h_{ij}(r) = -1$ for $r < \sigma_{ij}$ follows from $g_{ij}(r) = 0$ for $r < \sigma_{ij}$, which in turn occurs because the local density of j atoms $\rho_j g_{ij}(r)$ near an i-atom vanishes for $r < \sigma_{ij}$, the hard-sphere overlap distance.]

To proceed further we require an approximate form for $c_{ij}(r)$. The simplest guess is the ideal-gas approximation (4.18). This gives $c_{ij}(r) = 0$ for $r > \sigma_{ij}$ and $c_{ij}(r) = -1$ for $r < \sigma_{ij}$. The latter condition, however, is inconsistent with the OZ equation (4.13), given the rigorous constraint $h_{ij}(r) = -1$ for $r < \sigma_{ij}$. To remove this inconsistency, we instead choose

$$c_{ij}(r) = \begin{cases} ? & \text{for} \quad r < \sigma_{ij} \\ 0 & \text{for} \quad r > \sigma_{ij} \end{cases} \tag{4.26b}$$

Given Eqs. (4.26), the OZ equation (4.13) becomes a closed equation for an unknown matrix function $\mathbf{f}(r)$ where

$$f_{ij}(r) = \begin{cases} c_{ij}(r) & \text{for} \quad r < \sigma_{ij} \\ h_{ij}(r) & \text{for} \quad r > \sigma_{ij} \end{cases} \tag{4.27}$$

This equation has been solved *analytically* by Wertheim[42] and independently by Thiele for a pure hard-sphere fluid. A generalization to multicomponent hard-sphere fluids has been given by Lebowitz.[43]

Wertheim's result for $c(r)$ for the pure hard-sphere fluid is

$$c(r) = \begin{cases} \lambda_1 + 12\eta\lambda_2 y + 4\eta\lambda_1 y^3 & \text{for} \quad r < \sigma \\ 0 & \text{for} \quad r > \sigma \end{cases} \tag{4.28}$$

where

$$\lambda_1 = -[1 + 2\eta]^2 [1 - \eta]^{-4} \tag{4.29a}$$

$$\lambda_2 = [1 + \tfrac{1}{2}\eta]^2 [1 - \eta]^{-4} \tag{4.29b}$$

and where $y = r/2\sigma$ and $\eta = \frac{4}{3}\pi(\sigma/2)^3\rho$. Here η is the packing fraction (the fraction of the total volume occupied by hard spheres) of the fluid.

Given Eq. (4.28), the pair correlation function $h(r)$ may be determined numerically from the OZ equation. The calculated results for $h(r)$[44] show detailed oscillatory excluded-volume structure and are in reasonable agreement with results obtained from MD simulations.

The key point, for the present discussion, is that detailed N-body structure in $h(r)$ may be built up from a relatively simple form, Eq. (4.28), for $c(r)$.

4. The Percus–Yevick Approximation

We next briefly discuss one of the most popular integral-equation theories. This is the Percus–Yevick (PY) approximation. This approximation may be regarded as the following choice for $\mathbf{c}(r)$:

$$c_{ij}(r) = \left\{1 - \exp\left[\beta u_{ij}(r)\right]\right\}\left\{1 + h_{ij}(r)\right\} \tag{4.30}$$

Equations (4.13) and (4.30) provide a closed integral equation for $\mathbf{h}(r)$.

Using Eqs. (4.11) and (4.12), Eq. (4.30) may be recast as

$$c_{ij}(r) = \left\{\exp\left[-\beta u_{ij}(r)\right] - 1\right\} y_{ij}(r) \tag{4.31}$$

where

$$y_{ij}(r) = \exp\left[-\beta w_{ij}(r)\right] \tag{4.32}$$

Equation (4.31) shows that the PY approximation differs from our simple low-density model, Eq. (4.18), to the extent that the cavity function $w_{ij}(r)$ (the liquid-state correction to the gas-phase potential) differs from zero.

For the hard-sphere fluid, the PY approximation is identical to the model, Eq. (4.26b), developed earlier by intuition. This may be seen from Eq. (4.30). For $r > \sigma_{ij}$, $u_{ij}(r) = 0$ and thus $c_{ij}(r) = 0$. For $r < \sigma_{ij}$, $1 + h_{ij}(r) = 0$ and $\exp[\beta u_{ij}(r)] = \infty$. Thus $c_{ij}(r)$ ($= 0 \times \infty$) is undetermined for $r < \sigma_{ij}$. This is exactly Eq. (4.26b).

This concludes our introductory discussion of equilibrium liquid-state theory.

B. Some Topics from the Formal Theory of Irreversible Processes

We next present a concise development of certain basic methods and results from the modern theory of irreversible processes.[10-12]

We stress that the content of this section is on a very different footing from that of section IV.A. The central theorems of irreversibility theory follow rigorously from the basic laws of dynamics and equilibrium statistical mechanics. Thus these results reflect fundamental aspects of the N-body problem that are of far-reaching generality. The rigor and generality of the central results of irreversibility theory is part of the reason for their great conceptual and practical importance.

For simplicity of presentation, we shall often focus our general discussion by considering the classical dynamics of the monatomic liquid whose Hamiltonian is given by Eq. (4.1). We shall actually consider an ensemble of such liquids. Each member of the ensemble is characterized by the same macroscopic state T, ρ. The microscopic state at time t is specified by the phase point $p(t), q(t)$ of the liquid. This point differs for different members of the ensemble.

The ensemble as a whole is microscopically characterized by the normalized probability density for phase points. This is the phase-space PDF $f[p, q; t]$.

1. Time Evolution of Dynamical Variables

We next discuss the formal representation of the time evolution of dynamical variables (i.e., phase-point functions). First consider the time development of the phase point $p(t), q(t)$. Consider a particular member of the ensemble with initial phase point p_0, q_0. This point evolves in time according to Hamilton's equation, which are

$$\frac{d\mathbf{q}_\lambda(t)}{dt} = \frac{\partial H[p(t), q(t)]}{\partial \mathbf{p}_\lambda(t)} \tag{4.33a}$$

$$\frac{d\mathbf{p}_\lambda(t)}{dt} = - \frac{\partial H[p(t), q(t)]}{\partial \mathbf{q}_\lambda(t)} \qquad \lambda = 1, 2, \dots, N_s \tag{4.33b}$$

Next consider an arbitrary set of n dynamical variables $\mathbf{A}[p(t), q(t)] \equiv (A_1[p(t), q(t)], \dots, A_n[p(t), q(t)])$ of the liquid. We assume these variables have no explicit time dependence; that is, their development in time is completely determined by the motion of the phase point. Examples are the velocity $\dot{\mathbf{q}}_\lambda(t)$ of liquid atom λ, the Fourier components $p_\mathbf{q}(t) = \sum_{\lambda=1}^{N_s} \exp[i\mathbf{k} \cdot \mathbf{q}_\lambda(t)]\mathbf{p}_\lambda(t)$ of the momentum density in the liquid, and so on.

Consider the particular member of the ensemble with initial phase point p_0, q_0. Let $\mathbf{A}(t) \equiv \mathbf{A}[p(t), q(t)] = \mathbf{A}[t; p_0, q_0]$, and let $\mathbf{A} \equiv \mathbf{A}(t = 0) = \mathbf{A}[p_0, q_0]$. Since $\mathbf{A}(t)$ has no explicit time dependence, its equation of motion may be written as

$$\frac{d\mathbf{A}(t)}{dt} = \sum_{\lambda=1}^{N_s} \left\{ \frac{\partial \mathbf{A}(t)}{\partial \mathbf{q}_\lambda(t)} \cdot \frac{d\mathbf{q}_\lambda(t)}{dt} + \frac{\partial \mathbf{A}(t)}{\partial \mathbf{p}_\lambda(t)} \cdot \frac{d\mathbf{p}_\lambda(t)}{dt} \right\} \tag{4.34}$$

Using Eqs. (4.33), Eq. (4.34) may be written in Poisson-bracket form as

$$\frac{d\mathbf{A}(t)}{dt} = \sum_{\lambda=1}^{N_s} \left\{ \frac{\partial \mathbf{A}(t)}{\partial \vec{q}_\lambda(t)} \cdot \frac{\partial H(t)}{\partial \vec{p}_\lambda(t)} - \frac{\partial \mathbf{A}(t)}{\partial \vec{p}_\lambda(t)} \cdot \frac{\partial H(t)}{\partial \vec{q}_\lambda(t)} \right\}$$
$$\equiv \left[\mathbf{A}(t), H(t) \right]_{p(t), q(t)} \qquad (4.35a)$$

The Poisson bracket is, however, invariant under a canonical transformation of the phase space coordinates.[45] The transformation (inverse dynamical transformation) $\mathbf{p}_\lambda(t), \mathbf{q}_\lambda(t) \to \mathbf{p}_{\lambda 0}, \mathbf{q}_{\lambda 0}$ is such a canonical transformation.[46] Thus the Poisson bracket $[A(t), H(t)]_{p(t), q(t)}$ is equal to the Poisson bracket $[A(t), H(t)]_{p_0, q_0}$. Thus Eq. (4.35a) may be alternatively written as

$$\frac{d\mathbf{A}(t)}{dt} = \left[\mathbf{A}(t), H \right]_{p_0, q_0} \equiv \left[\mathbf{A}(t), H \right] \qquad (4.35b)$$

where the Poisson bracket [,] is defined by

$$[A, B] = \sum_{\lambda=1}^{N_s} \left\{ \frac{\partial A}{\partial \mathbf{q}_{\lambda 0}} \cdot \frac{\partial B}{\partial \mathbf{p}_{\lambda 0}} - \frac{\partial A}{\partial \mathbf{p}_{\lambda 0}} \cdot \frac{\partial B}{\partial \mathbf{q}_{\lambda 0}} \right\} \qquad (4.36)$$

2. Liouville Space

Equation (4.35b) may be written more compactly by defining the linear *Liouville operator L* via

$$iLX(t) = \left[X(t), H \right] \qquad (4.37)$$

Using the Liouville-operator notation, Eq. (4.35) may be rewritten as

$$\frac{d\mathbf{A}(t)}{dt} = iL\mathbf{A}(t) \qquad (4.38)$$

A key advantage of the Liouville-operator notation is that it emphasizes the formal analogy between nonequilibrium statistical mechanics and quantum dynamics. For example, Eq. (4.37) formally resembles the time-dependent Schrödinger equation

$$\frac{d|\psi(t)\rangle}{dt} = -iH|\psi(t)\rangle \qquad (4.39)$$

This formal analogy permits one to transfer many of the formal methods used in time-dependent quantum mechanics (e.g., perturbation theory,[47] projection operators,[48] Feynman graphs[49]) into nonequilibrium theory.

For example, just as the set of quantum wavefunctions may be regarded as vectors in an abstract Hilbert space, the set of complex-valued phase-point functions (dynamical variables) may be regarded as abstract vectors in *Liouville space*. Operators O in that space are objects that transform one phase-point function into another. Thus L is an operator in Liouville space.

Abstract inner products (X, Y^*) of pairs of Liouville-space vectors may be defined. The only restriction on the definition of an inner product is that it must satisfy the standard mathematical conditions

$$(X, Y^*) = (Y, X^*)^* \tag{4.40a}$$

$$\left(\sum_i c_i X_i, Y^* \right) = \sum_i c_i (X_i, Y^*) \tag{4.40b}$$

$$(X, X^*) \geqslant 0 \tag{4.40c}$$

The Hermitian adjoint $O^\dagger$ of an arbitrary Liouville-space operator O is defined for a *particular inner product* via

$$(X, OY^*) = (Y, O^\dagger X^*)^* \tag{4.41}$$

where X and Y are *arbitrary* phase-space functions.

An important inner product is the canonical-ensemble average of a pair of dynamical variables, defined by

$$\langle XY^* \rangle \equiv \int dp \, dq \, f_{CA}[p, q] X[p, q] Y^*[p, q] \tag{4.42}$$

A key point is that the Liouville operator is Hermitian if the inner product is $\langle \ \rangle$ [this is why the $i = \sqrt{-1}$ is included in the definition of L, Eq. (4.37)]; that is,

$$\langle XLY^* \rangle = \langle YL^\dagger X^* \rangle^* = \langle YLX^* \rangle^* \tag{4.43}$$

for arbitrary dynamical variables X, Y. Equation (4.43) is proven in Appendix B.

Equation (4.38) may be formally solved to yield

$$\mathbf{A}[p(t), q(t)] = \mathbf{A}(t) = e^{iLt}\mathbf{A} \tag{4.44}$$

Equation (4.44) despite its abstract appearance, is just a concise form for the power-series expansion of $\mathbf{A}(t)$ about $t = 0$. That is, since

$$\exp[iLt] \equiv \sum_{n=0}^{\infty} \frac{t^n}{n!} (iL)^n$$

and since [Eq. (4.38)]

$$(iL)^n \mathbf{A} = \left[\frac{d^n}{dt^n} \mathbf{A}(t) \right]_{t=0} \equiv \mathbf{A}^{(n)}$$

then

$$\exp[iLt]\mathbf{A} = \sum_{n=0}^{\infty} \frac{t^n}{n!} \mathbf{A}^{(n)} \tag{4.45}$$

which is the required power-series expansion.

Finally we note that the classical propagator

$$G(t) = \exp[iLt] \tag{4.46}$$

has formal properties identical to those of the quantum propagator. For example, $G(t)$ has the group property

$$G(t)G(t') = G(t + t') \tag{4.47a}$$

Moreover $G(t)$ is unitary if the inner product is the canonical-ensemble average $\langle \ \rangle$. This follows because of the Hermitian property (4.43). The mathematical statement of the unitarity property of $G(t)$ is

$$G^{\dagger}(t) = G^{-1}(t) = G(-t) \tag{4.47b}$$

3. Time Correlation Functions

Time correlation functions have been discussed from a physical viewpoint in Section III.C. We now give a more formal discussion.

Consider two typical members of the set of dynamical variables, $A_i(t)$ and $A_j(t)$. Their time correlation function is defined by

$$C_{ij}(t) = \langle \exp[iLt] A_i A_j^* \rangle$$
$$= \langle A_i(t) A_j^* \rangle \tag{4.48}$$

$C_{ij}(t)$ has the stationarity property [see Section III.C]:

$$C_{ij}(t) = \langle A_i(t + \tau) A_j^*(\tau) \rangle$$
$$= \langle A_i(t) A_j^* \rangle \tag{4.49}$$

for arbitrary τ. This follows from the group and unitary properties of $G(t)$

[Eqs. (4.47)], since

$$\langle A_i(t+\tau)A_j^*(\tau)\rangle = \langle A_i(t+\tau)G(\tau)A_j^*\rangle = \langle A_jG^\dagger(\tau)A_i^*(t+\tau)\rangle^*$$
$$= \langle A_jG(-\tau)A_i^*(t+\tau)\rangle^* = \langle A_jG(-\tau)G(t+\tau)A_i^*\rangle^*$$
$$= \langle A_jG(t)A_i^*\rangle^* = \langle A_jA_i^*(t)\rangle^* = \langle A_i(t)A_j^*\rangle.$$

An important consequence of Eq. (4.49) may be derived by setting $\tau = -t$ to yield

$$\langle A_i(t)A_j^*\rangle = \langle A_iA_j^*(-t)\rangle \tag{4.50a}$$

An immediate consequence of Eq. (4.50a) is the "dot interchange theorem"

$$\langle \dot{A}_i(t)A_j^*\rangle = -\langle A_i\dot{A}_j^*(-t)\rangle$$
$$= -\langle A_i(t)\dot{A}_j^*\rangle \tag{4.50b}$$

which we will use repeatedly below. A second consequence of Eq. (4.50) is that the *auto*correlation function of a *real* dynamical variable $R(t)$ is even:

$$\langle R(-t)R\rangle = \langle R(t)R\rangle \tag{4.51}$$

4.   The Liouville Equation

Our discussion so far has focused on the time development of dynamical variables. We next derive an equation of motion, the Liouville equation, for the time-dependent phase-space PDF $f[p,q;t]$ of the liquid.

First consider the particular member of the ensemble of liquids with initial phase point p_0, q_0. The PDF for this member is

$$f_{p_0,q_0}[p,q;t] = \delta[p-p(t)]\delta[q-q(t)] \tag{4.52}$$

where $p(t), q(t)$ is the phase-space trajectory that evolves from p_0, q_0:

$$\begin{pmatrix} p(t) \\ q(t) \end{pmatrix} = \exp[iLt]\begin{pmatrix} p_0 \\ q_0 \end{pmatrix} \tag{4.53}$$

To derive the Liouville equation for $f_{p_0,q_0}[p,q;t]$, differentiate Eq. (4.52) with respect to time. This yields

$$\frac{\partial f_{p_0,q_0}}{\partial t} = \sum_{\lambda=1}^{N_s}\left\{\frac{\partial f_{p_0,q_0}}{\partial \mathbf{p}_\lambda(t)}\cdot\frac{d\mathbf{p}_\lambda(t)}{dt} + \frac{\partial f_{p_0,q_0}}{\partial \mathbf{q}_\lambda(t)}\cdot\frac{d\mathbf{q}_\lambda(t)}{dt}\right\}$$

Because of Eq. (4.52), $p(t) = p$ and $q(t) = q$. Therefore

$$\frac{d\mathbf{p}_\lambda(t)}{dt} = \frac{d\mathbf{p}_\lambda}{dt} = \dot{\mathbf{p}}_\lambda \quad \text{and} \quad \frac{d\mathbf{q}_\lambda(t)}{dt} = \frac{d\mathbf{q}_\lambda}{dt} = \dot{\mathbf{q}}_\lambda$$

Also, since $f_{p_0, q_0}[p, q, t]$ is a function of $p - p(t)$ and $q - q(t)$, it follows that

$$\frac{\partial f_{p_0, q_0}}{\partial \mathbf{p}_\lambda(t)} = -\frac{\partial f_{p_0, q_0}}{\partial \mathbf{p}_\lambda} \quad \text{and} \quad \frac{\partial f_{p_0, q_0}}{\partial \mathbf{q}_\lambda(t)} = -\frac{\partial f_{p_0, q_0}}{\partial \vec{q}_\lambda}$$

Combining the last three equations yields

$$\frac{\partial f_{p_0, q_0}[p, q; t]}{\partial t} = -\sum_{\lambda=1}^{N_s} \left\{ \frac{\partial f_{p_0, q_0}[p, q; t]}{\partial \mathbf{p}_\lambda} \cdot \dot{\mathbf{p}}_\lambda + \frac{\partial f_{p_0, q_0}[p, q; t]}{\partial \mathbf{q}_\lambda} \cdot \dot{\mathbf{q}}_\lambda \right\}$$

Finally, using Hamilton's equations [Eqs. (4.33)] and the definitions of the Poisson bracket and the Liouville operator [Eqs. (4.36) and (4.37)], the above equation becomes

$$\frac{\partial f_{p_0, q_0}[p, q; t]}{\partial t} = -\left[f_{p_0, q_0}[p, q; t], H \right]$$

$$= -iL f_{p_0, q_0}[p, q; t] \qquad (4.54)$$

This is Liouville's equation derived for the special class of distribution functions for which the initial phase point is precisely specified, that is, δ-function phase-space PDFs.

Suppose instead that the initial distribution of phase points is

$$f[p, q; 0] = G[p, q], \qquad (4.55)$$

where $G[p, q]$ is an arbitrary phase space PDF. The initial distribution function (4.55) may be written as a superposition of initial phase-space δ-function PDFs $f_{p_0, q_0}[p, q; 0]$ via

$$f[p, q; 0] = \int dp_0 \, dq_0 \, G[p_0, q_0] f_{p_0, q_0}[p, q; 0] \qquad (4.56)$$

Since the initial phase points p_0, q_0 propagate independently via Eq. (4.53), the phase-space PDF that evolves from $f[p, q; 0]$ is

$$f[p, q; t] = \int dp_0 \, dq_0 \, G[p_0, q_0] f_{p_0, q_0}[p, q; t] \qquad (4.57)$$

Since the Liouville operator L in Eq. (4.54) is independent of the variables p_0 and q_0 and since L is linear, comparison of Eqs. (4.54) and (4.57) yields the general form of Liouville's equation,

$$\frac{\partial f[p,q;t]}{\partial t} = -iLf[p,q;t] \tag{4.58}$$

which is the basic result of this Section.

We next turn to a discussion of perturbation theory and linear-response theory. During this discussion we shall utilize the formal solution of the Liouville equation (4.58):

$$f[p,q;t] = \exp[-iLt]\,f[p,q;0]$$
$$= \exp[-iLt]\,G[p,q] \tag{4.59}$$

5. Perturbation Theory

Next assume that a weak perturbation $H^{(1)}(t)$ is added to the Hamiltonian H of Eq. (4.1). The Liouville operator of the perturbed system will be denoted by

$$i[L + L^{(1)}(t)]$$

where [cf. Eq. (4.37)]

$$iL^{(1)}(t)X = [X, H^{(1)}(t)]$$
$$= \sum_{\lambda=1}^{N_s} \left\{ \frac{\partial X}{\partial \mathbf{q}_\lambda} \cdot \frac{\partial H^{(1)}(t)}{\partial \mathbf{p}_\lambda} - \frac{\partial X}{\partial \mathbf{p}_\lambda} \cdot \frac{\partial H^{(1)}(t)}{\partial \mathbf{q}_\lambda} \right\} \tag{4.60}$$

We shall write the phase-space PDF in the presence of the perturbation as $f[p,q;t] + f^{(1)}[p,q;t]$. We assume that the unperturbed PDF $f[p,q;t]$ is known. Usually $f[p,q;t] = f_{CA}[p,q]$. The Liouville equation for the full PDF is

$$\frac{\partial}{\partial t}[f[p,q;t] + f^{(1)}[p,q;t]] = -i[L + L^{(1)}(t)][f[p,q;t] + f^{(1)}[p,q;t]]$$

To first order in the perturbation, the above equation simplifies to

$$\frac{\partial f^{(1)}[p,q;t]}{\partial t} = -iLf^{(1)}[p,q;t] - iL^{(1)}(t)f[p,q;t] \tag{4.61}$$

where we have used Eq. (4.58). The formal solution of Eq. (4.60) is [cf. Eq.

(4.59)]

$$f^{(1)}[p, q; t] = \exp[-iLt] f^{(1)}[p, q; 0]$$
$$+ \int_0^t d\tau \exp[-iL(t - \tau)] [-iL^{(1)}(\tau) f[p, q; \tau]]$$

(4.62)

Taking the time derivative of Eq. (4.62) shows that it is, indeed, the solution of Eq. (4.61).

6. Linear-Response Theory

We next outline Kubo's[10(a)] linear response theory. This theory was originally developed to provide a microscopic method for deriving phenomenological linear transport laws, like Ohm's law of conduction or Fick's law of particle diffusion, including rigorous statistical-mechanical expressions [see e.g. Eq. (4.77)] for the transport coefficients. The linear transport laws are statements of proportionality between a generalized current (e.g., heat flux) and a conjugate weak applied field, (e.g., temperature gradient). The transport coefficient is the proportionality factor. For example, the transport coefficient relating heat flux and temperature gradient is the thermal conductivity. Other familiar examples of transport coefficients include the diffusion coefficient, the electrical conductivity, and the shear and bulk viscosities.

The importance of linear-response theory is partly due to the fact that earlier theories for the transport coefficients were based on approximate kinetic equations, such as the Boltzmann equation,[50] and thus were of restricted validity. The universality of Kubo's formulas gives them a status in nonequilibrium theory somewhat analogous to the formulas of equilibrium statistical mechanics that connect the thermodynamic properties of a system to its partition function.

Transport coefficients may seem far removed from the main topic of this review—liquid-state reaction theory—but linear-response theory is, in fact, highly relevant. For example, it leads one naturally to the formal theory of generalized Brownian motion.

We begin our discussion of linear-response theory with a formal calculation of the electrical conductivity σ of an ionic fluid. To keep the treatment as simple as possible, suppose the liquid contains only *one* charged atom, atom μ with charge $Z_\mu e$. Thermal motion of atom μ then produces an instantaneous current

$$\mathbf{j}(t) = \frac{Z_\mu e}{m} \mathbf{p}_\mu(t) = \dot{\mathbf{M}}(t)$$

(4.63)

where $\mathbf{M}$ is the electric moment of the liquid. For this case

$$\mathbf{M} = Z_\mu e \mathbf{q}_\mu \tag{4.64}$$

Suppose the liquid is initially in thermal equilibrium. Then the ensemble-average current vanishes. Formally,

$$\langle \mathbf{j} \rangle = \int dp\, dq\, f_{CA}[p, q]\mathbf{j} = 0 \tag{4.65}$$

Now assume that a *weak* uniform electric field $\mathbf{E}(t)$ is turned on at $t = 0$. This perturbation throws the system slightly out of equilibrium. The phase-space PDF, for example, is modified to $f_{CA}[p, q] + f^{(1)}[p, q; t]$, where $f^{(1)}[p, q; t]$ is a nonequilibrium correction due to the field. Also, atom μ acquires a net drift leading to a nonvanishing average current $\langle \mathbf{j}(t) \rangle_{\mathbf{E}(t)}$ given by

$$\langle \mathbf{j}(t) \rangle_{\mathbf{E}(t)} = \int dp\, dq\, \left[f_{CA}[p, q] + f^{(1)}[p, q; t] \right] \mathbf{j}$$

$$= \int dp\, dq\, f^{(1)}[p, q; t]\mathbf{j} \tag{4.66}$$

where we have used Eq. (4.65).

Linear-response theory provides a formal algorithm based on Eq. (4.62) for calculating $\langle \mathbf{j}(t) \rangle_{\mathbf{E}(t)}$ to linear order in $\mathbf{E}(t)$. Since the proportionality factor between average current and field is the electrical conductivity σ, the linear-response algorithm provides a microscopic expression for the transport coefficient σ.

We begin the calculation by noting that the coupling of the field $\mathbf{E}(t)$ to the electric moment $\mathbf{M}$ of the liquid yields the following perturbation Hamiltonian:

$$H^{(1)}(t) = -\mathbf{M} \cdot \mathbf{E}(t) \tag{4.67}$$

The Liouville operator associated with $H^{(1)}(t)$ may be determined using Eqs. (4.60), (4.64), and (4.67). The result is

$$iL^{(1)}(t) = Z\mu e\, \mathbf{E}(t) \cdot \frac{\partial}{\partial \mathbf{p}_\mu} \tag{4.68}$$

We next determine $f^{(1)}[p, q; t]$ from Eq. (4.62). This equation simplifies to

$$f^{(1)}[p, q; t] = \int_0^t d\tau \exp[-iL(t - \tau)] \left[-iL^{(1)}(\tau) f_{CA}[p, q] \right] \tag{4.69}$$

since the liquid is in thermal equilibrium for $t \leqslant 0$. Thus the unperturbed PDF is $f[p, q; t] = f_{CA}[p, q]$ and the initial perturbation correction is $f^{(1)}[p, q; 0] = 0$.

The factor $-iL^{(1)}(\tau)f_{CA}[p, q]$ in Eq. (4.69) may be evaluated using Eqs. (4.1), (4.6), (4.64), and (4.68) to yield

$$-iL^{(1)}(\tau)f_{CA}[p, q] = \beta \mathbf{E}(\tau) \cdot \dot{\mathbf{M}} f_{CA}[p, q] \qquad (4.70)$$

Further [see Eq. (4.44)], $\exp[iLt] f_{CA}[p, q] = f_{CA}[p(t), q(t)] = f_{CA}[p, q]$, since $f_{CA}[p, q]$ is a function of $H[p, q]$, which is a constant of the unperturbed motion. Thus Eqs. (4.44) and (4.70) yield

$$\exp[-iL(t - \tau)]\left[-iL^{(1)}(\tau)f_{CA}[p, q]\right] = \beta \mathbf{E}(\tau) \cdot \dot{\mathbf{M}}[\tau - t] f_{CA}[p, q] \qquad (4.71)$$

Using this result in Eq. (4.69) yields our final result for $f^{(1)}[p, q; t]$:

$$f^{(1)}[p, q; t] = \beta f_{CA}[p, q] \int_0^t d\tau \mathbf{E}(\tau) \cdot \dot{\mathbf{M}}(\tau - t) \qquad (4.72)$$

Using Eq. (4.72) in Eq. (4.66) yields the following expression for the induced current:

$$\langle \mathbf{j}(t) \rangle_{\mathbf{E}(t)} = \beta \int_0^t d\tau \mathbf{E}(\tau) \cdot \int dp \, dq \, f_{CA}[p, q,] \dot{\mathbf{M}}(\tau - t)\mathbf{j}$$

$$= \beta \int_0^t d\tau \mathbf{E}(\tau) \cdot \langle \dot{\mathbf{M}}(\tau - t)\mathbf{j} \rangle \qquad (4.73)$$

Finally, using Eq. (4.50a), Eq. (4.73) may be rewritten as

$$\langle \mathbf{j}(t) \rangle_{\mathbf{E}(t)} = \beta \int_0^t d\tau \mathbf{E}(t - \tau) \cdot \langle \mathbf{j}(\tau)\dot{\mathbf{M}} \rangle \qquad (4.74)$$

This is the basic result of our linear response calculation.

Equation (4.74) way be rewritten in a number of different ways, which provide additional insight into its meaning. Using Eq. (4.63) and the fact that the liquid is isotropic permits one to rewrite Eq. (4.74) as

$$\langle \mathbf{j}(t) \rangle_{\mathbf{E}(t)} = \tfrac{1}{3}\beta \int_0^t d\tau \mathbf{E}(t - \tau)\langle \mathbf{j}(\tau) \cdot \mathbf{j} \rangle \qquad (4.75)$$

The Fourier–Laplace transform of Eq. (4.75) is

$$\langle \mathbf{j}(\omega) \rangle_{\mathbf{E}(t)} = \sigma(\omega)\hat{\mathbf{E}}(\omega) \qquad (4.76)$$

where

$$\sigma(\omega) \equiv \int_0^\infty \exp[i\omega t] \left[\tfrac{1}{3}\beta\langle \mathbf{j}(t)\cdot\mathbf{j}\rangle\right] dt \qquad (4.77)$$

Equation (4.76) shows that our linear-response calculation has provided a microscopic derivation of *Ohm's law*, the phenomenological linear relationship between induced current and applied electric field. Thus $\sigma(\omega)$ may be interpreted as the frequency-dependent electrical conductivity of the liquid, and Eq. (4.77) is thus a rigorous statistical-mechanical expression for this conductivity. A key feature of Eq. (4.77) is that it expresses a quantity, $\sigma(\omega)$, describing *nonequilibrium* response of the liquid in terms of a quantity, $\langle \mathbf{j}(t)\cdot\mathbf{j}\rangle$, describing the decay of *equilibrium* current fluctuations in the liquid. This feature is a characteristic of *Kubo formulas* for linear transport coefficients, of which Eq. (4.77) is a prototypical example. [The result that the response of the system to a weak applied field is determined by the properties of the system in the absence of the field is a simple consequence of perturbation theory and thus is familiar from other problems. The frequency-dependent polarizability of an atom, for example, is determined by the complete set of eigenfunctions and energy eigenvalues of the atom in the *absence* of an external field via a Kramers–Heisenberg dispersion expression. The Kubo formula, Eq. (4.76), is in fact a disguised form of such dispersion expression. This point has been mentioned by Kubo[10(a)] and developed in detail by Zwanzig.[51]]

Equation (4.75) may be rewritten in terms of the normalized VAF $\dot{\chi}(t)$ of atom μ. Denoting the force on atom μ by $\mathbf{F}(t) = Z_\mu e \mathbf{E}(t)$ and the velocity induced by $\mathbf{F}(t)$ by $\langle \dot{\mathbf{q}}_\mu(t)\rangle_{\mathbf{F}(t)} \equiv [Z_\mu e]^{-1}\langle \mathbf{j}(t)\rangle_{\mathbf{E}(t)}$, Eq. (4.75) may be rewritten using Eq. (4.63) as

$$\langle \dot{\mathbf{q}}_\mu(t)\rangle_{\mathbf{F}(t)} = m^{-1}\int_0^t \dot{\chi}(t-\tau)\mathbf{F}(\tau)\,d\tau \qquad (4.78)$$

where [cf. Eq. (3.15)]

$$\dot{\chi}(t) = \frac{m}{3k_B T}\langle \dot{\mathbf{q}}_\mu(t)\cdot\dot{\mathbf{q}}_\mu\rangle \qquad (4.79)$$

The physical interpretation of the VAF as the primary-system velocity with fluctuations removed was developed in Section III.C. Equation (4.77) provides additional insight into the meaning of the VAF. It shows that $\dot{\chi}(t)$ is the *response function* that governs velocity induction by a weak external field.

The linear-response calculation just presented may readily be generalized. Consider, for example, the problem of formally calculating the conductivity

$\sigma(\omega)$ of the ionic solution of Section III.A.3. Recall for this solution that there are n charged components. The number density and charge for the ith component are, respectively, ρ_i and $Z_i e$ [recall that $\sum_{i=1}^{n} Z_i = 0$]. The total number of ions of the ith type is $N_i = V\rho_i$, where V is the volume of the solution. By a straightforward generalization of the linear-response calculation just presented, one may show that Eqs. (4.74)–(4.77) still hold if $\mathbf{j}$ and $\mathbf{M}$ are properly reinterpreted. $\mathbf{M}$ must now be interpreted as the total ionic contribution to the electric moment of the solution, that is,

$$\mathbf{M} = \sum_{i=1}^{n} Z_i e \sum_{\mu=1}^{N_i} \mathbf{q}_{\mu}^{(i)} \tag{4.80}$$

where $\mathbf{q}_{\mu}^{(i)}$ is the coordinate of the μth ion of type i. The electric current $\mathbf{j}$ must similarly be reinterpreted as

$$\mathbf{j} = \dot{\mathbf{M}} = \sum_{i=1}^{n} \sum_{\mu=1}^{N_i} Z_i e \dot{\mathbf{q}}_{\mu}^{(i)} \tag{4.81}$$

Finally we consider the more general case of n real dynamical variables $\mathbf{A} = (A_1, \ldots, A_n)$ linearly coupled to a set of n weak external fields $\mathbf{E}_A(t)$ turned on at $t = 0$. Equation (4.67) generalizes to

$$H_A^{(1)}(t) = -\sum_{i=1}^{n} \lambda_i E_{Ai} A_i \tag{4.82}$$

where the λ_i are coupling constants analogous to $Z_i e$. Defining the generalized moment $\mathbf{M}_A$ by

$$M_{Ai} = \lambda_i A_i \tag{4.83}$$

Equation (4.82) may be rewritten in matrix notation as [cf. Eq. (4.67)]

$$H_A(t) = -\mathbf{M}_A^{\mathrm{T}} \cdot \mathbf{E}_A(t) \tag{4.84}$$

Next consider a set on n real dynamical variables $\mathbf{B}$. Assume that $\langle \mathbf{B} \rangle = 0$. (If $\langle \mathbf{B} \rangle \neq 0$, then we may redefine $\mathbf{B}$ by $\mathbf{B} \rightarrow \mathbf{B} - \langle \mathbf{B} \rangle$, so that $\langle \mathbf{B} \rangle = 0$.)

Denote by $\langle \mathbf{B}(t) \rangle_{\mathbf{E}_A(t)}$ the average value of $\mathbf{B}$ induced by $\mathbf{E}_A(t)$. Linear-response theory then yields for this quantity

$$\langle \mathbf{B}(t) \rangle_{\mathbf{E}_A(t)} = \beta \int_0^t \langle \mathbf{B}(\tau) \dot{\mathbf{M}}_A^{\mathrm{T}} \rangle \mathbf{E}_A(t - \tau) \, d\tau \tag{4.85}$$

An important special case of Eq. (4.85) occurs when $\mathbf{B}(t) = \mathbf{J}_A(t)$, where

$$\mathbf{J}_A(t) \equiv \dot{\mathbf{M}}_A(t) \tag{4.86}$$

is the generalized current associated with $A(t)$. Then Eq. (4.85) becomes

$$\langle \mathbf{J}_A(t) \rangle_{\mathbf{E}_A(t)} = \beta \int_0^t \langle \mathbf{J}_A(\tau) \mathbf{J}_A^T \rangle \mathbf{E}_A(t - \tau) \, d\tau \tag{4.87}$$

Notice that Eq. (4.75) is a special case of Eq. (4.87).
The Kubo formula associated with Eq. (4.87) is [cf. Eq. (4.77)]

$$\sigma_{AA}(\omega) = \int_0^\infty \exp[i\omega t] \langle \mathbf{J}_A(t) \mathbf{J}_A^T \rangle \, dt \tag{4.88}$$

where $\sigma_{AA}(\omega)$ is the frequency-dependent transport matrix that governs the induction of the current $\mathbf{J}_A(t)$ by the applied field $\mathbf{E}_A(t)$. Note that $\sigma_{AA}(\omega)$, in the general case, is an $n \times n$ matrix whose elements are 3×3 Cartesian tensors and is thus a $9n^2$-component quantity. Many of the $9n^2$ components are, however, identical, since the transport matrix has a number of rigorous symmetry properties. These properties have been worked out for the general (quantum-mechanical) case by Kubo. Here we limit ourselves to a discussion of the symmetry that holds if the variables $\mathbf{A}$ all have definite parities under the operation of time reversal. For this situation

$$TA_i = \varepsilon_i A_i \tag{4.89}$$

where T is the antiunitary[47] time-reversal operator and $\varepsilon_i = \pm 1$ is the time-reversal signature of A_i.

From Eq. (4.89) and the antiunitary property of T one may readily show that $\langle A_i(-t) A_j \rangle = \varepsilon_i \varepsilon_j \langle A_i(t) A_j \rangle$. Moreover, Eq. (4.50a) yields $\langle A_i(-t) A_j \rangle = \langle A_j(t) A_i \rangle$. Thus for real variables with definite time-reversal parity,

$$\langle A_i(t) A_j \rangle = \varepsilon_i \varepsilon_j \langle A_j(t) A_i \rangle \tag{4.90}$$

The "dot interchange theorem," Eq. (4.50b), further implies that $\langle \ddot{A}_i(t) A_j \rangle = -\langle \dot{A}_i(t) \dot{A}_j \rangle$ and $\langle \ddot{A}_j(t) A_i \rangle = -\langle \dot{A}_j(t) \dot{A}_i \rangle$. Thus, Eqs. (4.83), (4.86), and (4.90) imply that

$$\langle J_{A_i}(t) J_{A_j} \rangle = \varepsilon_i \varepsilon_j \langle J_{A_j}(t) J_{A_i} \rangle \tag{4.91}$$

Taking the Fourier–Laplace transform of Eq. (4.91) and then using Eq. (4.88) yields the symmetry relation

$$[\sigma_{AA}(\omega)]_{ij} = \varepsilon_i \varepsilon_j [\sigma_{AA}(\omega)]_{ji} \tag{4.92}$$

Equation (4.91) is an example of the celebrated *Onsager reciprocal relations*.[14]
    Finally we note that Eq. (4.87) may be recast using Eqs. (4.83) and (4.86)
as

$$\langle \dot{\mathbf{A}}(t) \rangle_{\mathbf{F}_A(t)} = \beta \int_0^t \dot{\chi}_{\dot{A}\dot{A}}(t - \tau) \langle \dot{\mathbf{A}}\dot{\mathbf{A}}^T \rangle \mathbf{F}_A(\tau)\, d\tau \tag{4.93}$$

where $[\mathbf{F}_A(t)]_i = \lambda_i [\mathbf{E}_A(t)]_i$ is the generalized force conjugate to the variable A_i, and where $\dot{\chi}_{\dot{A}\dot{A}}(t)$ is the normalized response matrix for the generalized velocity $\dot{\mathbf{A}}$, given by

$$\begin{aligned}
\dot{\chi}_{\dot{A}\dot{A}}(t) &= \langle \dot{\mathbf{A}}(t + \tau)\dot{\mathbf{A}}^T(\tau) \rangle \langle \dot{\mathbf{A}}\dot{\mathbf{A}}^T \rangle^{-1} \\
&= \langle \dot{\mathbf{A}}(t)\dot{\mathbf{A}}^T \rangle \langle \dot{\mathbf{A}}\dot{\mathbf{A}}^T \rangle^{-1}
\end{aligned} \tag{4.94}$$

Equations (4.93) and (4.94) may be regarded as generalizations of Eqs. (4.78) and (4.79).
    This concludes our simplified introductory discussion of linear-response theory.

7. Generalized Langevin Equations: The Response - Function Method

We next begin our development of the rigorous theory of generalized Langevin representations. A simplified discussion of such representations has already been given in Section 3. Generalized Langevin representations are usually developed via Mori's[11] projection-operator method. We instead begin with an alternative approach based on an analysis of the linear response functions, like $\dot{\chi}(t)$, just introduced. This linear-response approach will be fully developed in Section V.
    The response-function method[10(b)], although a perhaps bit less direct than the projection-operator method, more clearly brings out the physical, as opposed to the mathematical, basis for the existence of generalized Langevin representations. (That rigorous generalized Langevin representations, which closely resemble the phenomenological linear transport equations [Eq. (4.108) below], should exist is to us far from obvious. For example, the close parallelism between the rigorous equations (4.102) and (4.103) [see below] and the phenomenological equations (3.31), (3.25), and (3.19b) at first glance appears remarkable.) This physical basis is simply that if a subsystem of a thermal N-body system is slightly perturbed, it must eventually return to thermal equilibrium with the remainder of the system.

The response-function method also brings out an important related result. Namely, the essential content of a generalized Langevin representation is carried by its fluctuation–dissipation theorems. The rest is definition and notation.

To illustrate simply the response-function method, we derive the generalized Langevin representation for the velocity $v_0(t) = \dot{r}_0(t)$ of a particle of mass m translating in an isotropic liquid. This consists of the generalized Langevin equation and associated statistical theorems given in Eqs. (4.102) and (4.103) below.

To motivate the response-function method we begin with the *phenomenological* transport equation for the time dependence of the average velocity $\langle v_0(t) \rangle_{F(t)}$ of the particle, given that it is in an external force field $F(t)$. This may be developed from the Langevin equation (3.16) by suppressing the random-force term and adding a term $m^{-1}F(t)$ to account for the external force. This gives

$$\frac{d\langle v_0(t) \rangle_{F(t)}}{dt} = -\beta_1 \langle v_0(t) \rangle_{F(t)} + \frac{F(t)}{m} \tag{4.95}$$

We assume as in Section IV.B.6 that $F(t) = 0$ for $t < 0$. Thus $\langle v_0(t = 0) \rangle_{F(t)} = \langle v_0 \rangle_{\dot{q}q} = 0$. Then solving Eq. (4.95) yields

$$\langle v_0(t) \rangle_{F(t)} = m^{-1} \int_0^t \dot{\chi}_B(t - \tau)F(\tau)\,d\tau \tag{4.96}$$

where [see Eq. (3.18b)]

$$\dot{\chi}_B(t) = \exp[-\beta_1 t] \tag{4.97}$$

is the normalized VAF of a Brownian particle.

Notice the similarity in form between the rigorous linear-response result (4.78) and the phenomenological equation (4.96). This similarity suggests that one may derive a generalized Langevin equation for the atom from Eq. (4.78) by essentially inverting the steps that led from the Langevin equation (3.16) to Eq. (4.96). This strategy is the essence of the response-function method for generalized Langevin equations.

We begin with the Laplace-domain form of Eq. (4.97), which is

$$z\hat{\chi}_B(z) = [z + \beta_1]^{-1} \tag{4.98}$$

This suggests that the rigorous linear response function $\dot{\chi}(t)$ may analogously be represented as

$$z\hat{\chi}(z) = [z + \hat{\beta}_1(z)]^{-1} \tag{4.99}$$

This equation may be regarded as the *definition* of the friction kernel $\beta_1(t)$ in terms of the normalized VAF $\dot{\chi}(t)$; equivalently,

$$\beta_1(t) = \mathcal{L}^{-1}\hat{\beta}_1(z) \equiv \mathcal{L}^{-1}\left[\left[z\hat{\chi}(z)\right]^{-1} - z\right] \tag{4.100}$$

The time-domain form of Eq. (4.99) is

$$\frac{d\dot{\chi}(t)}{dt} = -\int_0^t \beta_1(t-\tau)\dot{\chi}(\tau)\,d\tau \tag{4.101}$$

We next introduce the generalized Langevin equation

$$\frac{d\mathbf{v}_0(t)}{dt} = -\int_0^t \beta_1(t-\tau)\mathbf{v}_0(\tau)\,d\tau + \mathbf{f}_1(t) \tag{4.102}$$

Eq. (4.102) may be regarded as the *definition* of the random force $\mathbf{f}_1(t)$ in terms of the atomic velocity $\mathbf{v}_0(t)$ and the normalized VAF $\dot{\chi}(t)$. Since Eq. (4.102) is merely a definition, it is so far devoid of content.

The content of (4.102) arises *solely* from the statistical theorems satisfied by $\mathbf{f}_1(t)$. These are the second fluctuation–dissipation theorem

$$\langle \mathbf{f}_1(t-\tau)\cdot\mathbf{f}_1(\tau)\rangle = \frac{3k_BT}{m}\beta_1(t) \tag{4.103a}$$

and the orthogonality relation

$$\langle \mathbf{f}_1(t)\cdot\mathbf{v}_0\rangle = 0 \tag{4.103b}$$

Thus to establish the generalized Langevin representation, we only need to prove Eqs. (4.103).

Equation (4.103b) is a corollary of the first fluctuation–dissipation theorem, Eq. (3.15), and the defining relation (4.101). To see this note that combining Eqs. (3.15) and (4.101) yields the relation

$$\langle \dot{\mathbf{v}}_0(t)\cdot\mathbf{v}_0\rangle = -\int_0^t \beta_1(t-\tau)\langle\mathbf{v}_0(\tau)\cdot\mathbf{v}_0\rangle\,d\tau = -\int_0^t \beta_1(\tau)\langle\mathbf{v}_0(t-\tau)\cdot\mathbf{v}_0\rangle\,d\tau \tag{4.104}$$

Next multiply Eq. (4.102) by $\langle\cdots\mathbf{v}_0\rangle$. This yields

$$\langle \dot{\mathbf{v}}_0(t)\cdot\mathbf{v}_0\rangle = -\int_0^t \beta_1(t-\tau)\langle\mathbf{v}_0(\tau)\cdot\mathbf{v}_0\rangle\,d\tau - \langle\mathbf{f}_1(t)\cdot\mathbf{v}_0\rangle \tag{4.105}$$

Comparison of Eqs. (4.104) and (4.105) establishes the orthogonality relation (4.103b).

Proof of the second fluctuation–dissipation theorem is a bit more complex. To keep the calculation simple, we only prove Eq. (4.103a) for the case $\tau = 0$; extension to $\tau \neq 0$ requires a lengthier but still straightforward calculation.

We begin with the relation

$$\mathbf{f}_1 = \dot{\mathbf{v}}_0$$

which follows from Eq. (4.102) evaluated at $t = 0$. Multiplying Eq. (4.102) by $\langle \cdots \mathbf{f}_1 \rangle$ and using $\mathbf{f}_1 = \dot{\mathbf{v}}_0$ yields

$$\langle \dot{\mathbf{v}}_0(t) \cdot \dot{\mathbf{v}}_0 \rangle = - \int_0^t \beta_1(t - \tau)\langle \mathbf{v}_0(\tau) \cdot \dot{\mathbf{v}}_0 \rangle + \langle \mathbf{f}_1(t) \cdot \mathbf{f}_1 \rangle \qquad (4.106)$$

Taking the time derivative of Eq. (4.104), and using Eq. (4.50b) yields

$$\langle \dot{\mathbf{v}}_0(t) \cdot \dot{\mathbf{v}}_0 \rangle = - \int_0^t \beta_1(t - \tau)\langle \mathbf{v}_0(\tau) \cdot \dot{\mathbf{v}}_0 \rangle \, d\tau + \beta_1(t)\langle v_0^2 \rangle \qquad (4.107)$$

Comparing Eq. (4.106) and (4.107) and using $\langle v_0^2 \rangle = 3k_B T / m$ establishes the second fluctuation–dissipation theorem.

Thus we have established the generalized Langevin representation (4.102), (4.103) for $\mathbf{v}_0(t)$.

The response-function method may readily be extended to develop a generalized Langevin representation for an arbitrary set of variables $\mathbf{A}(t)$. For this case, the phenomenological transport equation analogous to Eq. (4.95) is

$$\frac{d\langle \mathbf{A}(t) \rangle_{\mathbf{F}_A}}{dt} = - \mathbf{K}\langle \mathbf{A}\mathbf{A}^\mathsf{T} \rangle^{-1}\langle \mathbf{A}(t) \rangle_{\mathbf{F}_A} + \beta \langle \mathbf{A}\mathbf{A}^\mathsf{T} \rangle \mathbf{F}_A(t) \qquad (4.108)$$

where $\mathbf{K}$ is the internal[52] transport matrix. Defining $\lambda = \mathbf{K}\langle \mathbf{A}\mathbf{A}^\mathsf{T} \rangle^{-1}$, the solution to Eq. (4.108) for $\langle \mathbf{A}(t) \rangle_{\mathbf{F}_A}$ is

$$\langle \mathbf{A}(t) \rangle_{\mathbf{F}_A} = \beta \int_0^t \dot{\chi}_{AA}^{(0)}(t - \tau)\langle \mathbf{A}\mathbf{A}^\mathsf{T} \rangle \mathbf{F}_A(\tau) \, d\tau \qquad (4.109)$$

where the phenomenological response matrix is

$$\dot{\chi}_{AA}^{(0)}(t) = \exp[-\lambda t] \qquad (4.110)$$

The phenomenological and exact linear response relations (4.109) and (4.93) are identical in form [with the change in notation $\dot{\mathbf{A}} \to \mathbf{A}$ in Eq. (4.93)]. Thus the argument just presented to derive the generalized Langevin representation for $\mathbf{v}_0(t)$ may be extended to yield the following generalized Langevin representation for $\mathbf{A}(t)$:

$$\frac{d\mathbf{A}(t)}{dt} = i\hat{\omega}\mathbf{A}(t) - \int_0^t \mathbf{K}(t-\tau)\langle \mathbf{AA}^T \rangle^{-1}\mathbf{A}(\tau)\,d\tau + \mathbf{f}_A(t) \quad (4.111)$$

with associated statistical theorems

$$\langle \mathbf{f}_A(t+\tau)\mathbf{f}_A^T(\tau) \rangle = \mathbf{K}(t) \quad\quad (4.112a)$$

and

$$\langle \mathbf{f}_A(t)\mathbf{A}^T \rangle = \mathbf{0} \quad\quad (4.112b)$$

The *frequency matrix* $i\hat{\omega}$ may be evaluated in terms of equilibrium correlation functions as

$$i\hat{\omega} = \langle \dot{\mathbf{A}}\mathbf{A}^T \rangle \langle \mathbf{AA}^T \rangle^{-1} \quad\quad (4.113)$$

This equation may be derived by setting $t = 0$ in Eq. (4.111), multiplying by $\langle \cdots \mathbf{A}^T \rangle$, and using the orthogonality relation (4.112b).

Finally, Eqs. (4.111) and (4.112b) yield the following *memory-function* form for the equation of motion of the correlation function $\langle \mathbf{A}(t)\mathbf{A}^T \rangle$:

$$\frac{d\langle \mathbf{A}(t)\mathbf{A}^T \rangle}{dt} = i\hat{\omega}\langle \mathbf{A}(t)\mathbf{A}^T \rangle - \int_0^t \mathbf{K}(t-\tau)\langle \mathbf{AA}^T \rangle^{-1}\langle \mathbf{A}(\tau)\mathbf{A}^T \rangle\,d\tau \quad (4.114)$$

While the manipulations that have led to Eqs. (4.111)–(4.114) are simple, the results represent a major advance in nonequilibrium theory. This is because Eqs. (4.111) and (4.112) provide a rigorous microscopic foundation for the phenomenological linear transport equations and, in particular, give Onsager's general theory of linear transport[14] a firm statistical-mechanical basis.

We close this section by noting that if $\mathbf{A} \to \mathbf{v}_0$, Eqs. (4.111) and (4.112) reduce to Eqs. (4.102) and (4.103). This may not seem obvious, since the force term $i\hat{\omega}\mathbf{A}(t)$ in Eq. (4.111) [which governs the periodic motion $\mathbf{A}(t) = \exp[i\omega t]\mathbf{A}(0)$ that would hold if $\mathbf{f}(t)$ and hence $\mathbf{K}(t)$ vanished] has no analogue in Eq. (4.102). The term $i\hat{\omega}\mathbf{A}(t)$, however, vanishes if $\mathbf{A} \to \mathbf{v}_0$. This may be seen because $i\omega = \langle \dot{\mathbf{v}}_0 \cdot \mathbf{v}_0 \rangle \langle v_0^2 \rangle^{-1}$ [Eq. (4.113)] and because $\langle \dot{\mathbf{v}}_0 \cdot \mathbf{v}_0 \rangle = 0$, which follows from Eq. (4.90) and the fact that $\mathbf{v}_0$ and $\dot{\mathbf{v}}_0$ have opposite parity under time reversal, or alternatively from Eq. (4.50b) evaluated at $t = 0$.

8. Generalized Langevin Equations: The Projection-Operator Method

The generalized Langevin-representation (4.111)–(4.113) was first derived in a landmark paper by Mori.[11] The derivation is based on the representation of dynamical variables as abstract vectors in Liouville space [see Section IV.B.2]. The idea behind the derivation, illustrated in Fig. 11, provides a beautiful conceptual picture of how *irreversible* reduced dynamics can originate from fully reversible fundamental dynamics.

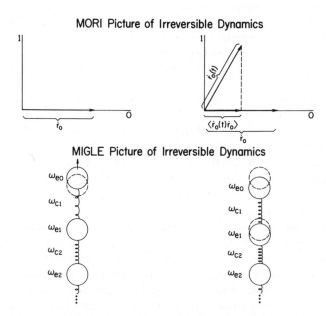

Fig. 11. Mori and MTGLE pictures of irreversible dynamics. The Mori and MTGLE representations of generalized Brownian-motion theory provide alternative and complementary conceptual pictures of how irreversible reduced dynamics originates from fully reversible fundamental dynamics. The condition for irreversibility is that the normalized VAF $\dot{\chi}(t) = \langle \dot{r}_0(t)\dot{r}_0 \rangle / \langle \dot{r}_0^2 \rangle$ must be a decaying function of time. According to the Mori theory $\dot{r}_0(t)$ is a Liouville-space vector. The norm of this vector is conserved, but memory of its initial Liouville-space direction 0 is lost as time progresses, since Liouville space has an infinite number of dimensions. Thus the Liouville-space projection of $\dot{r}_0(t)$ on the initial direction 0, which is $\langle \dot{r}_0(t)\dot{r}_0 \rangle$, is a decaying function of time. Hence $\dot{\chi}(t)$ is a decaying function of time. According to the MTGLE theory $\dot{r}_0(t)$ is the velocity of equivalent-chain atom 0. This velocity is dissipated by fictitious flow of the initial energy of chain atom 0, $\frac{1}{2}m\dot{r}_0^2$, into chain atoms 1,2,.... Hence $\dot{\chi}(t)$ is a decaying function of time.

Mori's development employs the technique of formal projection operation. Projection operators first entered irreversibility theory as a method for deriving generalized master equations from the Liouville equation (4.58). The first work along these lines is apparently due to Kohn and Luttinger,[53] in their quantum development of Bloch's theory of electron transport in solids, and also to Nakajima.[54]

The real value of projection operators in irreversibility theory, however, only became apparent with the work of Zwanzig,[12] who developed the method into a general approach to the master equation. The first rigorous development of the memory-function formalism [see Eqs. (4.101) and (4.114)] also apparently appeared in the papers of Zwanzig.[12]

Mori's treatment begins with the equation of motion, Eq. (4.38), for an arbitrary set of dynamical variables $\mathbf{A}(t)$. The Mori theory amounts to a rigorous transformation of the nonlinear N-body equation (4.38) into the linear generalized Langevin form (4.111). The simplicity, however, is only formal, since the full complexity of nonlinear N-body dynamics is carried by the Mori random force $\mathbf{f}_A(t)$. The key ideas of the Mori theory are:

1. The variables $\mathbf{A}(t)$, as discussed in Section III, may be decomposed into dissipative and stochastic components. We denote the dissipative component by $\mathbf{A}_P(t)$ and the stochastic component by $\mathbf{A}_Q(t)$.

2. The dissipative component is proportional to the time correlation function $\langle \mathbf{A}(t)\mathbf{A}^T \rangle$ (see Section III) and thus has a relatively simple time dependence. In particular, an effective equation of motion for $\mathbf{A}_P(t)$ of the simple memory-function type [see Eq. (4.114)] is expected to exist.

3. The complexity in the time dependence of $\mathbf{A}(t)$ is carried by $\mathbf{A}_Q(t)$. By analogy with the Langevin model, however, one expects this complexity may be dealt with implicitly by introducing a suitably defined random force. The effective equation of motion for $\mathbf{A}_Q(t)$, again by analogy with the Langevin model, is expected to include a random driving force, but otherwise to be identical to the equation for $\mathbf{A}_P(t)$.

We next implement these ideas. To minimize technical complication, we consider the dynamics of a single real variable $A(t)$. The equation of motion of $A(t)$ is

$$\frac{dA(t)}{dt} = iLA(t) \tag{4.115}$$

We introduce Liouville-space projection operators P and Q, which decompose $A(t)$ into dissipative components

$$A_P(t) = PA(t) \tag{4.116a}$$

and stochastic components

$$A_Q(t) = QA(t) \tag{4.116b}$$

Note that

$$A(t) = A_P(t) + A_Q(t) \tag{4.117}$$

since $P + Q = 1$. The projection operators are assumed to have the standard properties $P^2 = P$, $Q^2 = Q$, $PQ = 0$. Moreover, since $A_P(t)$ must be proportional to $\langle A(t)A \rangle$, P must project onto A; that is, $PA = A$. These restrictions imply that

$$PX = \langle XA \rangle \langle AA \rangle^{-1} A \tag{4.118a}$$

for arbitrary dynamical variables X. Thus $Q = 1 - P$ is given by

$$QX = X - \langle XA \rangle \langle AA \rangle^{-1} A \tag{4.118b}$$

Partitioning Eq. (4.115) yields

$$\frac{dA_P(t)}{dt} = PiLA_P(t) + PiLA_Q(t) \tag{4.119a}$$

and

$$\frac{dA_Q(t)}{dt} = QiLA_P(t) + QiLA_Q(t) \tag{4.119b}$$

Since $A_Q(t = 0) = QA = 0$, solving Eq. (4.119b) yields

$$A_Q(t) = \int_0^t \exp[QiL(t - \tau)] QiLA_P(\tau) \, d\tau \tag{4.120}$$

Substituting Eq. (4.120) into Eq. (4.119a) then yields

$$\frac{dA_P(t)}{dt} = PiLA_P(t) + \int_0^t PiL\exp[QiL(t - \tau)] QiLA_P(\tau) \, d\tau \tag{4.121}$$

We next rewrite Eqs. (4.120) and (4.121) in more explicit form using Eqs. (4.118). We begin by noting that

$$A_P(t) = \beta^{-1}\dot{\chi}_{AA}(t)A \qquad (4.122)$$

where

$$\dot{\chi}_{AA}(t) = \beta\langle A(t)A\rangle\langle AA\rangle^{-1} \qquad (4.123)$$

Thus $PiLA_P(t) = \beta^{-1}\dot{\chi}_{AA}(t)P\dot{A}$ is given by

$$PiLA_P(t) = i\hat{\omega}A_P(t) \qquad (4.124)$$

where

$$i\hat{\omega} = \langle \dot{A}A\rangle\langle AA\rangle^{-1} \qquad (4.125)$$

Similarly

$$\exp[QiL(t-\tau)]\,QiLA_P(\tau) = \beta^{-1}\dot{\chi}_{AA}(\tau)f_A(t-\tau) \qquad (4.126)$$

where the Mori random force is defined by

$$f_A(t) = \exp[QiLt]Q\dot{A} \qquad (4.127)$$

Notice that $f_A(t)$ remains in the Q-space for all t and is therefore orthogonal to all P-space vectors.

In particular,

$$\langle f_A(t)A\rangle = 0 \qquad (4.128a)$$

This is the orthogonality theorem satisfied by the Mori random force.

Finally, since

$$PiLf_A(t) = P\dot{f}_A(t) = \langle \dot{f}_A(t)A\rangle\langle AA\rangle^{-1}A$$
$$= -\langle f_A(t)Q\dot{A}\rangle\langle AA\rangle^{-1}A = -\langle f_A(t)f_A\rangle\langle AA\rangle^{-1}A$$

Eqs. (4.122) and (4.126) yield

$$PiL\exp[QiL(t-\tau)]\,QiLA_P(\tau) = -K(t-\tau)\langle AA\rangle^{-1}A_P(\tau) \qquad (4.129)$$

where

$$K(t) = \langle f_A(t+\tau)f_A^T(\tau)\rangle \qquad (4.128b)$$

Eq. (4.128b) is the second fluctuation–dissipation theorem satisfied by $f_A(t)$ [cf. Eq. (4.112a)].

Using Eqs. (4.124) and (4.129), the P-space equation (4.121) simplifies to

$$\frac{dA_P(t)}{dt} = i\hat{\omega}A_P(t) - \int_0^t K(t-\tau)\langle AA\rangle^{-1}A_P(\tau)\,d\tau \qquad (4.130a)$$

Using Eqs. (4.122) and (4.123) shows that Eq. (4.130a) is equivalent to the following memory-function [cf. Eq. (4.114)] equation of motion:

$$\frac{d}{dt}\chi_A(t) = i\hat{\omega}\chi_A(t) - \int_0^t K(t-\tau)\langle AA\rangle^{-1}\chi_A(\tau)\,d\tau \qquad (4.131)$$

Next, combining Eq. (4.120) and (4.126) yields

$$A_Q(t) = \beta^{-1}\int_0^t \chi_A(t-\tau)f_A(\tau)\,d\tau \qquad (4.132)$$

Differentiating Eq. (4.132) and using Eq. (4.131) shows that $A_Q(t)$ satisfies the effective equation of motion

$$\frac{dA_Q(t)}{dt} = i\hat{\omega}A_Q(t) - \int_0^t K(t-\tau)\langle AA\rangle^{-1}A_Q(\tau)\,d\tau + f_A(t) \qquad (4.130b)$$

Finally since $A_P(t) + A_Q(t) = A(t)$, combining Eqs. (4.130a) and (4.130b) yields the *Mori equation*

$$\frac{dA(t)}{dt} = i\hat{\omega}A(t) - \int_0^t K(t-\tau)\langle AA\rangle^{-1}A(\tau)\,d\tau + f_A(t) \qquad (4.133)$$

Equation (4.133), along with the statistical theorems (4.128), constitutes the Mori generalized Langevin equation for a single dynamical variable. For an arbitrary set of variables $\mathbf{A}(t)$, a generalization of the projection-operator reduction just presented yields the representation

$$\frac{d\mathbf{A}(t)}{dt} = i\hat{\boldsymbol{\omega}}\mathbf{A}(t) - \int_0^t \mathbf{K}(t-\tau)\langle \mathbf{AA}^T\rangle^{-1}\mathbf{A}(\tau)\,d\tau + \mathbf{f}_A(t) \qquad (4.134)$$

where

$$i\hat{\boldsymbol{\omega}} = \langle \dot{\mathbf{A}}\mathbf{A}^T\rangle\langle \mathbf{AA}^T\rangle^{-1} \qquad (4.135)$$

The Mori random force $\mathbf{f}_A(t)$ satisfies the statistical theorems

$$\langle \mathbf{f}_A(t + \tau)\mathbf{f}_A^{\mathsf{T}}(\tau)\rangle = K(t) \tag{4.136a}$$

and

$$\langle \mathbf{f}_A(t)\mathbf{A}^{\mathsf{T}}\rangle = 0 \tag{4.136b}$$

Notice that the above results are identical to those developed in Section IV.B.7 by the response-function method. One important new result, however, emerges from the projection-operator method. That is the following explicit result for $\mathbf{f}_A(t)$:

$$\mathbf{f}_A(t) = \exp[QiLt]\,Q\dot{\mathbf{A}} \tag{4.137}$$

where, as before, Q projects onto the subspace orthogonal to the variables $\mathbf{A}$. Equation (4.137) shows that the Mori random force evolves in time according to *fictitious* projected dynamics ($\exp[QLt]$) as opposed to physical dynamics ($\exp[iLt]$). This point has already been emphasized in Section III.F.1.

Finally we note that the *frequency matrix* $i\hat{\omega}$, the *thermodynamic derivative* $\langle \mathbf{A}\mathbf{A}^{\mathsf{T}}\rangle^{-1}$, and the *generalized transport matrix* $\mathbf{K}(t)$ appearing in the Mori equation (4.133) may be explicitly evaluated for particular choices of $\mathbf{A}$.[11] For example if $\mathbf{A}$ is taken as the momentum density of the monatomic liquid, the Mori equation reduces to a non-Markovian generalization of the linearized Navier–Stokes equation.[55]

We close this section by noting that the adiabatic generalized Langevin equation (3.48), discussed in Section III.F.3, may be derived from the Mori equation if we choose

$$\mathbf{A} = \begin{pmatrix} r_0 \\ \dot{r}_0 \end{pmatrix}$$

V.   RIGOROUS FORMULATION OF THE MTGLE REPRESENTATION OF IRREVERSIBLE DYNAMICS

The physics of subpicosecond-time-scale many-body dynamics, and the consequences of this physics for condensed-phase reaction dynamics, were discussed in Section III. The MTGLE representation of irreversible dynamics was also introduced in Section III via heuristic arguments based on the harmonic-chain model.

The MTGLE representation provides a natural formal framework for short-time-scale many-body dynamics. [Recall that the term short-time-scale dynamics refers to atomic motions that take place on a time scale shorter than the time required for relaxation of the local heatbath density.] The MTGLE representation thus complements the Mori representation, which provides a natural framework for the adiabatic-limit dynamics important for macroscopic transfort problems.

The MTGLE and Mori representations, moreover, provide alternative and complementary conceptual pictures of how irreversible reduced dynamics originates from fully reversible fundamental dynamics. The alternative pictures are illustrated in Fig. 11.

We next present a rigorous mathematical development of the basic theorems of the MTGLE representation. Our development is based on the methods of modern irreversibility theory reviewed in Section IV.B. We shall work, in particular, within the response-function approach to generalized Langevin representations outlined in Section IV.B.7.

The development in this section, since it is based on the methods of modern irreversibility theory, is admittedly abstract. Although some readers may find this repugnant, it is the abstract character of modern irreversibility theory that gives its basic theorems the generality and rigor that are the source of their real power.

To avoid technical complications, we develop the MTGLE representation for the dynamics of a single *real* variable $r_0(t)$. For convenience we shall refer to $r_0(t)$ as the coordinate of an atom of mass m. Its time derivative $\dot{r}_0(t)$ will correspondingly be referred to as the velocity of the atom. The quantity $r_0(t)$, however, may also be regarded as an arbitrary real dynamical variable.

Our development is based on the following mathematical assumptions:

1. We assume that motion in the underlying dynamical system is governed by a linear Liouville operator L; that is,

$$\frac{dA(t)}{dt} = iLA(t) \tag{5.1}$$

where $A(t)$ is an arbitrary system variable.

2. We assume that the statistics of the system is governed by an abstract inner product $\langle \ \rangle$ that obeys the standard mathematical rules [see Eqs. (4.40)]:

$$\langle XY^* \rangle = \langle YX^* \rangle^* \tag{5.2a}$$

$$\left\langle \sum_i C_i X_i Y^* \right\rangle = \sum_i C_i \langle X_i Y^* \rangle \tag{5.2b}$$

$$\langle XX^* \rangle \geqslant 0. \tag{5.2c}$$

We also choose the inner product so that the velocity autocorrelation function $\langle \dot{r}_0(t + \tau)\dot{r}_0(\tau)\rangle$ is real. Normally $\langle\ \rangle$ will be interpreted as a canonical-ensemble average. This interpretation, however, is not employed in the development of the MTGLE representation.

3. We assume the Liouville operator L is Hermitian if the inner product is $\langle\ \rangle$; that is [cf. Eq. (4.43)],

$$\langle XLY^*\rangle = \langle YLX^*\rangle^* \tag{5.3}$$

4. We assume the dynamics is smooth; that is, the time derivatives

$$r_0^{(p)}(t) \equiv \frac{d^p}{dt^p} r_0(t), \qquad p = 1, 2, \ldots$$

exist at all times t for every allowed trajectory $r_0(t) = \exp[iLt]\, r_0$. This smoothness assumption does not hold if the intermolecular potentials are singular—as is the case, for example, for the hard-sphere fluid. Thus the analysis presented below does not hold for such systems.

All subsequent results follow as rigorous consequences of the above assumptions and the additional assumption that all Fourier and Laplace transforms and spectral moments defined below exist.

A. The Normalized Velocity Autocorrelation Function

We define the normalized VAF of the atom by

$$\dot{\chi}(t) = \frac{\langle \dot{r}_0(t + \tau)\dot{r}_0(\tau)\rangle}{\langle \dot{r}_0^2\rangle} = \frac{\langle \dot{r}_0(t)\dot{r}_0\rangle}{\langle \dot{r}_0^2\rangle} \tag{5.4}$$

The stationarity property (5.4) follows from the Hermiticity of L [cf. Eq. (4.49)]. Setting $\tau = -t$ in Eq. (5.4) shows that $\dot{\chi}(t)$ is even in time [cf. Eq. (4.51)]. Moreover, by construction $\dot{\chi}(0) = 1$. Thus $\dot{\chi}(t)$ has the basic properties

$$\begin{aligned}
\dot{\chi}(-t) &= \dot{\chi}(t) \\
\dot{\chi}(0) &= 1, \qquad \chi(0) = 0
\end{aligned} \tag{5.5}$$

The spectral density $\rho(\omega)$ of $\dot{\chi}(t)$ will also play an important role in the subsequent development. This spectral density, defined by the cosine-transform pair

$$\begin{aligned}
\rho(\omega) &= \frac{2}{\pi} \int_0^\infty \cos \omega t\, \dot{\chi}(t)\, dt \\
&\equiv \frac{2}{\pi} \mathrm{Re}\left\{ \lim_{\varepsilon \to 0} \int_0^\infty \exp[-(i\omega + \varepsilon)t]\, \dot{\chi}(t)\, dt \right\}
\end{aligned} \tag{5.6a}$$

and

$$\dot{\chi}(t) = \int_0^\infty \cos \omega t \, \rho(\omega) \, d\omega \tag{5.6b}$$

has the following basic properties:

$$\rho(-\omega) = \rho(\omega) \tag{5.7a}$$

$$\int_0^\infty \rho(\omega) \, d\omega = 1 \tag{5.7b}$$

$$\rho(\omega) \geqslant 0 \tag{5.7c}$$

Equations (5.7a, b) follow directly from Eqs. (5.5) and (5.6). The positivity of $\rho(\omega)$, Eq. (5.7c), may be proven by considering the eigenvalue problem

$$L\phi_\lambda = \omega_\lambda \phi_\lambda \tag{5.8}$$

Since L is Hermitian, its eigenfunctions ϕ_λ form a complete orthonormal set and its eigenvalues ω_λ are real. These properties permit one to express $\rho(\omega)$ $(\omega \geqslant 0)$ as[‡]

$$\rho(\omega) = \frac{2}{\langle \dot{r}_0^2 \rangle} \sum_\lambda |\langle \phi_\lambda \dot{r}_0 \rangle|^2 \delta(\omega - \omega_\lambda) \tag{5.9}$$

Eq. (5.9) shows that $\rho(\omega) \geqslant 0$.

All subsequent results of Section V follows from the properties (5.5) and (5.7). These properties in turn follow from the assumption that L is Hermitian if the inner product is $\langle \ \rangle$.

B. MTGLE Hierarchy for Response Functions

The mathematical structure of the MTGLE theory is based on an infinite hierarchy of effective equations of motion for the response functions $\theta_1(t), \theta_2(t), \ldots$ and associated coordinate variables $R_1(t), R_2(t), \ldots$ that characterize the dynamics of an infinite sequence of abstract heatbaths. [The approximate solvation-shell interpretation of these heatbaths for the case of solute motion in liquids has been discussed in Section III.] We next develop the hierarchy for response functions.

1. Einstein Frequency and Coupling Constant

We begin by introducing the Einstein frequency ω_{e0} of the atom and the coupling constant ω_{c1}^2 that couples it to the first abstract heatbath. These are

---

[‡]Eq. (5.9) may be proven as follows. We begin with Eq. (5.4) written as $\dot{\chi}(t) = \langle \dot{r}_0^2 \rangle^{-1} \langle \dot{r}_0 \exp[-iLt] \dot{r}_0 \rangle$. Using Eq. (5.8), one may write $\exp[-iLt] = \Sigma_\lambda \exp[-i\omega_\lambda t] |\phi_\lambda\rangle\langle\phi_\lambda|$. Thus $\dot{\chi}(t) = \langle \dot{r}_0^2 \rangle^{-1} \Sigma_\lambda |\langle \phi_\lambda \dot{r}_0 \rangle|^2 \exp[-i\omega_\lambda t]$. Combining this result with Eq. (5.6a) yields Eq. (5.9).

defined in terms of the second and fourth moments of the spectral density via

$$\omega_{e0}^2 = \int_0^\infty \omega^2 \rho(\omega)\, d\omega \geqslant 0 \qquad (5.10a)$$

and

$$\omega_{c1}^4 = \int_0^\infty \left[\omega^2 - \omega_{e0}^2\right]^2 \rho(\omega)\, d\omega \geqslant 0 \qquad (5.10b)$$

Comparing Eqs. (5.6b) and (5.10a) shows that the short-time expansion of $\dot{\chi}(t)$ is of the form

$$\dot{\chi}(t) = 1 - \tfrac{1}{2}\omega_{e0}^2 t^2 + \cdots$$
$$= \cos \omega_{e0} t, \qquad \text{(short times)} \qquad (5.11)$$

The short-time limit of $\dot{\chi}(t)$, which is $\cos \omega_{e0} t$, is identical to the normalized VAF of an isolated or Einstein harmonic oscillator. (Hence the designation "Einstein frequency" for ω_{e0}.) The physical basis of this fundamental point (e.g., solute cage oscillations) has been extensively discussed in Section III.

The Einstein limit breaks down at longer times due to coupling, via ω_{c1}^2, of the atom to the heatbath. We next construct a representation for $\dot{\chi}(t)$ that explicitly displays the Einstein limit and also rigorously accounts for its breakdown.

2. The Normalized Heatbath Response Function $\dot{\theta}_1(t)$

This representation, written in the Laplace domain, is

$$\hat{\chi}(z) = \left[z^2 + \omega_{e0}^2 - \omega_{c1}^4 \hat{\theta}_1(z)\right]^{-1} \qquad (5.12a)$$

This equation may be regarded as the *definition* of the heatbath response function $\dot{\theta}_1(t) = \mathcal{L}^{-1}[z\hat{\theta}_1(z)]$ in terms of the normalized VAF $\dot{\chi}(t)$. Thus Eq. (5.12a) is in the spirit of the response-function approach [see Section IV.B.7 and cf. Eqs. (4.99) and (5.12a)] to the construction of generalized Langevin representations. Comparison of Eqs. (5.11) and (5.12a), moreover, shows that the heatbath response function $\dot{\theta}_1(t)$ rigorously accounts for the breakdown of the Einstein limit: if $\hat{\theta}_1(z)$ vanished, then Eq., (5.12a) would be the Laplace transform of Eq. (5.11).

The time-domain form of Eq. (5.12a) is the following effective equation of motion for $\dot{\chi}(t)$:

$$\ddot{\chi}(t) = -\omega_{e0}^2 \chi(t) + \omega_{c1}^4 \int_0^t \chi(t-\tau)\theta_1(\tau)\, d\tau \qquad (5.12b)$$

From Eqs. (5.10) and (5.12b) it is straightforward[‡] to show that $\dot{\theta}_1(t)$ has basic properties identical to those of $\dot{\chi}(t)$, that is [cf. Eq. (5.5)],

$$\dot{\theta}_1(-t) = \dot{\theta}_1(t)$$

$$\dot{\theta}_1(0) = 1, \qquad \theta_1(0) = 0$$

(5.13)

We next define the spectral density $\sigma_1(\omega)$ of $\dot{\theta}_1(t)$ by a cosine-transform pair analogous to Eqs. (5.6):

$$\sigma_1(\omega) = \frac{2}{\pi} \int_0^\infty \cos \omega t \, \dot{\theta}_1(t)$$

$$\equiv \frac{2}{\pi} \operatorname{Re} \left\{ \lim_{\varepsilon \to 0} \int_0^\infty \exp[-(i\omega + \varepsilon)t] \, \dot{\theta}_1(t) \, dt \right\} \qquad (5.14a)$$

and

$$\dot{\theta}_1(t) = \int_0^\infty \cos \omega t \, \sigma_1(\omega) \, d\omega \qquad (5.14b)$$

The spectral density $\sigma_1(\omega)$ has basic properties identical to those of $\rho(\omega)$, that is [cf. Eqs. (5.7)],

$$\sigma_1(-\omega) = \sigma_1(\omega) \qquad (5.15a)$$

$$\int_0^\infty \sigma_1(\omega) \, d\omega = 1 \qquad (5.15b)$$

$$\sigma_1(\omega) \geqslant 0 \qquad (5.15c)$$

Equations (5.15a, b) follow directly from Eqs. (5.13) and (5.14). The positivity of $\sigma_1(\omega)$ follows from the relation[§] $\sigma_1(\omega) = \lim_{\varepsilon \to 0} \{ \omega_{cl}^{-4} \rho(\omega) / |\hat{\chi}(i\omega + \varepsilon)|^2 \}$. Since $\rho(\omega)$ and ω_{cl}^4 are positive, this relation shows that $\sigma_1(\omega)$ is also positive.

[‡] This may be proven, for example, by taking successive time derivatives of Eq. (5.12b), setting $t = 0$, and using Eq. (5.5). This procedure shows that all even derivatives of $\theta_1(t)$, when evaluated at $t = 0$, vanish. This shows that $\theta_1(t)$ is an odd function of time and thus $\dot{\theta}_1(t)$ is an even function; that is, $\dot{\theta}_1(-t) = \dot{\theta}_1(t)$. To prove $\dot{\theta}_1(0) = 1$, differentiate Eq. (5.12b) three times. Using Eq. (5.5) and setting $t = 0$, this yields $\chi^{(5)}(0) = -\omega_{e0}^2 \chi^{(3)}(0) + \omega_{cl}^4 \dot{\theta}_1(0)$. However, Eqs. (5.10) and (5.6b) show that $\omega_{cl}^4 = \chi^{(5)}(0) + \omega_{e0}^2 \chi^{(3)}(0)$. Thus $\dot{\theta}_1(0) = 1$. Thus we have proven Eq. (5.13).

[§] This relation may be proven from $\omega_{cl}^4 \hat{\theta}_1(z) = \hat{\chi}^{-1}(z) - z^2 - \omega_{e0}^2$, which is a rearrangement of Eq. (5.12a), and from the facts that $\rho(\omega) = (2\omega/\pi) \times \operatorname{Im} \hat{\chi}(i\omega + \varepsilon)$ and $\sigma_1(\omega) = (2\omega/\pi) \operatorname{Im} \hat{\theta}_1(i\omega + \varepsilon)$, which follow from Eqs. (5.6a) and (5.14a).

3. Response-Function Hierarchy

We have just seen that $\dot{\theta}_1(t)$ and $\sigma_1(\omega)$ have basic properties identical to those of $\dot{\chi}(t)$ and $\rho(\omega)$. This indicates that one may recursively continue the procedure of Section V.B.2 and thus define an infinite sequence of abstract response functions $\dot{\theta}_2(t), \dot{\theta}_3(t), \ldots$. All response functions have basic properties identical to those of $\dot{\chi}(t)$. All satisfy effective equations of motion identical in form to Eq. (5.12). The general pth ($p = 1, 2, \ldots$) heatbath response function has the basic properties

$$\dot{\theta}_p(-t) = \dot{\theta}_p(t)$$
$$\dot{\theta}_p(0) = 1, \qquad \theta_p(0) = 0 \tag{5.16}$$

The spectral density $\sigma_p(\omega)$ of $\dot{\theta}_p(t)$ is defined by the cosine-transform pair

$$\sigma_p(\omega) = \frac{2}{\pi} \int_0^\infty \cos \omega t \, \dot{\theta}_p(t) \, dt \tag{5.17a}$$

$$\dot{\theta}_p(t) = \int_0^\infty \cos \omega t \, \sigma_p(\omega) \, d\omega \tag{5.17b}$$

The spectral density $\sigma_p(\omega)$ has basic properties identical to those of $\rho(\omega)$:

$$\sigma_p(-\omega) = \sigma_p(\omega) \tag{5.18a}$$

$$\int_0^\infty \sigma_p(\omega) \, d\omega = 1 \tag{5.18b}$$

$$\sigma_p(\omega) \geqslant 0 \tag{5.18c}$$

The Einstein frequency of the pth heatbath and the coupling constant link-ing the pth and ($p + 1$)th heatbaths are defined by relations analogous to Eqs. (5.10):

$$\omega_{ep}^2 = \int_0^\infty \omega^2 \sigma_p(\omega) \, d\omega \geqslant 0 \tag{5.19a}$$

$$\omega_{cp+1}^2 = \int_0^\infty \left[\omega^2 - \omega_{ep}^2\right]^2 \sigma_p(\omega) \, d\omega \geqslant 0 \tag{5.19b}$$

As discussed in Section III.F.4, all information about heatbath influence on atomic dynamics is carried by the complete set of MTGLE parameters $\{\omega_{ep}^2, \omega_{cp+1}^2\}$, $p = 0, 1, \ldots, \infty$. An algorithm for constructing these parameters from $\dot{\chi}(t)$ is given in Appendix A.

The response functions of the $(p+1)$th and pth heatbaths are related by defining equations analogous to Eqs. (5.12), namely,

$$\hat{\theta}_p(z) = \left[z^2 + \omega_{ep}^2 - \omega_{cp+1}^4 \hat{\theta}_{p+1}(z)\right]^{-1} \qquad (5.20a)$$

or equivalently, in the time domain,

$$\ddot{\theta}_p(t) = -\omega_{ep}^2 \theta_p(t) + \omega_{cp+1}^4 \int_0^t \theta_p(t-\tau)\theta_{p+1}(\tau)\, d\tau \qquad (5.20b)$$

Notice that Eqs. (5.12) and (5.20) constitute a recursive sequence that successively *define* the heatbath response functions $\dot{\theta}_1(t), \dot{\theta}_2(t), \ldots$ in terms of $\dot{\chi}(t)$. That is, Eqs. (5.12) define $\dot{\theta}_1(t)$ in terms of $\dot{\chi}(t)$, Eqs. (5.20) for $p = 1$ define $\dot{\theta}_2(t)$ in terms of $\dot{\theta}_1(t)$ and hence $\dot{\chi}(t)$, and so on.

4. Continued-Fraction Representations

It is now straightforward to derive a rigorous continued-fraction representation for $\hat{\chi}(z)$. For example, specializing Eq. (5.20a) to the case $p = 1$ yields

$$\hat{\theta}_1(z) = \left[z^2 + \omega_{e1}^2 - \omega_{c2}^4 \hat{\theta}_2(z)\right]^{-1}$$

Using the above equation in Eq. (5.12a) to eliminate $\hat{\theta}_1(z)$ yields the continued fraction

$$\hat{\chi}(z) = \cfrac{1}{z^2 + \omega_{e0}^2 - \cfrac{\omega_{c1}^4}{z^2 + \omega_{e1}^2 - \omega_{c2}^4 \hat{\theta}_2(z)}} \qquad (5.21)$$

The response function $\hat{\theta}_2(z)$ may similarly be eliminated from Eq. (5.21) to yield a continued fraction involving $\hat{\theta}_3(z)$. Continuing this procedure indefinitely yields the following infinite-continued-fraction representation of $\hat{\chi}(z)$:

$$\hat{\chi}(z) = \cfrac{1}{z^2 + \omega_{e0}^2 - \cfrac{\omega_{c1}^4}{z^2 + \omega_{e1}^2 - \cfrac{\omega_{c2}^4}{z^2 + \omega_{e2}^2 - \cfrac{\omega_{c3}^4}{\cdots}}}} \qquad (5.22)$$

A corresponding infinite-continued-fraction representation of $\hat{\theta}_p(z)$ for $p =$

$1, 2, 3, \ldots$ may be similarly derived from Eq. (5.20a):

$$\hat{\theta}_p(z) = \cfrac{1}{z^2 + \omega_{ep}^2 - \cfrac{\omega_{cp+1}^4}{z^2 + \omega_{ep+1}^2 - \cfrac{\omega_{cp+2}^4}{z^2 + \omega_{ep+2}^2 - \cfrac{\omega_{cp+3}^4}{\cdots}}}} \tag{5.23}$$

C. Equivalent-Harmonic-Chain Representation for Response Functions

A central feature of the MTGLE representation discussed in Section III.F is that the temporal development of the response functions $\chi(t), \theta_1(t), \ldots$ and the associated dynamical variables $r_0(t), R_1(t), \ldots$ may be rigorously represented as trajectories of fictitious atoms belonging to abstract harmonic chains. The chain parameters are the set of Einstein frequencies and coupling constants just defined.

The atomic coordinate $r_0(t)$ (which, we recall, can be an *arbitrary* real dynamical variable) and its associated VAF $\dot{\chi}(t)$ are trajectories of the terminal chain atom 0. The abstract heatbath coordinate $R_p(t)$ and its associated VAF $\dot{\theta}_p(t)$ are trajectories of chain atom p in the pth clamped chain.

1. Chain Representation of the Normalized VAF $\dot{\chi}(t)$

We first derive the chain representation for $\dot{\chi}(t)$. We define the coordinates $\rho_0(t)$ of chain atom 0 and $\rho_1(t)$ of chain atom 1 by

$$\rho_0(t) = \chi(t) \tag{5.24a}$$

and

$$\rho_1(t) = \omega_{c1}^2 \int_0^t \chi(t-\tau)\theta_1(\tau)\,d\tau$$

$$= \omega_{c1}^2 \int_0^t \theta_1(t-\tau)\rho_0(\tau)\,d\tau \tag{5.24b}$$

Given the definitions of Eq. (5.24), the effective equation of motion for $\chi(t)$, Eq. (5.12b), becomes the following equation of motion for $\rho_0(t)$

$$\ddot{\rho}_0(t) = -\omega_{e0}^2\rho_0(t) + \omega_{c1}^2\rho_1(t) \tag{5.25a}$$

The equation of motion for $\rho_1(t)$ is, correspondingly,

$$\ddot{\rho}_1(t) = -\omega_{e1}^2\rho_1(t) + \omega_{c1}^2\rho_0(t) + \omega_{c1}^4\int_0^t \theta_2(t-\tau)\rho_1(\tau)\,d\tau \tag{5.25b}$$

Equation (5.25b) may be derived as follows. The Laplace transform of Eq. (5.24b) is

$$\hat{\rho}_1(z) = \omega_{c1}^2 \hat{\theta}_1(z) \hat{\rho}_0(z).$$

Using Eq. (5.20a), for $p = 1$, the above equation may be rewritten as

$$\left[z^2 + \omega_{e1}^2 - \omega_{c2}^4 \hat{\theta}_2(z) \right] \hat{\rho}_1(z) = \omega_{c1}^2 \hat{\rho}_0(z).$$

The inverse Laplace transform of the above equation is Eq. (5.25b).

The equations (5.25b) constitute the equivalent-chain equations for the case $N = 1$. The general $(N + 1)$-atom chain equations $N = 1, 2, \ldots$ may be developed by continuing the procedure just described. This procedure yields

$$\ddot{\rho}_0(t) = -\omega_{e0}^2 \rho_0(t) + \omega_{c1}^2 \rho_1(t)$$

$$\ddot{\rho}_1(t) = -\omega_{e1}^2 \rho_1(t) + \omega_{c1}^2 \rho_0(t) + \omega_{c2}^2 \rho_2(t)$$

$$\vdots$$

$$\ddot{\rho}_N(t) = -\omega_{eN}^2 \rho_N(t) + \omega_{cN}^2 \rho_{N-1}(t) + \omega_{cN+1}^4 \int_0^t \theta_{N+1}(t - \tau) \rho_N(\tau) \, d\tau$$

$$(5.26)$$

The qth chain-atom coordinate $\rho_q(t)$ is defined in terms of the $(q - 1)$th coordinate by

$$\rho_q(t) = \omega_{cq}^2 \int_0^t \theta_q(t - \tau) \rho_{q-1}(\tau) \, d\tau \qquad (5.24c)$$

Equation (5.24) provides a recursive sequence that successively defines the coordinates $\rho_q(t)$, $q = 0, 1, 2, \ldots$, in terms of the normalized VAF $\dot{\chi}(t)$.

The initial conditions of the chain atoms are as follows. Since $\rho_0(t) = \chi(t)$, Eq. (5.5) yields that $\rho_0 = 0$ and $\dot{\rho}_0 = 1$. Moreover, since $\theta_q(0) = 0$ [Eq. (5.16)], Eq. (5.27) gives $\rho_q = \dot{\rho}_q = 0$, $q \geqslant 1$. These considerations show that $\dot{\chi}(t)$ is the following particular velocity of chain atom 0:

$$\dot{\chi}(t) = \rho_0(t) \qquad (5.27a)$$

given the initial conditions

$$\dot{\rho}_0 = 1, \qquad \rho_0 = 0$$
$$\dot{\rho}_q = \rho_q = 0, \qquad q \geqslant 1 \qquad (5.27b)$$

We next pass to the limit of an infinite chain. For this case the chain equations of motion (5.26) may be written in matrix notation as

$$\ddot{\rho}(t) = -\omega^2 \rho(t) \tag{5.28}$$

where

$$\rho(t) = \begin{pmatrix} \rho_0(t) \\ \rho_1(t) \\ \vdots \end{pmatrix} \tag{5.29}$$

is an infinite vector of chain-atom coordinates and where ω^2 is the following infinite symmetric dynamical matrix [cf. Eq. (3.2a)]:

$$\omega^2 = \begin{bmatrix} \omega_{e0}^2 & -\omega_{c1}^2 & 0 & 0 & 0 & \cdots \\ -\omega_{c1}^2 & \omega_{e1}^2 & -\omega_{c2}^2 & 0 & 0 & \cdots \\ 0 & -\omega_{c2}^2 & \omega_{e2}^2 & -\omega_{c3}^2 & 0 & \cdots \\ \vdots & \vdots & \vdots & \vdots & \vdots & \end{bmatrix} \tag{5.30}$$

The normalized VAF $\dot{\chi}(t)$ is the velocity $\dot{\rho}_0(t)$ of atom 0 in the infinite chain, given the initial conditions (5.27b). Moreover, for *arbitrary* chain-atom initial conditions, solving Eq. (5.28) yields the following expression for the velocity of chain atom 0:

$$\dot{\rho}_0(t) = \sum_{q=0}^{\infty} \{ [\cos \omega t]_{0q} \dot{\rho}_q - [\omega \sin \omega t]_{0q} \rho_q \} \tag{5.31}$$

Thus using the special initial conditions (5.27b) in Eq. (5.31) yields the following rigorous representation of $\dot{\chi}(t)$ in terms of the dynamical matrix ω^2 [cf. Eq. (3.12)]:

$$\dot{\chi}(t) = [\cos \omega t]_{00} \tag{5.32}$$

We next note that the normal-mode frequencies ω_λ [$\lambda = 0, 1, \ldots$] of the infinite chain may be determined by diagonalizing the symmetric matrix ω^2 via an orthogonal matrix $\mathbf{U}$; that is,

$$\omega_D^2 = \mathbf{U}^T \omega^2 \mathbf{U} \tag{5.33}$$

where

$$\mathbf{U}^\mathsf{T}\mathbf{U} = \mathbf{U}\mathbf{U}^\mathsf{T} = 1 \tag{5.34}$$

and where

$$\omega_D^2 = \begin{bmatrix} \omega_0^2 & 0 & 0 & \cdots \\ 0 & \omega_1^2 & 0 & \cdots \\ 0 & 0 & \omega_2^2 & \cdots \\ \vdots & \vdots & \vdots & \end{bmatrix} \tag{5.35}$$

is the dynamical matrix in the normal-mode representation.

The normal-mode frequencies are just the eigenvalues ω_λ of the Liouville operator L belonging to eigenfunctions ϕ_λ with nonzero Liouville space projections $\langle \dot{r}_0 \phi_\lambda \rangle$ onto $\dot{r}_0$. This may be proven by noting that $\dot{\chi}(t) = [\cos \omega t]_{00} = [\mathbf{U} \cos \omega_D t \mathbf{U}^\mathsf{T}]_{00}$, or equivalently,

$$\dot{\chi}(t) = \sum_{\lambda = 0}^{\infty} U_{0\lambda} U_{0\lambda} \cos \omega_\lambda t \tag{5.36}$$

which is a new rigorous representation of $\dot{\chi}(t)$. Taking the cosine transform of Eq. (5.36) yields the following rigorous normal-mode representation of the spectral density ($\omega \geqslant 0$):

$$\rho(\omega) = \sum_{\lambda = 0}^{\infty} U_{0\lambda} U_{0\lambda} \delta(\omega - \omega_\lambda) \tag{5.37}$$

Comparing Eqs. (5.9) and (5.37) shows that the infinite-chain normal-mode frequencies ω_λ are the eigenvalues of the Liouville operator L.

2. Chain Representation of the Heatbath Response Functions $\dot{\theta}_p(t)$

As discussed in Section III.F, the heatbath response function $\dot{\theta}_p(t)$ has a representation in the pth clamped equivalent chain analogous to that of $\dot{\chi}(t)$ in the full chain. The proofs of the basic results parallel those of Section V.C.1 and thus will be omitted.

The basic results are as follows. The pth infinite clamped chain is governed by the dynamical matrix

$$[\omega^{(p)}]^2 = \begin{bmatrix} \omega_{ep}^2 & -\omega_{cp+1}^2 & 0 & 0 & \cdots \\ -\omega_{cp+1}^2 & \omega_{ep+1}^2 & -\omega_{cp+2}^2 & 0 & \cdots \\ 0 & -\omega_{cp+2}^2 & \omega_{ep+2}^2 & -\omega_{cp+3}^2 & \cdots \\ \vdots & \vdots & \vdots & \vdots & \end{bmatrix} \tag{5.38}$$

which is obtained by striking out the first p columns and rows of the dynamical matrix ω^2 [Eq. (5.30)] of the full chain. Note that ω^2 and $[\omega^{(p)}]^2$ are identical in form to the dynamical matrices (3.2), for the full and clamped *physical* chain. The coordinates of the pth clamped chain will be denoted collectively by

$$\rho^{(p)}(t) = \begin{pmatrix} \rho_p^{(p)}(t) \\ \rho_p^{(p+1)}(t) \\ \vdots \end{pmatrix} \tag{5.39}$$

These are defined recursively in terms of $\theta_p(t)$ via relations analogous to Eqs. (5.24) and (5.27), namely,

$$\rho_p^{(p)}(t) = \theta_p(t) \tag{5.40a}$$

and

$$\rho_q^{(p)}(t) = \omega_{cq}^2 \int_0^t \theta_q(t-\tau) \rho_{q-1}^{(p)}(\tau) \, d\tau, \qquad q \geqslant p+1 \tag{5.40b}$$

The chain coordinates satisfy the following infinite-matrix equation of motion [cf. Eq. (5.28)]:

$$\ddot{\rho}^{(p)}(t) = -[\omega^{(p)}]^2 \rho^{(p)}(t) \tag{5.41}$$

The chain-trajectory representation of $\dot{\theta}_p(t)$ is

$$\dot{\theta}_p(t) = \dot{\rho}_p^{(p)}(t) \tag{5.42a}$$

given the initial conditions

$$\begin{aligned} \dot{\rho}_p^{(p)} &= 1, \qquad \rho_p^{(p)} = 0 \\ \dot{\rho}_q^{(p)} &= \rho_q^{(p)} = 0, \qquad q \geqslant p+1 \end{aligned} \tag{5.42b}$$

The dynamical matrix representation of $\dot{\theta}_p(t)$ is

$$\dot{\theta}_p(t) = [\cos \omega^{(p)} t]_{pp} \tag{5.43}$$

The normal-mode frequencies of the pth clamped chain may be found by diagonalizing $[\omega^{(p)}]^2$ via an orthogonal transformation $\mathbf{U}^{(p)}$:

$$[\omega_D^{(p)}]^2 = [\mathbf{U}^{(p)}]^T [\omega^{(p)}]^2 \mathbf{U}^{(p)} \tag{5.44}$$

where

$$[\mathbf{U}^{(p)}]^{\mathsf{T}}\mathbf{U}^{(p)} = \mathbf{U}^{(p)}[\mathbf{U}^{(p)}]^{\mathsf{T}} \tag{5.45}$$

and where

$$[\omega_D^{(p)}]^2 = \begin{bmatrix} [\omega_p^{(p)}]^2 & 0 & 0 & \cdots \\ 0 & [\omega_{p+1}^{(p)}]^2 & 0 & \cdots \\ 0 & 0 & [\omega_{p+2}^{(p)}]^2 & \cdots \\ \vdots & \vdots & \vdots & \end{bmatrix} \tag{5.46}$$

The normal-mode representations of $\dot{\theta}_p(t)$ and its spectral density $\sigma_p(\omega)$ are

$$\dot{\theta}_p(t) = \sum_{\lambda=p}^{\infty} U_{p\lambda}^{(p)} U_{p\lambda}^{(p)} \cos \omega_\lambda^{(p)} t \tag{5.47}$$

and

$$\sigma_p(\omega) = \sum_{\lambda=p}^{\infty} U_{p\lambda}^{(p)} U_{p\lambda}^{(p)} \delta[\omega - \omega_\lambda^{(p)}] \tag{5.48}$$

D.   Molecular-Time-Scale Generalized Langevin Equations

We next turn to the MTGLE hierarchy for the dynamical variables $r_0(t)$, $R_1(t)$,.... This is described by the following infinite sequence of MTGLEs [cf. Eqs. (3.37) and (3.43)]:

$$\ddot{r}_0(t) = -\omega_{e0}^2 r_0(t) + \omega_{c1}^4 \int_0^t \theta_1(t-\tau) r_0(\tau)\, d\tau + \omega_{c1}^2 R_1(t)$$

$$\ddot{R}_1(t) = -\omega_{e1}^2 R_1(t) + \omega_{c2}^4 \int_0^t \theta_2(t-\tau) R_1(\tau)\, d\tau + \omega_{c2}^2 R_2(t)$$

$$\vdots$$

$$\ddot{R}_p(t) = -\omega_{ep}^2 R_p(t) + \omega_{cp+1}^4 \int_0^t \theta_p(t-\tau) R_p(\tau)\, d\tau + \omega_{cp+1}^2 R_{p+1}(t)$$

$$\vdots \tag{5.49}$$

In the spirit of Eq. (4.102), the equations (5.49) may be regarded as recursive *defining* relations for the MTGLE random forces $\omega_{c1}^2 R_1(t),\ldots,$

$\omega_{cp+1}^2 R_{p+1}(t)$, ar equivalently the abstract heatbath coordinates $R_1(t)$, $R_2(t),\ldots,R_{p+1}(t)$, in terms of the atomic coordinate $r_0(t)$. For example, the first equation of (5.49) *defines* $R_1(t)$ in terms of $r_0(t)$, the second defines $R_2(t)$ in terms of $R_1(t)$ and hence $r_0(t)$, and so on.

A key point concerning (5.49) is that the pth heatbath coordinate variable is proportional to the random force acting on the $(p-1)$th heatbath variable.

Finally we note that the equations (5.49) are, so far, simply definitions and are thus devoid of physical content. Their content arises from the fluctuation–dissipation and orthogonality theorems developed below [Eqs. (5.77)].

E. Equivalent-Chain Representation for the Coordinate Variables

The main results of this section are rigorous equivalent-chain representations for $r_0(t)$, $R_1(t),\ldots$. The atomic coordinate $r_0(t)$ is recast as the trajectory of chain atom 0 in the full chain; the pth abstract heatbath coordinate $R_p(t)$ is recast as the trajectory of chain atom p in the pth clamped chain. The normal modes of the chains are then examined.

1. Chain Representation for $r_0(t)$

The chain representation for $r_0(t)$ may be derived from the MTGLE hierarchy (5.49). Since the derivation parallels the development of the chain representation for $\dot{\chi}(t)$, we skip the details and pass to the main results.

The $(N+1)$-atom chain representation for $r_0(t)$ is (note that N can equal $1,2,3,\ldots$)

$$\ddot{r}_0(t) = -\omega_{e0}^2 r_0(t) + \omega_{c1}^2 r_1(t)$$
$$\ddot{r}_1(t) = -\omega_{e1}^2 r_1(t) + \omega_{c1}^2 r_0(t) + \omega_{c2}^2 r_2(t)$$
$$\vdots \qquad\qquad\qquad\qquad\qquad\qquad (5.50)$$
$$\ddot{r}_N(t) = -\omega_{eN}^2 r_N(t) + \omega_{cN}^2 r_{N-1}(t)$$
$$\qquad + \omega_{cN+1}^4 \int_0^t \theta_{N+1}(t-\tau) r_N(\tau)\, d\tau + \omega_{cN+1}^2 R_{N+1}(t)$$

The coordinate $r_0(t)$ is the atomic coordinate. The coordinates $r_1(t)$, $r_2(t),\ldots$ are mathematical constructs defined recursively in terms of $r_0(t)$ and $\dot{\chi}(t)$ via the sequence

$$r_p(t) = R_p(t) + \omega_{cp}^2 \int_0^t \theta_p(t-\tau) r_{p-1}(\tau)\, d\tau_1, \qquad p \geq 1 \qquad (5.51)$$

Since $\theta_p(0) = 0$, Eq. (5.51) shows that the initial conditions of atom p in the full chain $[r_p(t)]$ and atom p in the pth clamped chain $[R_p(t)]$ are identical; that is,

$$\dot{r}_p = \dot{R}_p \text{ and } r_p = R_p, \quad p \geqslant 1 \tag{5.52}$$

Finally, passing to the limit of an infinite chain $[N \to \infty]$, Eq. (5.50) may be rewritten in matrix notation as

$$\ddot{r}(t) = -\omega^2 r(t) \tag{5.53}$$

where ω^2 is defined by Eq. (5.30) and where $r(t) = (r_0(t), r_1(t), \ldots)$. Solving Eq. (5.53) yields the following expansion for the velocity:

$$\dot{r}_0(t) = \sum_{q=0}^{\infty} \{[\cos \omega t]_{0q} \dot{r}_p - [\omega \sin \omega t]_{0q} r_q\} \tag{5.54}$$

2. Normal Modes and Eigenfunctions of the Liouville Operator

The normal modes of the full chain which we denote collectively by $\zeta(t) \equiv (\zeta_0(t), \zeta_1(t), \ldots)$ are determined in terms of the atom coordinates $r(t)$ by the orthogonal transformation

$$\zeta(t) = \mathbf{U}^T r(t) \tag{5.55a}$$

The inverse of Eq. (5.55a) is

$$r(t) = \mathbf{U}\zeta(t) \tag{5.55b}$$

where, we recall, $\mathbf{U}$ is the matrix that diagonalizes ω^2 to produce ω_D^2, the matrix of normal-mode frequencies ω_λ. The normal-mode coordinates obey equations of motion of the form $\ddot{\zeta}_\lambda(t) = -\omega_\lambda^2 \zeta_\lambda(t)$. Thus

$$\zeta_\lambda(t) = \cos \omega_\lambda t \zeta_\lambda + \omega_\lambda^{-1} \sin \omega_\lambda t \dot{\zeta}_\lambda \tag{5.56}$$

Equation (5.55b) and (5.56) yield the following mode expansion for the atomic velocity $\dot{r}_0(t)$:

$$\dot{r}_0(t) = \sum_{\lambda=0}^{\infty} U_{0\lambda}\{\cos \omega_\lambda t \dot{\zeta}_\lambda - \omega_\lambda \sin \omega_\lambda t \zeta_\lambda\} \tag{5.57}$$

The normal-mode coordinates and velocities $\zeta_\lambda, \dot{\zeta}_\lambda$ are closely related to the eigenfunctions ϕ_λ of the Liouville operator L [see Eq. (5.8)]. This is expected, since the normal-mode frequencies are the eigenvalues of L. The relationship between ϕ_λ and $\zeta_\lambda, \dot{\zeta}_\lambda$ follows by comparing Eq. (5.56) with the following result for the time dependence of $\phi_\lambda(t)$:

$$\phi_\lambda(t) = \exp[i\omega_\lambda t]\,\phi_\lambda \tag{5.58}$$

which follows from $\dot{\phi}_\lambda(t) = iL\phi_\lambda(t) = i\omega_\lambda\phi_\lambda(t)$. This comparison yields

$$\phi_\lambda = N_\lambda[\zeta_\lambda - i\omega_\lambda^{-1}\dot{\zeta}_\lambda] \tag{5.59}$$

where N_λ is a normalization factor that depends on the normalization chosen for ϕ_λ.[56]

These considerations show that the MTGLE theory provides a formal algorithm for constructing the eigenfunctions of L. Alternatively one may regard the MTGLE dynamical matrix ω^2 as a tridiagonal representation of the eigenvalue spectrum of L.

3. The Clamped Equivalent Chain

As mentioned, the heatbath variable $R_p(t)$ is the trajectory of atom p in the pth clamped chain. The development of this result parallels the derivation of the chain representation for $r_0(t)$ just presented. Thus we skip the derivation and pass directly to the main results.

The matrix equation of motion for the infinite pth clamped chain is [cf. Eq. (5.53)]

$$\ddot{r}^{(p)}(t) = -[\omega^{(p)}]^2 r^{(p)}(t) \tag{5.60}$$

where $[\omega^{(p)}]^2$ is defined by Eq. (5.38). The coordinates of the pth clamped chain $r^{(p)}(t) = (r_p^{(p)}(t), r_{p+1}^{(p)}(t), \ldots)$ are defined recursively in terms of $R_p(t)$ via [cf. Eqs. (5.51)]

$$r_p^{(p)}(t) = R_p(t)$$

$$r_q^{(p)}(t) = R_q(t) + \omega_{cq}^2 \int_0^t \theta_q(t-\tau) r_{q-1}^{(p)}(\tau)\,d\tau, \qquad q \geqslant p+1 \tag{5.61}$$

Equation (5.52) and (5.61) yield the following relations between initial conditions:

$$\dot{r}_q^{(p)} = \dot{R}_q = r_q \quad \text{and} \quad r_q^{(p)} = R_q = r_q, \qquad q \geqslant p \tag{5.62}$$

Note that $\dot{r}_q^{(p)}$ is the initial velocity of atom q in the pth clamped chain, $\dot{R}_q$ is its initial velocity in the qth clamped chain, and $\dot{r}_q$ is its initial velocity in the full chain. Equation (5.62) states that the initial velocities in the various chains are identical, as are the corresponding initial coordinates.

Solution of Eq. (5.60) gives an atom expansion of $\dot{R}_p(t)$ analogous to Eq. (5.54). This is [since $r_p^{(p)}(t) = R_p(t)$ by Eq. (5.61)]

$$\dot{R}_p(t) = \sum_{q=p}^{\infty} \{[\cos \omega^{(p)} t]_{pq} \dot{r}_q^{(p)} - [\omega^{(p)} \sin \omega^{(p)} t]_{pq} r_q^{(p)}\} \qquad (5.63)$$

The normal modes of the clamped chain $\zeta^{(p)}(t) \equiv (\zeta_p^{(p)}(t), \zeta_{p+1}^{(p)}(t), \ldots)$ are related to the atomic coordinates $r^{(p)}(t)$ by the transform pair [cf. Eqs. (5.55)]

$$\zeta^{(p)}(t) = [\mathbf{U}^{(p)}]^T r^{(p)}(t) \qquad (5.64a)$$

$$r^{(p)}(t) = \mathbf{U}^{(p)} \zeta^{(p)}(t) \qquad (5.64b)$$

Recall that the normal mode frequencies $\omega_\lambda^{(p)}$ of the clamped chain are the elements of the diagonal matrix $\omega_\mathcal{B}^{(p)}$, where $[\omega_\mathcal{B}^{(p)}]^2 = [\mathbf{U}^{(p)}]^T [\omega^{(p)}]^2 \mathbf{U}^{(p)}$. Thus $\ddot{\zeta}_\lambda^{(p)}(t) = -[\omega_\lambda^{(p)}]^2 \zeta_\lambda^{(p)}(t)$, so that

$$\zeta_\lambda^{(p)}(t) = \cos \omega_\lambda^{(p)} t \, \zeta_\lambda^{(p)} + \omega_\lambda^{-1} \sin \omega_\lambda^{(p)} t \, \dot{\zeta}_\lambda^{(p)} \qquad (5.65)$$

Equations (5.64) and (5.65) thus yield the following normal-mode expansion for $\dot{R}_p(t)$:

$$\dot{R}_p(t) = \sum_{\lambda=p}^{\infty} U_{p\lambda}^{(p)} \{\cos \omega_\lambda^{(p)} t \, \dot{\zeta}_\lambda^{(p)} - \omega_\lambda^{(p)} \sin \omega_\lambda^{(p)} t \, \zeta_\lambda^{(p)}\} \qquad (5.66)$$

Finally we note that we may define orthonormal eigenfunction $\phi_\lambda^{(p)}$ for the clamped chain via relations analogous to Eq. (5.8). A *fictitious* Liouville operator may then be defined by

$$L^{(p)} = \sum_\lambda \omega_\lambda^{(p)} |\phi_\lambda^{(p)}\rangle\langle\phi_\lambda^{(p)}| \qquad (5.67)$$

This fictitious Liouville operator governs motion in the pth clamped chain. In particular the time dependence of the pth MTGLE random force is given by

$$\omega_{cp}^2 R_p(t) = \exp[iL^{(p)} t] \left[\omega_{cp}^2 R_p(t=0)\right] \qquad (5.68)$$

This in both the MTGLE and Mori representations the random forces may be regarded as evolving according to *fictitious* [cf. Eqs. (4.137) and (5.68)], as opposed to physical [$\exp(iLt)$], dynamics.

F. Basic Statistical Theorems

We have just shown that the abstract dynamics defined by the Liouville operator L may be rigorously recast as equivalent-harmonic-chain dynamics. This suggests that the abstract statistics governed by the inner product $\langle \ \rangle$ might resemble physical harmonic-chain statistics.

We next show that the *binary* statistical properties of the equivalent chain are identical to the physical statistical properties of real chains belonging to *classical canonical ensembles*. The proof of this assertion, given elsewhere,[21(b)] follows from the orthonormality and completeness of the eigenfunctions ϕ_λ of the Hermitian Liouville operator L.

1. Mode Statistics

The formal statistical properties of the normal-mode phase-space coordinates $p_\lambda = m\dot{\zeta}_\lambda$ and ζ_λ are

$$\frac{1}{2m}\langle p_\lambda p_{\lambda'}\rangle = \tfrac{1}{2}m\langle \dot{r}_0\rangle\delta_{\lambda\lambda} \tag{5.69a}$$

$$\langle p_\lambda p_{\lambda'}\rangle = 0 \tag{5.69b}$$

$$\tfrac{1}{2}m\omega_\lambda^2\langle \zeta_\lambda \zeta_{\lambda'}\rangle = \tfrac{1}{2}m\langle \dot{r}_0^2\rangle\delta_{\lambda\lambda} \tag{5.69c}$$

If one defines a formal temperature T by

$$\tfrac{1}{2}m\langle \dot{r}_0^2\rangle = \tfrac{1}{2}k_B T \tag{5.70}$$

then Eqs. (5.69) are identical to the classical Boltzmann statistical properties of the modes of physical chains. Notice that the formal temperature T is the physical Kelvin temperature if: (i) r_0 is the coordinate of a physical atom of mass m (recall that r_0 can be an arbitrary real phase-point function); (ii) the abstract inner product $\langle \ \rangle$ is interpreted as a classical Boltzmann average.

2. Atom Statistics

The statistical theorems for the chain-atom phase-space coordinates $p_q = m\dot{r}_q$ and r_q may be immediately obtained from Eqs. (5.69) and the fact that the atom and mode representations are connected by the orthogonal transformation **U**. Equations (5.33), (5.55), and (5.69) yield

$$\frac{1}{2m}\langle p_q p_{q'}\rangle = \tfrac{1}{2}m\langle \dot{r}_0^2\rangle\delta_{qq'} \tag{5.71a}$$

$$\langle p_q r_q\rangle = 0 \tag{5.71b}$$

$$\frac{1}{2m}\langle r_q r_{q'}\rangle = \tfrac{1}{2}m\langle \dot{r}_0^2\rangle[\omega^{-2}]_{qq'} \tag{5.71c}$$

These equations with Eq. (5.70) are identical to the physical-harmonic-chain statistical properties presented in Section III.A, Eqs. (3.5b,c,d).

G.   Fluctuation–Dissipation and Orthogonality Theorems for the Inner Product $\langle \ \rangle$

The fluctuation–dissipation and orthogonality theorems which hold for the inner product $\langle \ \rangle$ are

$$\theta_p(t) = \frac{\langle \dot{R}_p(t)\dot{R}_p \rangle}{\langle \dot{r}_0^2 \rangle} \tag{5.72a}$$

and

$$\langle R_p(t)\dot{R}_{p-1} \rangle = 0 \tag{5.72b}$$

They follow immediately from the statistical theorems (5.71) and the following expansions, which follow from Eqs. (5.62) and (5.63):

$$R_p(t) = \sum_{q=p}^{\infty} \left\{ [\cos \omega^{(p)}t]_{pq}r_q + [(\omega^{(p)})^{-1}\sin \omega^{(p)}t]_{pq}\dot{r}_q \right\} \tag{5.73a}$$

and

$$\dot{R}_p(t) = \sum_{q=p}^{\infty} \left\{ [\cos \omega^{(p)}t]_{pq}\dot{r}_q - [\omega^{(p)}\sin \omega^{(p)}t]_{pq}r_q \right\} \tag{5.73b}$$

In particular, to prove Eq. (5.72b) we note that by Eqs. (5.62) and (5.73a),

$$\langle R_p(t)\dot{R}_{p-1} \rangle = \langle R_p(t)\dot{r}_{p-1} \rangle$$

$$= \sum_{q=p}^{\infty} \left\{ [\cos \omega^{(p)}t]_{pq}\langle r_q\dot{r}_{p-1} \rangle + [(\omega^{(p)})^{-1}\sin \omega^{(p)}t]_{pq}\langle \dot{r}_q\dot{r}_{p-1} \rangle \right\}.$$

Equations (5.71) show that $\sum_{q=p}^{\infty}\{ \ \}=0$, which proves Eq. (5.72b).

Equation (5.72a) may be proven analogously from Eqs. (5.73b) and (5.62). These equations give $\langle \dot{R}_p(t)\dot{R}_p \rangle = \langle \dot{R}_p(t)\dot{r}_p \rangle = \sum_{q=p}^{\infty}\{[\cos \omega^{(p)}t]_{pq}\langle \dot{r}_q\dot{r}_p \rangle - [\omega^{(p)}\sin \omega^{(p)}t]_{pq}\langle r_q\dot{r}_p \rangle\}$. Using Eqs. (5.71), this sum simplifies to $\langle \dot{R}_p(t)\dot{R}_p \rangle = \langle \dot{r}_0^2 \rangle[\cos \omega^{(p)}t]_{pp}$. Equation (5.43) shows that this result is equivalent to the fluctuation–dissipation theorem (5.72a).

The proofs just given for the fluctuation–dissipation and orthogonality theorems show that these basic theorems are consequences of the orthonormality of the eigenfunctions ϕ_λ of the Liouville operator L. This is because the statistical theorems (5.70) are immediate consequences of this orthonormality.

Recall that the fluctuation–dissipation and orthogonality theorems originally emerged from purely phenomenological considerations (the Onsager fluctuation-regression hypothesis and the requirement that a prepared nonequilibrium primary-system ensemble relax to an equilibrium ensemble as $t \to \infty$) during our discussion of the Langevin model in Section III.D. The present analysis helps clarify the deep origin of these theorems in the mathematical structure of fundamental dynamics.

H.   The Inner Product $\langle\ \rangle_{r_0, r_1, \ldots, r_{p-1}}$

The pth fluctuation–dissipation theorem in the form given in Eq. (5.73a) suffers from a flaw that has both practical and theoretical implications. This flaw is that the time correlation function $\langle \dot{R}_p(t+\tau)R_p(\tau)\rangle$ *non*stationary; that is, $\langle \dot{R}_p(t+\tau)\dot{R}_p(\tau)\rangle \neq \langle \dot{R}_p(t)\dot{R}_p\rangle$ for $\tau \neq 0$. This may be proven by constructing $\langle \dot{R}_p(t+\tau)\dot{R}_p(\tau)\rangle$ using Eqs. (5.71) and (5.73b).

The practical implication of this flaw is that Eq. (5.72a) cannot be used in stochastic calculations. This is because the stationarity property is explicitly used in stochastic calculations. This point was illustrated in our construction of the PDF $P(\mathbf{r}_0(t))$ [Eq. (3.26)] for Brownian velocities in Section III.D.

From the theoretical standpoint, the nonstationary character of $\langle \dot{R}_p(t+\tau)\dot{R}_p(\tau)\rangle$ breaks the symmetry, repeatedly emphasized, between the atomic velocity $\dot{r}_0(t)$ and the abstract heatbath velocity $\dot{R}_p(t)$. This is because the *first* fluctuation–dissipation theorem, Eq. (5.4), involves the *stationary* time correlation function $\langle r_0(t+\tau)r_0(\tau)\rangle$.

The nonstationary character of $\langle R_p(t+\tau)R_p(\tau)\rangle$ originates because the time development of $R_p(t)$ is generated by $\exp[iL^{(p)}t]$ [Eq. (5.68)] rather than by $\exp[iLt]$. The Liouville operator $L^{(p)}$ [Eq. (4.67)] is *not*, however, Hermitian if the inner product is $\langle\ \rangle$. Thus our earlier proof of the stationarity property of time correlation functions [see Eq. (4.49)], based on the Hermiticity of L, breaks down.

We next define a *clamped-chain* abstract inner product $\langle\ \rangle_{r_0, r_1, \ldots, r_{p-1}}$. The key point is that for this inner product the Liouville operator $L^{(p)}$ is Hermitian. The new inner product is defined *only* for Liouville space vectors $X^{(p)}$ and $Y^{(p)}$ that lie in the *subspace* spanned by the eigenfunctions $\phi_\lambda^{(p)}$ of $L^{(p)}$.

The inner product $\langle\ \rangle_{r_0, r_1, \ldots, r_{p-1}}$ has the standard mathematical properties given for $\langle\ \rangle$ in Eqs. (5.2). Moreover, as mentioned, $L^{(p)}$ is Hermitian for this inner product; that is [cf. Eq. (5.3)],

$$\langle X^{(p)}L^{(p)}Y^{(p)*}\rangle_{r_0, r_1, \ldots, r_{p-1}} = \langle Y^{(p)}L^{(p)}X^{(p)*}\rangle^*_{r_0, r_1, \ldots, r_{p-1}} \qquad (5.74)$$

For the case of physical chains, the inner product $\langle\ \rangle_{r_0, r_1, \ldots, r_{p-1}}$ has been defined in Section III.A as an average over the classical Boltzmann distribution function $f_{CA}^{(p)}$ of the pth clamped chain. We employ here a more general definition. This definition is expressed in terms of the normal-mode phase-space coordinates, $p_\lambda^{(p)} = m\dot{\zeta}_\lambda(p)$ and $\zeta_\lambda^{(p)}$, via $(\lambda, \lambda' = p, p+1, \ldots)$

$$\frac{1}{2m}\langle p_\lambda^{(p)} p_{\lambda'}^{(p)}\rangle_{r_0, r_1, \ldots, r_{p-1}} = \tfrac{1}{2}\langle \dot{r}_0^2\rangle\delta_{\lambda\lambda'} \tag{5.75a}$$

$$\langle p_\lambda^{(p)}\zeta_{\lambda'}^{(p)}\rangle_{r_0, r_1 \cdots r_{p-1}} = 0 \tag{5.75b}$$

$$\tfrac{1}{2}m\left[\omega_\lambda^{(p)}\right]^2\langle \zeta_\lambda^{(p)}\zeta_{\lambda'}^{(p)}\rangle_{r_0, r_1 \cdots r_{p-1}} = \tfrac{1}{2}m\langle \dot{r}_0^2\rangle\delta_{\lambda\lambda'} \tag{5.75c}$$

These equations define $\langle\ \rangle_{r_0, r_1 \cdots r_{p-1}}$, for the following reason. The phase-space coordinates $p_\lambda^{(p)}, \zeta_\lambda^{(p)}$ span the same subspace as the eigenfunctions $\phi_\lambda^{(p)}$. This is because $\phi_\lambda^{(p)}$ is a linear combination of $p_\lambda^{(p)}$ and $\zeta_\lambda^{(p)}$, as discussed above. Thus all variables $X^{(p)}, Y^{(p)}$ confined to the subspace may be expanded as linear combinations of $p_\lambda^{(p)}$ and $\zeta_\lambda^{(p)}$. Thus all possible subspace inner products $\langle X^{(p)}Y^{(p)*}\rangle_{r_0, r_1 \cdots r_{p-1}}$ may be computed as linear combinations of basic inner products in Eqs. (5.75). This is why Eqs. (5.75) *define* the abstract inner product $\langle\ \rangle_{r_0, r_1, \ldots, r_{p-1}}$.

Using the transformation relation $\zeta^{(p)}(t) = [\mathbf{U}^{(p)}]^{\mathsf{T}} r^{(p)}(t)$ [Eq. (5.64a)] and the fact that $\dot{r}_q^{(p)} = \dot{r}_q$ and $r_q^{(p)} = r_q$ [Eq. (5.62)], Eqs. (5.75) may be transformed to the following statistical relations for the chain-atom phase-space coordinates p_q, r_q $[q = p, p+1, \ldots]$:

$$\frac{1}{2m}\langle p_q p_{q'}\rangle_{r_0, r_1, \ldots, r_{p-1}} = \tfrac{1}{2}m\langle \dot{r}_0^2\rangle\delta_{qq'} \tag{5.76a}$$

$$\langle r_q p_{q'}\rangle_{r_0, r_1, \ldots, r_{p-1}} = 0 \tag{5.76b}$$

$$\tfrac{1}{2}m\langle r_q r_{q'}\rangle_{r_0, r_1, \ldots, r_{p-1}} = \tfrac{1}{2}m\langle \dot{r}_0^2\rangle\left[\left(\omega^{(p)}\right)^{-2}\right]_{qq'} \tag{5.76c}$$

Note that given Eq. (5.70), Eqs. (5.76) are identical to the physical clamped-chain thermal averages of Eqs. (3.6b, c, d). This shows that our abstract definition of $\langle\ \rangle_{r_0, r_1, \ldots, r_{p-1}}$ Eqs. (5.75), reduces to the simple definition of Section III.A for the case that the underlying dynamical system is a thermal physical harmonic chain.

I. Fluctuation–Dissipation and Orthogonality Theorems for the Inner Product $\langle \ \rangle_{r_0, r_1, \ldots, r_{p-1}}$

The fluctuation–dissipation and orthogonality theorems that hold for the clamped inner products are

$$\theta_p(t) = \frac{\langle \dot{R}_p(t+\tau)\dot{R}_p(\tau)\rangle_{r_0, r_1, \ldots, r_{p-1}}}{\langle \dot{r}_0^2\rangle}$$

$$= \frac{\langle \dot{R}_p(t)\dot{R}_p\rangle_{r_0, r_1, \ldots, r_{p-1}}}{\langle \dot{r}_0^2\rangle} \tag{5.77a}$$

and

$$\langle R_p(t)\dot{R}_{p-1}\rangle_{r_0, r_1, \ldots, r_{p-2}} = 0 \tag{5.77b}$$

These results have already been presented for the chain model in Eqs. (3.46) and (3.47). An important special case ($p = 0$) of Eq. (5.77b) is the orthogonality of the atomic coordinate $r_0(t)$ to $\dot{R}_1$:

$$\langle r_0(t)\dot{R}_1\rangle = 0 \tag{5.77c}$$

Given Eqs. (5.77), the complete symmetry between $r_0(t)$ and $\dot{\chi}(t)$ in the full chain and $R_p(t)$ and $\dot{\theta}_p(t)$ in the pth clamped chain is established.

The proof of Eq. (5.77a) is easily accomplished by explicitly constructing $\langle \dot{R}_p(t+\tau)\dot{R}_p(\tau)\rangle_{r_0, r_1, \ldots, r_{p-1}}$ using Eqs. (5.73b), (5.76), and the relation $\dot{\theta}_p(t) = [\cos \omega^{(p)}t]_{pp}$ [Eq. (5.43)].

The proof of Eq. (5.77b) follows immediately from the MTGLE for $R_{p-1}(t)$ [Eq. (5.49)]:

$$\ddot{R}_{p-1}(t) = -\omega_{e\,p-1}^2 R_{p-1}(t) + \omega_{cp}^4 \int_0^t \theta_{p-1}(t-\tau)R_{p-1}(\tau)\,d\tau + \omega_{cp}^2 R_p(t) \tag{5.78}$$

and the effective equation of motion for the associated response function $\theta_{p-1}(t)$ [Eq. (5.20b)]:

$$\ddot{\theta}_{p-1}(t) = -\omega_{e\,p-1}^2 \theta_{p-1}(t) + \omega_{cp}^4 \int_0^t \theta_{p-1}(t-\tau)\theta_p(\tau)\,d\tau \tag{5.79}$$

Multiplying Eq. (5.78) by $\langle \cdots \dot{R}_{p-1}\rangle_{r_0, r_1, \ldots, r_{p-2}}$ and using the fluctuation–

dissipation theorem (5.77a) yields

$$\ddot{\theta}_{p-1}(t) = -\omega_{e\,p-1}^2 \theta_{p-1}(t)$$
$$+ \omega_{cp}^2 \int_0^t \theta_{p-1}(t-\tau) \theta_p(\tau)\, d\tau + \omega_{cp}^2 \langle R_p(t) \dot{R}_{p-1} \rangle_{r_0,\,r_1,\dots,r_{p-2}}$$

(5.80)

Comparison of Eqs. (5.79) and (5.80) establishes the orthogonality theorem (5.77b).

Notice that our proof of the orthogonality theorem is based on the fluctuation–dissipation theorem. Thus only the fluctuation–dissipation theorem is fundamental; the orthogonality theorem is a derived result.

This shows that the *full* content of the MTGLE representation is carried by the first, second,... fluctuation-dissipation theorems [Eqs. (5.4) and (5.77a)]. All the rest is notation and definitions.

J. Formal Equipartition Theorems

The formal statistical properties of the terminal atoms in the zeroth, first,..., pth clamped chains, $r_0, R_1, \dots, R_p$, are identical to the classical Boltzmann equipartition theorems for a sequence of *isolated* harmonic oscillators with frequencies $\Omega_0, \Omega_1, \dots, \Omega_p$.

The frequency Ω_0 is the adiabatic frequency of the atom discussed in Section III.3. This is defined in Eq. (3.50). An equivalent alternative definition is

$$\Omega_0^{-2} = \hat{\chi}(z=0) = [\omega^{-2}]_{00}$$

(5.81a)

Here the first equality follows from Eq. (3.56b); the second equality follows by taking the Laplace transform of Eq. (5.32) and then setting $z = 0$.

The adiabatic frequency of the pth heatbath Ω_p may be analogously defined as

$$\Omega_p^{-2} = \hat{\theta}_p(z=0) = [\omega^{(p)}]_{pp}^{-2}$$

(5.81b)

The formal equipartition theorems satisfied by the atomic coordinate are

$$\tfrac{1}{2} m \langle \dot{r}_0^2 \rangle = \tfrac{1}{2} k_B T$$

(5.82a)

$$\langle \dot{r}_0 r_0 \rangle = 0$$

(5.82b)

$$\tfrac{1}{2} m \Omega_0^2 \langle r_0^2 \rangle = \tfrac{1}{2} k_B T$$

(5.82c)

where the temperature T is defined in Eq. (5.70). The formal equipartition theorems satisfied by the heatbath coordinates are, correspondingly,

$$\tfrac{1}{2}m\langle \dot{R}_p^2\rangle_{r_0,r_1,\ldots,r_{p-1}} = \tfrac{1}{2}k_B T \tag{5.83a}$$

$$\langle \dot{R}_p R_p\rangle_{r_0,r_1,\ldots,r_{p-1}} = 0 \tag{5.83b}$$

$$\tfrac{1}{2}m\,\Omega_p^2\langle R_p^2\rangle_{r_0,r_1,\ldots,r_{p-1}} = \tfrac{1}{2}k_B T \tag{5.83c}$$

These statistical theorems are immediate consequences[‡] of the fluctuation–dissipation theorems Eqs. (5.4) and (5.77a) and also Eqs. (5.81).

This concludes our discussion of the basic theorems of the MTGLE representation of irreversible dynamics.

VI.   A THEORY FOR LIQUID-STATE CHEMICAL-REACTION DYNAMICS

We develop in this section an approximate theory of liquid-state chemical-reaction dynamics based on the rigorous MTGLE formalism of Section V. This theory has already been sketched in Section III.J.

The key result of our development is an approximate reduced equation of motion, Eq. (6.15) below, for reagents interacting in liquid solution. This equation accounts for the effects of reagent configuration-dependent caging through *matrix* equivalent-chain parameters $\{\omega_{ep}^2(r_0), \omega_{cp+1}^2(r_0)\}, p = 0, 1, \ldots$, which depend explicitly on the instantaneous reagent configuration point r_0.

The idea behind the treatment [see Section III.J] is as follows. The systematic force acting on the reagents is approximated as the reaction force that the local mean-field solvent density exerts in response to *small forced oscillations* $\Delta r_0(t)$ of the reagents about the configuration point r_0. This reaction force is calculated via an application of linear-response theory [see Sections IV.B.5 and IV.B.6] in which $\Delta r_0(t)$ is the smallness parameter.

The random force acting on the reagents is approximated as the force due to thermal fluctuations of the local solvent density about its equilibrium value, conditional on the assumption that the reagents are *clamped* at point r_0. The systematic and random forces determined by this procedure depend

---

[‡]The equipartition theorems of Eqs. (5.83), for example, may be proven from the $(p+1)$th fluctuation–dissipation theorem (5.77a), the definition of the formal temperature [Eq. (5.70)], the basic properties of $\dot{\theta}_p(t)$ [Eq. (5.16)], and the definition of Ω_p^2 [Eq. (5.81b)]. The proofs are as follows. Setting $t = \tau = 0$ in Eq. (5.77a) and using Eqs. (5.16) and (5.70) yields Eq. (5.83a). Integrating Eq. (5.77a) once with respect to time and setting $t = \tau = 0$ yields Eq. (5.83b). Integrating Eq. (5.77a) again with respect to time, using the fact that $\int_0^\infty \theta_p(\tau)\,d\tau = \dot{\theta}_p(z=0)$, and then using Eq. (5.81b) yields Eq. (5.83c).

on the space–time local solvent density fluctuations about the clamped reagents. For this reason we will refer to the procedure as the *clamping approximation*.

The clamping approximation is somewhat analogous to the gas-phase Born–Oppenheimer approximation. It leads, for example, to the concept of reagent-configuration-dependent response-function surfaces, just as the Born–Oppenheimer approximation leads to the concept of potential-energy surfaces.

The plan of this section is as follows. For clarity we first present, without derivation, the basic results of our theory. This is done is Section VI.A. In Section VI.B we present the derivation, which is based on the linear response theory reviewed Sections IV.B.5 and IV.B.6

A. Basic Results for Interacting Reagents

We assume n reagent atoms with masses

$$\mathbf{m} = \begin{pmatrix} m_1 & 0 & 0 & \cdots & 0 \\ 0 & m_2 & 0 & \cdots & 0 \\ 0 & 0 & m_3 & \cdots & 0 \\ \vdots & \vdots & \vdots & & \vdots \\ 0 & 0 & 0 & \cdots & m_n \end{pmatrix} \tag{6.1}$$

are dissolved in a pure monatomic solvent composed of N_s solvent atoms, each of mass M. (The assumption of a pure monatomic solvent is made to avoid complexity of notation; generalization of the notation to molecular or mixed solvents is straightforward.)

Collectively denote the solvent coordinates and momenta by

$$q = \begin{pmatrix} \mathbf{q}_1 \\ \mathbf{q}_2 \\ \vdots \\ \mathbf{q}_{N_s} \end{pmatrix} \quad \text{and} \quad p = M\dot{q} = \begin{pmatrix} \mathbf{p}_1 \\ \mathbf{p}_2 \\ \vdots \\ \mathbf{p}_{N_s} \end{pmatrix} \tag{6.2a}$$

The coordinates and momenta of the reagents will be correspondingly denoted by

$$r_0 = \begin{pmatrix} \mathbf{r}_{01} \\ \mathbf{r}_{02} \\ \vdots \\ \mathbf{r}_{0n} \end{pmatrix} \quad \text{and} \quad \mathbf{m}\dot{r}_0 = \begin{pmatrix} m_1\dot{\mathbf{r}}_{01} \\ m_2\dot{\mathbf{r}}_{02} \\ \vdots \\ m_n\dot{\mathbf{r}}_{0n} \end{pmatrix} \tag{6.2b}$$

1. Results for the Matrix MTGLE Quantities

The reduced dynamics of an *isolated* solute atom is governed by the quantities ω_{e0}^2, ω_{c1}^4, $\theta_1(t)$, and $\mathbf{R}_1(t)$. For the case of interacting reagents, the isolated-solute-atom quantities generalize to quantities $[\omega_{e0}^2(r_0)$, etc.] that depend on the reagent configuration point r_0.

The r_0-dependent MTGLE quantities are $n \times n$ matrices whose elements are three-dimensional Cartesian tensors. They are thus $3n \times 3n = 9n^2$-component quantities. We shall show below that the r_0-dependent quantities are symmetric matrices and thus only $\frac{1}{2}3n(3n + 1)$ of their elements are distinct. For particular systems, molecular symmetry often leads to a further great reduction in complexity [see Section VII].

The matrix MTGLE quantities are defined as thermal averages over the solvent phase-space coordinates p, q, conditional on the assumption that the reagents are clamped at configuration point r_0. Such thermal averages will be denoted by the symbol $\langle \ \rangle_{r_0}$. The thermal averages involve the following forces:

$$\mathcal{F}_i^\alpha(r_0) \equiv \mathcal{F}_i^\alpha = \alpha\text{th Cartesian component of the solvent force acting}$$
on the ith reagent atom

$$\mathcal{F}_{0i}^\alpha(t; r_0) = \mathcal{F}_{0i}^\alpha(t) = \alpha\text{th Cartesian component of the solvent force acting}$$
on the ith reagent atom conditional on the assumption that the reagents are clamped at configuration point r_0

$$\tilde{\mathcal{F}}_{0i}^\alpha(t) \equiv \mathcal{F}_{0i}^\alpha(t) - \langle \mathcal{F}_i^\alpha \rangle_{r_0} = \alpha\text{th component of the } \textit{fluctuating force} \text{ acting on reagent } i \text{ at time } t.$$

More precise definitions of the above forces will be given in Section VI.B.

The results for the matrix components of $\omega_{e0}^2(r_0)$ and the like are as follows ($i, j,\dots$ are solute-atom indices and $\alpha, \beta,\dots$ are Cartesian indices):

$$\left[\omega_{e0}^2(r_0)\right]_{ij}^{\alpha\beta} = -m_i^{-1/2}\left\langle \frac{\partial}{\partial r_{0i}^\alpha}\mathcal{F}_j^\beta \right\rangle_{r_0} m_j^{-1/2} \tag{6.3a}$$

$$\left[\omega_{c1}^4(r_0)\right]_{ij}^{\alpha\beta} = (k_BT)^{-1}m_i^{-1/2}\langle \tilde{\mathcal{F}}_{0i}^\alpha\tilde{\mathcal{F}}_{0j}^\beta\rangle_{r_0}m_j^{-1/2} \tag{6.3b}$$

$$\left[\omega_{c1}^2(r_0)\theta_1(t; r_0)\left[\omega_{c1}^2(r_0)\right]^T\right]_{ij}^{\alpha\beta} = (k_BT)^{-1}m_i^{-1/2}$$
$$\times \langle \dot{\tilde{\mathcal{F}}}_{0i}^\alpha(t + \tau)\dot{\tilde{\mathcal{F}}}_{0j}^\beta(\tau)\rangle_{r_0}m_j^{-1/2} \tag{6.3c}$$

$$\left[\omega_{c1}^2(r_0)\mathbf{m}^{1/2}R_1(t; r_0)\right]_i^\alpha = m_i^{-1/2}\dot{\tilde{\mathcal{F}}}_{0i}^\alpha(t) \quad \text{(clamping approximation)} \tag{6.3d}$$

The physical interpretation of these matrices has been discussed in Section III.J.

Ww further note that while the matrix $\omega_{cl}^4(r_0)$ is uniquely defined by Eq. (6.3b), its matrix square root $\omega_{cl}^2(r_0)$ requires further specification. We define $\omega_{cl}^2(r_0)$ by the conditions:

1. $\omega_{cl}^2(r_0)[\omega_{cl}^2(r_0)]^T = \omega_{cl}^4(r_0)$;
2. $\omega_{cl}^2(r_0)$ is a *lower triangular* matrix.

This is the so-called Choleski prescription[57] for matrix square roots. It will be employed for all matrix square roots employed below.

A key point is that the heatbath response matrix $\dot{\theta}_1(t; r_0)$ and the (vector) heatbath dynamical variable $R_1(t; r_0)$ are related by a second fluctuation–dissipation theorem analogous to Eq. (3.41). Namely, comparing Eqs. (6.3c, d) yields

$$\left[\dot{\theta}_1(t; r_0)\right]_{ij}^{\alpha\beta} = \frac{(m_i m_j)^{1/2}}{k_B T}\langle \dot{R}_{1i}^{\alpha}[t + \tau; r_0]\dot{R}_{1j}^{\beta}[\tau; r_0]\rangle_{r_0} \qquad (6.4)$$

The information in $\omega_{cl}^2(r_0)\dot{\theta}_1(t; r_0)[\omega_{cl}^2(r_0)]^T$ may also be expressed in terms of the r_0-dependent friction kernel $\beta_1(t; r_0)$. This is defined by analogy to the isolated-solute-atom friction kernel by [cf. Eq. (3.49)]

$$\beta_1(t; r_0) = \int_t^\infty \omega_{cl}^2(r_0)\theta_1(\tau; r_0)[\omega_{cl}^2(r_0)]^T d\tau \qquad (6.5)$$

Comparing Eqs. (6.3c) and (6.5) shows that the matrix components of the friction kernel are given by

$$\left[\beta_1(t; r_0)\right]_{ij}^{\alpha\beta} = (k_B T)^{-1} m_i^{-1/2}\langle \tilde{\mathscr{F}}_{0i}^{\alpha}(t + \tau)\tilde{\mathscr{F}}_{0j}^{\beta}(\tau)\rangle_{r_0} m_j^{-1/2} \qquad (6.6)$$

2. Symmetry Properties of the MTGLE Matrices

We next prove that the matrices $\omega_{e0}^2(r_0)$, $\omega_{cl}^4(r_0)$, $\dot{\theta}_1(t; r_0)$, and $\beta_1(t; r_0)$ are fully symmetric, that is, they are symmetric in both the solute-atom and Cartesian indices.

The symmetry of $\beta_1(t; r_0)$ follows from the following facts: (i) The solvent forces $\mathscr{F}_{0i}^{\alpha}(t)$ exerted on the solutes all have identical (even) parity under the operation of time reversal. Thus [cf. Eq. (4.90)]

$$\left[\beta_1(t; r_0)\right]_{ij}^{\alpha\beta} = \left[\beta_1(-t; r_0)\right]_{ij}^{\alpha\beta} \qquad (6.7a)$$

(ii) The stationarity of $\langle \tilde{F}_{0i}^{\alpha}(t + \tau)\tilde{F}_{0j}^{\beta}(\tau)\rangle_{r_0}$ implies that [set $\tau = -t$ in

Eq. (6.6)]

$$[\beta_1(t; r_0)]_{ij}^{\alpha\beta} = [\beta_1(-t; r_0)]_{ji}^{\beta\alpha} \tag{6.7b}$$

Comparing Eqs. (6.7a) and (6.7b) shows that

$$[\beta_1(t; r_0)]_{ij}^{\alpha\beta} = [\beta_1(t; r_0)]_{ji}^{\beta\alpha} \tag{6.8}$$

or equivalently

$$\beta_1(t; r_0) = \beta_1^{\mathsf{T}}(t; r_0) \tag{6.9a}$$

Thus we have shown that $\beta_1(t; r_0)$ is a symmetric matrix. Notice that our argument is identical to that used to prove Onsager's reciprocal relations in Section IV.B.6.

The symmetry of $\omega_{cl}^4(r_0)$ and $\dot\theta_1(t; r_0)$ follows immediately from their definitions, Eqs. (6.3b,c), and the symmetry of $\beta_1(t; r_0)$. Thus

$$\omega_{cl}^4(r_0) = \left[\omega_{cl}^4(r_0)\right]^{\mathsf{T}} \tag{6.9b}$$

and

$$\dot\theta_1(t; r_0) = \dot\theta_1^{\mathsf{T}}(t; r_0) \tag{6.9c}$$

Finally, the symmetry of $\omega_{e0}^2(r_0)$ follows immediately from its definition, Eq. (6.3a), and the fact that

$$\mathscr{F}_j^\beta = -\frac{\partial U_{VU}(q, r_0)}{\partial r_{0j}^\beta}$$

where $U_{VU}(q, r_0)$ is the solvent–reagent potential-energy function. Thus

$$\omega_{e0}^2(r_0) = \left[\omega_{e0}^2(r_0)\right]^{\mathsf{T}} \tag{6.9d}$$

3. MTGLE *Hierarchy for Interacting Reagents*

In Section V we showed that a rigorous infinite hierarchy of response functions $\dot\theta_1(t), \dot\theta_2(t), \ldots$ and associated heatbath variables $R_1(t), R_2(t), \ldots$ may be developed from a small number of mathematical properties of $\dot\theta_1(t)$ and its spectral density $\sigma_1(\omega)$. These properties are given in Eqs. (5.13) and (5.15).

An analogous hierarchy may be developed for interacting reagents. This follows because $\dot{\theta}_1(t; r_0)$ and its matrix spectral density $\sigma_1(\omega; r_0)$ have basic properties analogous to those of $\dot{\theta}_1(t)$ and $\sigma_1(\omega)$.

For example, it is straightforward to show from Eq. (6.3b,c) and the symmetry relation (6.9c) that

$$\dot{\theta}_1(-t; r_0) = \dot{\theta}_1(t; r_0) = \dot{\theta}_1^T(t; r_0)$$

$$\dot{\theta}_1(0; r_0) = 1, \qquad \theta_1(0; r_0) = 0 \tag{6.10}$$

This is the analogue of Eq. (5.13).

The detailed development of the MTGLE hierarchy for interacting reagents[21(c)] parallels the development of Section V and thus will be omitted. The main results of the hierarchy are as follows.

1. One may define an infinite set of $3n \times 3n$-component matrix MTGLE parameters $\{\omega_{ep}^2(r_0), \omega_{c\,p+1}^2(r_0)\}$, $p = 1, 2, \ldots$, from the coefficients of the power-series expansion of $\dot{\theta}_1(t; r_0)$ in time. The procedure, which has been worked out in detail elsewhere,[58] is a matrix generalization of the algorithm of Appendix A.

2. One may define an infinite sequence of $3n \times 3n$-component matrix heatbath response functions $\dot{\theta}_p(t; r_0)$, $p = 2, 3, \ldots$, from $\dot{\theta}_1(t; r_0)$ via matrix defining relations analogous to Eqs. (5.20).

3. One may recursively define an infinite sequence of $3n$-component vector heatbath dynamical variables $R_p(t; r_0)$, $p = 2, 3, \ldots$, from $R_1(t; r_0)$ via defining relations analogous to Eqs. (5.49). Qualitatively, these heatbath dynamical variables describe density fluctuations in the pth solvation shell of the common cage surrounding the reagents.

4. One may recursively define a set of $3n$-component vector equivalent-chain dynamical variables $\rho_p(t; r_0) \equiv \rho_p(t)$, $p = 1, 2, \ldots$, from $\dot{\theta}(t; r_0)$ by defining relations[21(c)] somewhat analogous to Eqs. (5.24). These variables permit the construction of a matrix equivalent-chain representation for response functions, Eq. (6.12) below.

5. Correspondingly, one may recursively define a set of $3n$-component vector equivalent-chain dynamical variables $r_p(t; r_0) \equiv r_p(t)$, $p = 1, 2, \ldots$, from $R_1(t; r_0)$ via defining relations analogous to Eqs. (5.51). These variables permit the construction of a matrix equivalent-chain equation for the dynamics of interacting reagents, Eq. (6.15) below, which provides a basic framework for liquid-state chemical-reaction dynamics.

Finally we note that the heatbath response functions $\dot{\theta}_p(t; r_0)$ and the corresponding dynamical variables $\dot{R}_p(t; r_0)$ are related by the fluctuation–dissipation theorem sipation theorem

$$\left[\dot{\theta}_p(t; r_0)\right]_{ij}^{\alpha\beta} = \frac{(m_i m_j)^{1/2}}{k_B T} \left\langle \dot{R}_{pi}^{\alpha}(t+\tau; r_0) \dot{R}_{pj}^{\beta}(\tau; r_0) \right\rangle_{r_0, r_1, \ldots, r_{p-1}} \quad (6.11)$$

This is the analogue of the $(p+1)$th fluctuation–dissipation theorem of the rigorous MTGLE formalism given in Eq. (5.77a). [Note that $\langle \ \rangle_{r_0, r_1, \ldots, r_{p-1}}$ implies physical clamping of the reagents at configuration point r_0 and mathematical clamping (see Section V.H) of the equivalent-chain coordinates $r_1, r_2, \ldots, r_{p-1}$.]

4. Equivalent-Chain Representation for Response Functions

An N-"atom" equivalent-chain representation for response functions $\theta_p(t; r_0)$, $p = 1, 2, \ldots$, may be derived[21(c)] from the hierarchy just discussed via a matrix analogue of the procedure presented in Section V.C. This matrix chain representation is exact within the clamping approximation. The chain representation is really an infinite class of representation, since N can equal $2, 3, \ldots$.

The general N-atom chain representation, in matrix notation, is

$$\ddot{\rho}_1(t) = -\omega_{e1}^2(r_0)\rho_1(t) + \omega_{c2}^2(r_0)\rho_2(t)$$

$$\ddot{\rho}_2(t) = -\omega_{e2}^2(r_0)\rho_2(t) + \left[\omega_{c2}^2(r_0)\right]^{\mathsf{T}}\rho_1(t) + \omega_{c3}^2(r_0)\rho_3(t)$$

$$\vdots$$

$$\ddot{\rho}_N(t) = -\omega_{eN}^2(r_0)\rho_N(t) + \left[\omega_{cN}^2(r_0)\right]^{\mathsf{T}}\rho_{N-1}(t)$$
$$+ \int_0^t \omega_{cN+1}^2(r_0)\theta_{N+1}(t-\tau; r_0)\left[\omega_{cN+1}^2(r_0)\right]^{\mathsf{T}}\rho_N(\tau)\, d\tau \quad (6.12)$$

The chain-atom coordinates $\rho_q(t)$, $q = 1, 2, \ldots, N$, are $3n$-component vectors as discussed above.

The response functions $\theta_p(t; r_0)$ may be calculated from particular chain trajectories via prescriptions that are matrix analogues of the theorems of Section V.C. The matrix elements of the basic response function $\theta_1(t; r_0)$ may be calculated as particular trajectories of chain atom 1 in the full equivalent chain. The matrix elements of the derived response functions $\theta_p(t; r_0)$, $p \geq 2$, are corresponding trajectories of chain atom p in the $(p-1)$th clamped chain [i.e., the chain in which $\rho_1(t) = \rho_2(t) = \rho_{p-1}(t) = 0$.]

We shall require in Section VII.D the chain trajectory theorem for the basic response function $\theta_1(t; r_0)$. This is

$$\left[\theta_1(t; r_0)\right]_{ij}^{\alpha\beta} = \left[\rho_1(t)\right]_i^{\alpha} \quad (6.13)$$

given the initial conditions

$$[\dot{\rho}_1(0)]_k^\gamma = \delta_{kj}\delta_{\gamma\beta}$$

$$\dot{\rho}_{q+1}(0) = \rho_q(0) = 0, \qquad q \geqslant 1 \tag{6.14}$$

5. Equation of Motion for Interacting Reagents

The general $(N+1)$-atom equivalent-chain equation of motion for interacting reagents, in matrix notation, is [cf. Eq. (5.50)]

$$\mathbf{m}\ddot{r}_0(t) = \langle F \rangle_{r_0} - \mathbf{m}^{1/2}\omega_{e0}^2[r_0(t)]\mathbf{m}^{1/2}r_0(t) + \mathbf{m}^{1/2}\omega_{c1}^2[r_0(t)]\mathbf{m}^{1/2}r_1(t)$$

$$\mathbf{m}\ddot{r}_1(t) = -\mathbf{m}^{1/2}\omega_{e1}^2[r_0(t)]\mathbf{m}^{1/2}r_1(t) + \mathbf{m}^{1/2}\left[\omega_{c1}^2[r_0(t)]\right]^{\mathsf{T}}\mathbf{m}^{1/2}r_0(t)$$

$$+ \mathbf{m}^{1/2}\omega_{c2}^2[r_0(t)]\mathbf{m}^{1/2}r_2(t)$$

$$\vdots$$

$$\mathbf{m}\ddot{r}_N(t) = -\mathbf{m}^{1/2}\omega_{eN}^2[r_0(t)]\mathbf{m}^{1/2}r_N(t) + \mathbf{m}^{1/2}\left[\omega_{cN}^2[r_0(t)]\right]^{\mathsf{T}}\mathbf{m}^{1/2}r_{N-1}(t)$$

$$+ \int_0^t \mathbf{m}^{1/2}\omega_{cN+1}^2[r_0(t)]\theta_{N+1}[t-\tau; r_0(t)]\left[\omega_{cN+1}[r_0(t)]\right]^{\mathsf{T}}$$

$$\times \mathbf{m}^{1/2}r_N(\tau)\, d\tau + \mathbf{m}^{1/2}\omega_{cN+1}^2[r_0(t)]\mathbf{m}^{1/2}R_{N+1}[t; r_0(t)] \tag{6.15}$$

Note that $\langle F \rangle_{r_0}$ is the $3n$-dimensional vector of mean forces acting on the reagents when they are clamped at configuration point r_0. The component $\langle F_i^\alpha \rangle_{r_0}$, which is the αth Cartesian component of the mean force acting on the ith reagent atom, may be decomposed into a liquid-state contribution $\langle \mathcal{F}_i^\alpha \rangle_{r_0}$ and a gas-phase contribution $F_{gi}^\alpha(r_0) = -\partial U_{UU}(r_0)/\partial r_{0i}^\alpha$, where $U_{UU}(r_0)$ is the gas-phase-potential energy function.

Finally, not that while the equivalent-chain representation for response functions [Eq. (6.12)] is rigorous within the clamping approximation, the equation of motion for interacting reagents [Eq. (6.15)] requires an additional approximation. This is that the equilibrium solvent density $\rho[r_0(t)]$ is slowly varying on time scales comparable to a solvent-atom velocity relaxation time (~ 1 psec). This approximation is reasonable: (i) if the reagent motion is sufficiently slow; (ii) if the density $\rho[r_0]$ is a sufficiently slowly varying function of the configuration point r_0. The condition that $\rho[r_0(t)]$ is slowly varying is often reasonable at liquid-state densities, since large-scale reagent configurational changes are hindered by the solvent cage and are thus expected to occur on time scales slower than that for velocity reversal.

6. Low-Frequency Parameters

The practical value of the equivalent-chain equation for interacting reagents [Eq. (6.15)] is that it may be truncated to obtain a physically realistic and computationally practical effective few-body equation of motion for interacting reagents. The truncation procedure is a matrix generalization of the heatbath modeling method sketched in Section III.G. It thus qualitatively corresponds to treating the influence of the first few (usually one or two) solvation shells of the common cage on the reaction in detail while modeling the influence of the solvent external to these shells by that of a bulk energy reservoir.

Numerical application of truncated matrix chain models to the modeling of matrix r_0-dependent correlation functions [Section VII] and to chemical reaction dynamics [Section VIII] will be presented below.

In order to truncate the equivalent-chain representations (6.12) and (6.15) [see, for example, Eq. (7.17) below], we require the low-frequency parameters $\langle \Omega_p^2(r_0), \beta_{p+1}(r_0) \rangle$, $p = 0, 1, \ldots$. These are $3n \times 3n$ matrix generalizations of the corresponding quantities of the rigorous MTGLE formalism.

The adiabatic frequency matrix of the reagents $\Omega_0(r_0)$ is defined by analogy to Eq. (3.50) via

$$\Omega_0^2(r_0) = \omega_{e0}^2(r_0) - \beta_1(t = 0; r_0) \tag{6.16a}$$

The matrix $\Omega_0^2(r_0)$ has a simple physical interpretation. Suppose the reagents are displaced by a small amount $\Delta r_0 = (\Delta \mathbf{r}_{01}, \ldots, \Delta \mathbf{r}_{0n})$ at time $t = 0$. Then $-\mathbf{m}^{1/2} \Omega_0^2(r_0) \mathbf{m}^{1/2} \cdot \Delta r_0$ is the *fully relaxed* restoring force exerted on the reagents by the local solvent density, that is, the restoring force at $t = \infty$. [The quantity $-\mathbf{m}^{1/2} \omega_{e0}^2(r_0) \mathbf{m}^{1/2} \cdot \Delta r_0$ is in contrast the *instantaneous* restoring force exerted on the displaced reagents by the local solvent density.]

Given this interpretation, it is straightforward to derive [21(c)] the following alternative form for $\Omega_0^2(r_0)$:

$$\left[\Omega_0^2(r_0)\right]_{ij}^{\alpha\beta} = -m_i^{1/2} \frac{\partial \langle \mathcal{F}_j^\beta \rangle_{r_0}}{\partial r_{0i}^\alpha} m_j^{1/2} \tag{6.16b}$$

For isolated solutes $\langle \mathcal{F} \rangle_{r_0} = 0$. Thus $\Omega_0^2 = 0$ for isolated solutes. This result was derived from a different point of view in Section III [see Eq. (3.55)].

We next turn to the friction matrix $\beta_1(r_0)$. This is defined by analogy to Eq. (3.54) as

$$\beta_1(r_0) = \lim_{z \to 0} \int_0^\infty e^{-zt} \beta_1(t; r_0) \, dt \tag{6.17}$$

To develop expressions for $\Omega_p^2(r_0)$, $p \geqslant 1$, and $\beta_{p+1}(r_0)$, $p \geqslant 1$, we require the $(p+1)$th friction kernel $\beta_{p+1}(t; r_0)$. This is defined in analogy to Eq. (6.5) as

$$\beta_{p+1}(t; r_0) = \int_t^\infty \omega_{cp+1}^2(r_0)\theta_{p+1}(\tau; r_0)\left[\omega_{cp+1}^2(r_0)\right]^\mathsf{T} d\tau \qquad (6.18)$$

Then $\Omega_p^2(r_0)$ and $\beta_{p+1}(r_0)$ are defined by analogy to Eqs. (6.16a) and (6.17) as

$$\Omega_p^2(r_0) = \omega_{ep}^2(r_0) - \beta_{p+1}(t = 0; r_0) \qquad (6.19)$$

and

$$\beta_{p+1}(r_0) = \lim_{z \to 0} \int_0^\infty e^{-zt}\beta_{p+1}(t; r_0)\, dt \qquad (6.20)$$

The parameters $\{\Omega_p^2(r_0), \beta_{p+1}(r_0)\}$ for $p \geqslant 1$ may also be explicitly constructed from the equivalent-chain parameters $\{\omega_{ep}^2(r_0), \omega_{cp+1}^2(r_0)\}$ and the basic low-frequency parameters $\Omega_0^2(r_0)$ and $\beta_1(r_0)$. This construction is based on the definitions in Eqs. (6.19) and (6.20) and the results of the MTGLE hierarchy. The procedure is a matrix generalization of the algorithm for constructing Ω_p^2 and β_{p+1} given in Appendix A. This matrix generalization has been described in detail elsewhere.[58]

B.   Development of Basic Results

We next present the derivations of the basic results of Section VI.A [Eqs. (6.3)]. We begin with a fuller specification of the notation used in Section VI.A. This notation will be employed throughout the remainder of this chapter.

1.   Notation

We assume that interactions in the solution are governed by the following potential-energy functions:

$$U_{VV}(q) = \text{solvent potential energy}$$

$$U_{VU}(q, r_0) = \text{solute–solvent potential energy}$$

$$U_{UU}(r_0) = \text{solute–solute potential energy}$$

where we recall that $q = (\mathbf{q}_1, \ldots, \mathbf{q}_{N_s})$ are the solvent coordinates and $r_0 = (\mathbf{r}_{01}, \ldots, \mathbf{r}_{0n})$ are the reagent coordinates.

The formal linear-response development of this section holds whether or not the potential-energy functions are pairwise additive. In the practical calculations of Sections VII and VIII we assume, however, that the potential functions are pairwise additive.

We also require the gas-phase force $F_g[r_0] \equiv F_g$ and the instantaneous solvent force $\mathcal{F}(r_0) \equiv \mathcal{F}$ acting on the reagents. These quantities are n-component vectors whose elements are three-dimensional Cartesian vectors, and are thus $3n$-component quantities. The elements of F_g and $\mathcal{F}$ are

$$F_{gi}^\alpha = - \frac{\partial U_{UU}(r_0)}{\partial r_{0i}^\alpha} \tag{6.21a}$$

and

$$\mathcal{F}_i^\alpha = - \frac{\partial U_{VU}(q, r_0)}{\partial r_{0i}^\alpha} \tag{6.21b}$$

The total force on the reagents will be denoted by the n-component vector $F \equiv F[q, r_0]$ whose elements are

$$F_i^\alpha = F_{gi}^\alpha + \mathcal{F}_i^\alpha \tag{6.21c}$$

The classical Hamiltonian H of the solution is

$$H = \sum_{\lambda=1}^{N_s} \frac{p_\lambda^2}{2M} + \frac{1}{2} \sum_{i=1}^{n} m_i \dot{r}_{0i}^2 + U_{VV}(q) + U_{VU}(q, r_0) + U_{UU}(r_0) \tag{6.22a}$$

We also require the classical Hamiltonian H_0 of the solvent conditional on the assumption that the reagent atoms are clamped at point r_0. This is

$$H_0 = \sum_{\lambda=1}^{N_s} \frac{p_\lambda^2}{2M} + U_{VV}(q) + U_{VU}(q, r_0) \tag{6.22b}$$

The corresponding canonical ensemble distribution functions are

$$f_{CA}[p, q; m\dot{r}_0, r_0] = Z^{-1}(\beta) \exp[-\beta H] \tag{6.23a}$$

and

$$f_{CA}^{(0)}[p, q] = Z_0^{-1}(\beta) \exp[-\beta H_0] \tag{6.23b}$$

In accord with our earlier notation, $\langle \ \rangle$ and $\langle \ \rangle_{r_0}$ will respectively denote averages over $f_{CA}[p, q; p_0, q_0]$ and $f_{CA}^{(0)}[p, q]$.

We shall later require the local equilibrium density of solvent atoms given that the reagents are fixed at configuration point r_0. We denote this quantity by $\rho^{(n+1)}(\mathbf{q}, r_0)$. As the solvent coordinate $\mathbf{q} \to \infty$, we have $\rho^{(n+1)}(\mathbf{q}; r_0) \to \rho$, the bulk density. By analogy to the pair distribution function and the pair potential of the mean force discussed in Section IV.A.1, we define the $(n + 1)$-particle solute–solvent distribution function $g^{(n+1)}(\mathbf{q}; r_0)$ and the corresponding mean-force potential $W^{(n+1)}(\mathbf{q}; r_0)$ by

$$\rho^{(n+1)}(\mathbf{q}; r_0) = \rho g^{(n+1)}(\mathbf{q}; r_0) = \rho \exp\left[-\beta W^{(n+1)}(\mathbf{q}; r_0)\right] \quad (6.24)$$

We may also define a n-particle reagent potential of the mean force, $W^{(n)}(r_0)$. This may be defined in terms of the mean force acting on the reagents,

$$\langle F \rangle_{r_0} = F_g + \langle \mathcal{F} \rangle_{r_0} \quad (6.25)$$

by

$$\langle F_i^\alpha \rangle_{r_0} = -\frac{\partial W^{(n)}(r_0)}{\partial r_{0i}^\alpha} \quad (6.26)$$

We may define by analogy to Eq. (4.4) an n-particle solute cavity functions $w^{(n)}(r_0)$. This function, which is the liquid-state contribution to $W^{(n)}(r_0)$, is defined by

$$W^{(n)}(r_0) = U_{UU}(r_0) + w^{(n)}(r_0) \quad (6.27)$$

Note that the liquid-state contribution to the mean force is given by

$$\langle \mathcal{F} \rangle_{r_0} = -\frac{\partial w^{(n)}(r_0)}{\partial r_0} \quad (6.28)$$

We next turn to quantities which describe solution dynamics given that the reagents are clamped at configuration point r_0. The Liouville operator of the solvent for this case is

$$iL_0 = \sum_{\lambda=1}^{N_s} \left\{ \frac{\mathbf{p}_\lambda}{M} \cdot \frac{\partial}{\partial \mathbf{q}_\lambda} - \frac{\partial}{\partial \mathbf{q}_\lambda}[U_{VV}(q) + U_{VU}(q, r_0)] \cdot \frac{\partial}{\partial \mathbf{p}_\lambda} \right\} \quad (6.29)$$

The value of an arbitrary solvent dynamical variable A at time t given that the reagents are clamped will be denoted by $A_0(t) \equiv A_0(t; r_0)$. Explicitly

$$A_0(t) = \exp[iL_0 t] A \tag{6.30}$$

Note that $A_0(t = 0) = A$ and also that $\dot{A}_0 = iL_0 A$.

Finally we note that the Liouville operator L_0 is Hermitian if the averaging process is $\langle \; \rangle_{r_0}$. Thus time correlation functions of arbitrary solvent dynamical variables $A_0(t)$ and $B_0(t)$ have the following stationary property:

$$\langle A_0(t + \tau) B_0(\tau) \rangle_{r_0} = \langle A_0(t) B_0 \rangle_{r_0} \tag{6.31}$$

2.   Linear-Response Calculation

We next present the linear-response development of the results for the r_0-dependent MTGLE quantities given in Eqs. (6.3). We assume the reagents are initially clamped at configuration point r_0. The initial phase-space PDF is thus $f_{CA}^{(0)}(p, q)$ given in Eq. (6.23b).

At time $t = 0$, the reagents begin to execute *small forced oscillations* $\Delta r_0(t)$ $= (\Delta \mathbf{r}_{01}(t), \ldots, \Delta \mathbf{r}_{0n}(t))$ about configuration point r_0. The Liouville operator of the solvent at time $t > 0$, which we denote by $L(t)$, is determined by replacing $U_{VU}(q, r_0)$ with $U_{VU}[q, r_0 + \Delta r_0(t)]$ in Eq. (6.29). That is,

$$iL(t) = \sum_{\lambda=1}^{N_s} \left\{ \frac{\mathbf{p}_\lambda}{M} \cdot \frac{\partial}{\partial \mathbf{q}_\lambda} - \frac{\partial}{\partial \mathbf{q}_\lambda} [U_{VV}(q) + U_{VU}[q, r_0 + \Delta r_0(t)]] \cdot \frac{\partial}{\partial \mathbf{p}_\lambda} \right\} \tag{6.32}$$

To linear order in $\Delta r_0(t)$, however,

$$U_{VU}[q, r_0 + \Delta r_0(t)] = U_{VU}[q, r_0] + \sum_{i=1}^{n} \sum_{\alpha=1}^{3} \frac{\partial}{\partial r_{0i}^\alpha} U_{VU}[q, r_0] \Delta r_{0i}^\alpha(t)$$

$$= U_{VU}(q, r_0) - \sum_{i=1}^{n} \sum_{\alpha=1}^{3} \mathscr{F}_i^\alpha \Delta r_{0i}^\alpha(t) \tag{6.33}$$

where we have used the definition of $\mathscr{F}_i^\alpha$, Eq. (6.21b). Finally, in matrix notation,

$$\sum_{i=1}^{n} \sum_{\alpha=1}^{3} \mathscr{F}_i^\alpha \Delta r_{0i}^\alpha(t) = \mathscr{F}^{\mathrm{T}} \cdot \Delta r_0(t) \tag{6.34}$$

Combining Eqs. (6.32)–(6.34) gives $iL(t) = iL_0 + iL^{(1)}(t)$, where

$$iL^{(1)}(t) = \sum_{\lambda=1}^{N_s} \frac{\partial}{\partial \mathbf{q}_\lambda} \left[\mathscr{F}^T \cdot \Delta r_0(t) \right] \cdot \frac{\partial}{\partial \mathbf{p}_\lambda} \tag{6.35}$$

The solvent phase-space PDF at time $t > 0$, $f[p, q; t]$, is modified from its initial value $f_{CA}^{(0)}[p, q]$ by the perturbation $iL^{(1)}(t)$. To linear order in $iL^{(1)}(t)$, $f[p, q; t]$ may be calculated using the linear-response-theory result (4.69). This yields

$$f[p, q; t] = f_{CA}^{(0)}[p, q]$$
$$+ \int_0^t d\tau \exp[-iL_0(t - \tau)] \left[-iL^{(1)}(\tau) f_{CA}^{(0)}(p, q) \right] \tag{6.36}$$

Equation (6.36) may be simplified in the following manner. Combining Eqs. (6.22b), (6.23b), and (6.35) yields

$$-iL^{(1)}(t) f_{CA}^{(0)}(p, q) = \frac{1}{k_B T} f_{CA}^{(0)}(p, q) \sum_{\lambda=1}^{N_s} \frac{\mathbf{p}_\lambda}{M} \cdot \frac{\partial}{\partial \mathbf{q}_\lambda} \left[\mathscr{F}^T \cdot \Delta r_0(t) \right]$$

Equation (6.29), however, yields that

$$\dot{\mathscr{F}}_0^T = iL_0 \mathscr{F}^T = \sum_{\lambda=1}^{N_s} \frac{\mathbf{p}_\lambda}{M} \cdot \frac{\partial \mathscr{F}^T}{\partial \mathbf{q}_\lambda}$$

Comparing the above two equations yields

$$-iL^{(1)}(t) f_{CA}^{(0)}(p, q) = \frac{1}{k_B T} f_{CA}^{(0)}(p, q) \dot{\mathscr{F}}_0^T \cdot \Delta r_0(t) \tag{6.37}$$

Combining Eqs. (6.36) and (6.37) and using Eq. (6.30) yields the following result for the phase-space PDF $f[p, q; t]$:

$$f[p, q; t] = f_{CA}^{(0)}[p, q] \left\{ 1 + \frac{1}{k_B T} \int_0^t \dot{\mathscr{F}}_0(\tau - t) \cdot \Delta r_0(\tau) \, d\tau \right\} \tag{6.38}$$

Within our linear response model the liquid-state part of the systematic force acting on the reagents is the average induced force $\langle \mathscr{F} \rangle_t$, which is given by

$$\langle \mathcal{F} \rangle_t = \int dp\, dq\, f[\,p, q; t]\, \mathcal{F}[\,q, r_0 + \Delta r_0(t)]$$

$$= \langle \mathcal{F} \rangle_{r_0} + \int dp\, dq\, f[\,p, q; t]\{\mathcal{F}[\,q, r_0 + \Delta r_0(t)] - \langle \mathcal{F} \rangle_{r_0}\}$$

Combining the above equation and Eq. (6.38) and dropping terms higher than linear order in $\Delta r_0(t)$ yields (matrix notation) the following result for $\langle \mathcal{F} \rangle_t$:

$$\langle \mathcal{F} \rangle_t = \langle \mathcal{F} \rangle_{r_0} + \left\langle \frac{\partial \mathcal{F}^{\mathsf{T}}}{\partial r_0} \right\rangle_{r_0} \cdot \Delta r_0(t)$$

$$+ \frac{1}{k_B T} \int_0^t \langle \tilde{\mathcal{F}}_0 \dot{\tilde{\mathcal{F}}}_0^{\mathsf{T}}(\tau - t) \rangle_{r_0} \cdot \Delta r_0(\tau)\, d\tau$$

where $\tilde{\mathcal{F}}_0(t) = \mathcal{F}_0(t) - \langle \mathcal{F} \rangle_{r_0}$.

Using Eq. (4.50b), which applies because of the stationarity property (6.31), the above equation may be rewritten as

$$\langle \mathcal{F} \rangle_t = \langle \mathcal{F} \rangle_{r_0} + \left\langle \frac{\partial \mathcal{F}^{\mathsf{T}}}{\partial r_0} \right\rangle_{r_0} \cdot \Delta r_0(t)$$

$$- \frac{1}{k_B T} \int_0^t \langle \dot{\tilde{\mathcal{F}}}_0(t - \tau) \tilde{\mathcal{F}}_0^{\mathsf{T}} \rangle_{r_0} \cdot \Delta r_0(\tau)\, d\tau \qquad (6.39)$$

This equation gives the liquid-state contribution to the systematic part of the force acting on the oscillating reagents. The total force $F(t)$ exerted on the reagents additionally includes the gas-phase contribution $F_g[r_0 + \Delta r_0(t)]$ and the random force $\tilde{\mathcal{F}}_0(t)$. Thus within our linear-response approximation $F(t)$ may be decomposed as [cf. Eq. (3.78)]

$$F(t) = F_g[r_0 + \Delta r_0(t)] + \langle \mathcal{F} \rangle_{r_0}$$

$$\times \left\langle \frac{\partial \mathcal{F}^{\mathsf{T}}}{\partial r_0} \right\rangle_{r_0} \cdot \Delta r_0(t) - \beta \int_0^t \langle \dot{\tilde{\mathcal{F}}}_0(t - \tau) \tilde{\mathcal{F}}_0^{\mathsf{T}} \rangle_{r_0} \cdot \Delta r_0(\tau)\, d\tau + \tilde{\mathcal{F}}_0(t)$$

$$(6.40)$$

In Section III.J, however, we indicated how MTGLE quantities for interacting reagents may be defined. The method of definition is based on the approximate density-response interpretations of the MTGLE quantities for isolated solutes and the fact that these interpretations for isolated solutes (spherical local solvent densities) are readily generalized to interacting reagents ("ellipsoidal" local solvent densities). The density-response interpretations lead to the following *definition* of the MTGLE quantities:

$$F(t) = F_g[r_0 + \Delta r_0(t)] + \langle \mathcal{F} \rangle_{r_0} - \mathbf{m}^{1/2}\omega_{e0}^2(r_0)\mathbf{m}^{1/2} \cdot \Delta r_0(t)$$

$$+ \int_0^t \mathbf{m}^{1/2}\omega_{c1}^2(r_0)\theta_1(t-\tau; r_0)[\omega_{c1}^2(r_0)]^T \mathbf{m}^{1/2} \cdot \Delta r_0(\tau) \, d\tau$$

$$+ \mathbf{m}^{1/2}\omega_{c1}^2(r_0)\mathbf{m}^{1/2}R_1(t; r_0) \tag{6.41}$$

Comparing Eqs. (6.40) and (6.41) yields the results of Eq. (6.3).

3.   Isolated-Solute-Atom Limit

The extension of the rigorous formalism of Section V into a model for liquid-solution kinetics is based on the clamping approximation.

We next present an analytical test of the validity of the clamping approximation. A numerical test will be presented in Section VII. What we do here is compare the clamping approximation results for ω_{e0}^2 and ω_{c1}^2 with the corresponding exact results.

The clamping-approximation results are obtained by specializing Eqs. (6.3a,b) to the isolated-solute-atom case. This gives

$$\omega_{e0}^2 = -(3m)^{-1}\left\langle \frac{\partial}{\partial \mathbf{r}_0} \cdot \mathcal{F} \right\rangle_{\mathbf{r}_0} \tag{6.42}$$

and

$$\omega_{c1}^4 = (3mk_BT)^{-1}\langle \dot{\mathcal{F}}_0^2 \rangle_{\mathbf{r}_0} \quad \text{(clamping approximation)} \tag{6.43}$$

The clamping-approximation result for ω_{e0}^2 may be recast as

$$\omega_{e0}^2 = (-3m)^{-1}\left\langle \frac{\partial}{\partial \mathbf{r}_0} \cdot \mathcal{F} \right\rangle$$

$$= (3mk_BT)^{-1}\langle \mathcal{F}^2 \rangle \quad \text{(clamping approximation)} \tag{6.44a}$$

The first equality in Eq. (6.44a) follows from the isotropy of the liquid [which permits one to replace $\langle \ \rangle_{r_0}$ by $\langle \ \rangle$ in Eq. (6.44a)]. The second equality follows from the virial theorem.[59]

The rigorous result for ω_{e0}^2 follows from the fact that $\omega_{e0}^2 = -\ddot{\chi}(0) = -m(3k_B T)^{-1}\langle \ddot{\mathbf{r}}_0 \cdot \mathbf{r}_0 \rangle = m(3k_B T)^{-1}\langle \dot{r}_0^2 \rangle$. However, for an isolated solute $m\ddot{\mathbf{r}}_0 = \mathscr{F}$. Thus

$$\omega_{e0}^2 = (3mK_B T)^{-1}\langle \mathscr{F}^2 \rangle \qquad \text{(exact)} \qquad (6.44b)$$

Comparison of the clamping approximation for ω_{e0}^2 [Eq. (6.44a)] and the rigorous result for ω_{e0}^2 [Eq. (6.44b)] shows that the clamping approximation is actually exact. (This has already been mentioned in Section III.J.)

An analogous reduction of the clamping-approximation and exact results for ω_{cl}^4 yields

$$\omega_{cl}^4 = \frac{1}{3mM} \sum_{\lambda=1}^{N_s} \left\langle \frac{\partial \mathscr{F}}{\partial \mathbf{q}_\lambda} : \frac{\partial \mathscr{F}}{\partial \mathbf{q}_\lambda} \right\rangle \qquad \text{(clamping approximation)} \quad (6.45a)$$

and

$$\omega_{cl}^4 = \frac{1}{3mM} \left[\sum_{\lambda=1}^{N_s} \left\langle \frac{\partial \mathscr{F}}{\partial \mathbf{q}_\lambda} : \frac{\partial \mathscr{F}}{\partial \mathbf{q}_\lambda} \right\rangle + \frac{M}{m} \left\{ \left\langle \frac{\partial \mathscr{F}}{\partial \mathbf{r}_0} : \frac{\partial \mathscr{F}}{\partial \mathbf{r}_0} \right\rangle - \frac{1}{3} \left\langle \frac{\partial \mathscr{F}}{\partial \mathbf{r}_0} \right\rangle^2 \right\} \right]$$

$$\text{(exact)} \quad (6.45b)$$

Equations (6.45) show that the clamping approximation and exact results for ω_{cl}^4 differ by the term containing the braces in Eq. (6.45b). This term is guaranteed to be small only if: (i) the mass ratio $M/m \ll 1$; (ii) one may approximately break the average in the term $\langle \partial \mathscr{F}/\partial \mathbf{r}_0 \rangle : (\partial \mathscr{F}/\partial \mathbf{r}_0) \rangle$ to yield

$$\left\langle \frac{\partial \mathscr{F}}{\partial \mathbf{r}_0} : \frac{\partial \mathscr{F}}{\partial \mathbf{r}_0} \right\rangle \simeq \frac{1}{3} \left\langle \frac{\partial \mathscr{F}}{\partial \mathbf{r}_0} \right\rangle^2 .$$

This last condition holds rigorously for solutes linearly coupled to perfectly harmonic solvents.

VII. CALCULATION AND MODELING OF r_0-DEPENDENT CORRELATION FUNCTIONS

We next present four topics that are important for the practical implementation of the model for liquid-solution reactions developed in Section VI: (i) We present a numerical test of the clamping approximation for the case of an isolated solute atom (Section VII.A). (ii) We present molecular-dynamics (MD) calculations of the r_0-dependent quantities of Eqs. (6.3) for the case of diatomic solutes interacting in Lennard–Jones solvents (Section VII.B). (iii) We develop simplified methods for the calculation of the r_0-dependent quantities. These methods permit one to bypass MD (Section VII.C). (iv) We develop truncated chain models and test their numerical accuracy by comparing matrix response functions for the models with the corresponding exact response functions (Section VII.D). This test is a matrix analogue of the tests shown in Fig. 5.

A. Test of the Clamping Approximation for Isolated Solute Atoms

We have performed MD simulations for four model Lennard–Jones solutions consisting of a single solute atom (mass m) moving in an otherwise pure solvent (mass M). The Lennard–Jones parameters, solute and solvent masses, and thermodynamic states for the four solutions are given in Table I.

The simulations were designed to provide a numerical test of the clamping approximations. The results for ω_{e0}^2, ω_{c1}^2, $\beta_1(t)$, and β_1 for freely moving and for clamped solutes were compared. Details of the MD simulation methods have been presented elsewhere.[23(a)]

Results of our MD test of the clamping approximation for isolated solutes are summarized in Table II and Fig. 12. Table II gives a comparison of the exact results for ω_{e0}^2 and ω_{c1}^4 [computed, respectively, from Eqs. (6.42) and (6.45b)] and the corresponding clamping-approximation results [computed, respectively, from Eqs. (6.44a) and (6.45a)].

TABLE I
Isolated-Solute-Atom Solutions[a]

Model system	Solute-atom mass m (g/mol)	σ_{12} (Å)	ε_{12} (K)	T^*	ρ^*
1	39.95	3.4	120	0.92 ± 0.2	0.75
2	199.75	3.4	120	0.89 ± 0.4	0.75
3	39.95	3.4	240	0.83 ± 0.3	0.75
4	39.95	3.4	360	0.86 ± 0.2	0.75

[a]All potential-energy functions are of the Lennard–Jones form $U_{ij}(r) = 4\varepsilon_{ij}[(\sigma_{ij}/r)^{12} - (\sigma_{ij}/r)^6]$, where $U_{11}(r)$ is the solute–solute potential and $U_{12}(r)$ is the solute–solvent potential. Table gives solute masses, solute–solvent Lennard–Jones parameters, and solvent thermodynamic states (Lennard–Jones units are used) for all four solutions. The solvent atomic mass, potential-energy cutoff used in MD simulations[23(a)], and solvent–solvent Lennard–Jones parameters are, respectively, $M = 39.95$ g/mol, cutoff = 7.5 Å, $\varepsilon_{11} = 120$ K, and $\sigma_{11} = 3.4$ Å, for all four solutions. A total of 48 particles were used for all cases. A test of the use of small numbers of particles is given in Ref. 23(a).

TABLE II
Molecular-Dynamics Test of the Clamping Approximation
for the Isolated-Solute-Atom Model Solutions of Table I

Model system	ω_{e0}^2 (psec^{-2})	ω_{c1}^4 (psec^{-4})	β_1 (psec^{-1})
1 (unclamped)	51.0 ±2.5	6850± 340	5.67±1.00
1 (clamped)	47.4 ±4.0	3860± 440	5.75±0.73
2 (unclamped)	10.55±0.5	801± 24	1.81±0.23
2 (clamped)	9.55±0.8	772± 87	1.15±0.15
3 (unclamped)	77.9 ±3.9	11,060± 860	8.67±0.45
3 (clamped)	78.3 ±3.9	7,370± 780	5.88±0.33
4 (unclamped)	108.4 ±6.9	17,660±2120	8.09±1.03
4 (clamped)	103.7 ±4.4	10,180±2120	9.52±2.02

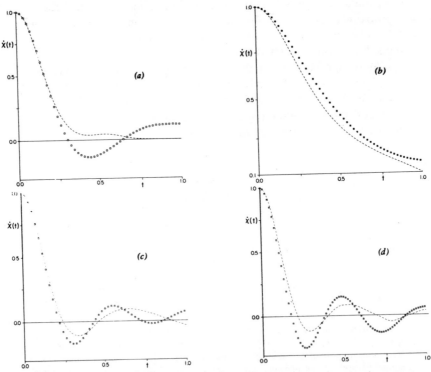

Fig. 12. Test of the clamping approximation for isolated-solute-atom model solutions. Clamped (squares) and unclamped (dashes) normalized velocity autocorrelation functions $\dot{\chi}(t)$ vs. time in picoseconds for (a) model system 1, (b) model system 2, (c) model system 3, (d) model system 4. Masses, Lennard–Jones potential parameters, and thermodynamic states for the isolated-solute-atom model systems are given in Table I.

The results for model systems 1, 3, and 4 provide a reasonably stringent test of the clamping approximation, since for these cases the solute/solvent mass ratio $m/M = 1$. (Recall that the clamping approximation is exact for liquids only if $m/M \to \infty$.) The results of Table II show that for all model systems the exact and clamping-approximation results for ω_{e0}^2 agree to within statistical error. This provides a useful check on the calculations, since, as shown in Section VI.B.3, the clamping approximation is exact for ω_{e0}^2.

The errors due to clamping are, however, significant in ω_{c1}^4 for the mass-ratio-unity model systems. The values of ω_{c1}^4 calculated from the clamping approximation are too small by $\simeq 25$–40%. The clamping-approximation result for ω_{c1}^4 is much better for mass ratio 5 (model system 2). The clamping-approximation result agrees with the corresponding exact result to within statistical error. A key point is that despite the quantitative discrepancies the clamping approximation reproduces the qualitative variation of ω_{c1}^4 over an order-of-magnitude range (cf. the results for model systems 2 and 3 in Table II).

The magnitudes and variation of the solute friction coefficient β_1, calculated from $\beta_1 = \int_0^\infty \beta_1(t)\, dt$, are surprisingly well reproduced by clamping. The clamped and unclamped function coefficients for model system 2, for example, are respectively 1.8 ± 0.2 and 1.2 ± 0.2 psec^{-1}, while the corresponding results for model system 4 are 8.1 ± 1.0 and 9.5 ± 2.0 psec^{-1}.

A further test of the clamping approximation is made in Fig. 12, where we compare clamped and unclamped normalized VAFs for the four model systems. The clamping approximation reproduces the chemically important short-time-scale behavior of the autocorrelation functions as well as their qualitative features in all four cases. Note that, in particular, clamping is able to reproduce the transition from very weak caging (model system 2) to fairly strong caging in a qualitatively correct manner.

In summary, our MD results for single solute atoms suggest that while the clamping approximation is inadequate for precision liquid-state work of the traditional type,[7-9] it may well be adequate for the purpose of modeling many-body systems at a chemically useful level.

B. Calculation of the r_0-Dependent Correlation Functions

We next present numerical calculations of the r_0-dependent quantities ω_{e0}^2, $\omega_{c1}^4(r_0)$, and $\beta_1(t; r_0)$, which are required to construct the matrix equivalent-chain equations (6.12) and (6.15). We explicitly treat the simplest case of homonuclear diatomic solutes interacting in pure Lennard–Jones solvents. This is the case relevant for I_2 photodissociation dynamics discussed in Section VIII. For this case the calculated quantities depend only on the solute–solute internuclear separation R.

We consider, in particular, four model Lennard–Jones solutions. Model solution 1 consists of two argon solute atoms dissolved in the Levesque–Verlet[8] model for liquid argon. The results for this rather artificial solution are included because they result clearly illustrate the features that govern the R-dependence of the MTGLE parameters. Model solution 2 is designed to roughly simulate the interaction of two iodine atoms dissolved in liquid carbon tetrachloride. Model solutions 3 and 4 are designed to roughly simulate interacting iodine atoms dissolved in ethane in two different thermodynamic states. The solute and solvent masses, Lennard–Jones potential parameters, and thermodynamics states for the four model solutions are listed in Table III.

We shall regard the MTGLE matrices as functions of the reduced internuclear separation $x = R/\sigma$ (σ = Lennard–Jones diameter of an argon atom) for model solution 1, and $x = R/R_{eq}$ ($R_{eq} = 2.667$ Å $= I_2$ bond length) for the remaining solutions.

Full molecular-dynamics calculations of the x-dependent quantities are presented for model solution 2 [I_2 in CCl_4]. Simplified calculations of the MTGLE matrices based on the superposition approximation (SA) are presented for all four model solutions. These approximate results are then tested against MD results for model solution 2.

We begin by presenting formulae, based on the SA, which permit construction of the MTGLE matrices $\omega_{e0}^2(r_0)$, $\omega_{c1}^4(r_0)$, and $\Omega_0^2(r_0)$ from solute–solvent and solute–solute equilibrium *pair* distribution functions. These pair

TABLE III
Interacting-Solute-Atom Model Solutions[a]

Model solution	Solute mass m (g/mol)	Solvent mass M (g/mol)	σ_{11} (Å)	ε_{11} (K)	σ_{12} (Å)	ε_{12} (K)	T (K)	ρ (Å^{-3})
1. Ar_2 in liquid Ar	39.95	39.95	3.4	120	3.4	120	105	1.9×10^{-2}
2. I_2 in liquid CCl_4	126.9	153.8	5.27	373	4.96	274	305	6.3×10^{-3}
3. I_2 in dense gaseous C_2H_6	126.9	30.0	4.38	236	4.53	218	314	5.2×10^{-3}
4. I_2 in liquid C_2H_6	126.9	30.0	4.38	236	4.53	218	298	9.0×10^{-3}

[a]All potential-energy functions are of the Lennard–Jones type (see Table I). For MD simulations 88 solvent particles were employed for internuclear separations $R < 2.20 R_{eq}$, while 103 solvent particles were employed for $R \geqslant 2.20 R_{eq}$ (Note that $R_{eq} = 2.667$ Å $= I_2$ bond length). The potential-energy cutoff was taken as 12.0 Å for all MD simulations on model solution 2.

distributions, for the case of Lennard–Jones liquid, may be conveniently and accurately constructed as solutions to the Percus–Yevick equation Eq. (4.13) with the closure (4.30).

1. Approximate Formulae for $\omega_{e0}^2(r_0)$, $\omega_{c1}^4(r_0)$, and $\Omega_0^2(r_0)$

We next specialize our general results for $\omega_{e0}^2(r_0)$ and $\omega_{c1}^4(r_0)$—Eqs. (6.3a, b)—to the case of pairwise additive solute–solvent forces. For this case the solute–solvent potential-energy function may be decomposed as

$$U_{VU}(q, r_0) = \sum_{\lambda=1}^{N_s} \sum_{j=1}^{n} u_{uv}^{(j)}\left[|\mathbf{r}_{0j} - \mathbf{q}_\lambda|\right] \qquad (7.1)$$

where $u_{uv}^{(j)}(r)$ is the pair potential coupling a solvent molecule with solute atom j. Given Eq. (7.1), the total solvent force on reagent atom j may be written as

$$\mathscr{F}_j = \sum_{\lambda=1}^{N_s} \mathscr{F}_j[\mathbf{q}_\lambda] \qquad (7.2)$$

where $\mathscr{F}_j[\mathbf{q}_\lambda]$ is the force on solute atom j due to solvent atom λ. Note that

$$\mathscr{F}_j[\mathbf{q}_\lambda] = -\frac{\partial}{\partial \mathbf{r}_{0j}} u_{uv}^{(j)}\left[|\mathbf{r}_{0j} - \mathbf{q}_\lambda|\right] \qquad (7.3)$$

Given this assumption of pairwise additivity, it is straightforward to derive the following *exact* expressions for $\omega_{e0}^2(r_0)$ and $\omega_{c1}^4(r_0)$:

$$\left[\omega_{e0}^2(r_0)\right]_{ij}^{\alpha\beta} = -(m_i m_j)^{-1/2} \delta_{ij} \int d\mathbf{q} \frac{\partial \mathscr{F}_i^\beta(\mathbf{q})}{\partial r_{0i}^\alpha} \rho^{(n+1)}(\mathbf{q}; r_0) \qquad (7.4)$$

and

$$\left[\omega_{c1}^4(r_0)\right]_{ij}^{\alpha\beta} = \left[(m_i m_j)^{1/2} M\right]^{-1} \int d\mathbf{q} \frac{\partial \mathscr{F}_i^\alpha(\mathbf{q})}{\partial \mathbf{q}} \cdot \frac{\partial \mathscr{F}_j^\beta(\mathbf{q})}{\partial \mathbf{q}} \rho^{(n+1)}(\mathbf{q}, r_0) \qquad (7.5)$$

Also, combining Eq. (6.16b) and Eq. (6.28) yields the following result for the matrix components of $\Omega_0^2(r_0)$:

$$\left[\Omega_0^2(r_0)\right]_{ij}^{\alpha\beta} = +(m_i m_j)^{-1/2} \frac{\partial^2 w^{(n)}(r_0)}{\partial r_{0i}^\alpha \partial r_{0j}^\beta} \qquad (7.6)$$

The SA for $\rho^{(n+1)}(\mathbf{q}; r_0)$ amounts to replacing the $(n+1)$-atom solute reagent potential of mean force $W^{(n+1)}(\mathbf{q}; r_0)$ [Eq. (6.24)] with the following superposition of solute–solvent pair potentials of the mean force:

$$W^{(n+1)}(\mathbf{q}; r_0) \simeq \sum_{j=1}^{n} W_j^{(2)}\big[|\mathbf{q} - \mathbf{r}_{0j}|\big] \tag{7.7}$$

Using Eq. (4.12), Eq. (7.7) is clearly equivalent to the following approximation for $\rho^{(n+1)}(\mathbf{q}; r_0)$:

$$\rho^{(n+1)}(\mathbf{q}; r_0) \simeq \rho g_1\big[|\mathbf{r}_{01} - \mathbf{q}|\big]\cdots g_n\big[|\mathbf{r}_{0n} - \mathbf{q}|\big] \tag{7.8}$$

where $g_i[|\mathbf{r}_{0i} - \mathbf{q}|]$ is the equilibrium pair distribution function between a solvent atom and solute atom i. Using Eq. (7.8) in Eqs. (7.4) and (7.5), one may determine $\omega_{e0}^2(r_0)$ and $\omega_{c1}^4(r_0)$ directly from the equilibrium solvent–solute pair distribution function.

An analogous superposition approximation may be made for the solute cavity function $w^{(n)}(r_0)$. Namely, one may approximate

$$w^{(n)}(r_0) \simeq \sum_{i>j} w_{ij}^{(2)}\big[|\mathbf{r}_{0i} - \mathbf{r}_{0j}|\big] \tag{7.9}$$

The solute–solute pair cavity functions $w_{ij}^{(2)}(r)$ may be calculated from the corresponding pair distribution functions and the gas-phase pair potential functions using Eq. (4.11). Thus, given Eq. (7.6), the matrix $\Omega_0^2(r_0)$ may be constructed within the SA from solute–solute pair distribution functions.

Summarizing, within the SA the basic MTGLE matrices $\omega_{e0}^2(r_0)$, $\omega_{c1}^4(r_0)$, and $\Omega_0^2(r_0)$ may be calculated from the solute–solute and solute–solvent equilibrium pair distribution functions. For Lennard–Jones solutions these may be conveniently and accurately constructed from the Percus–Yevick equation. For the present case of homonuclear diatomic solutes immersed in pure solvents, only two distribution functions need to be calculated.

We next discuss the structure of the MTGLE matrices for the case of homonuclear diatomic solutes. Recall that, in general, these matrices have $\frac{1}{2}3n(3n+1)$ distinct components. For the present case ($n=2$) one expects, *a priori*, 21 distinct components. As we now discuss, symmetry considerably reduces this complexity.

2. Structure of the MTGLE Matrices for Homonuclear Diatomic Solutes

The structure of the MTGLE matrices for homonuclear diatomic solutes is simplest in a Cartesian reference system whose origin bisects and whose

z-axis runs through the bond axis.[60] For this choice the quantities $F \equiv \omega_{e0}^2(x)$, $\omega_{c1}^4, (x)$, or $\Omega_0^2(x)$ have the following matrix structure:

$$
F =
\begin{bmatrix}
\begin{array}{ccc|ccc}
 & 1 & & & 2 & \\
z & y & x & z & y & x \\
[F]_\parallel^D & 0 & 0 & [F]_\parallel^{OD} & 0 & 0 \\
0 & [F]_\perp^D & 0 & 0 & [F]_\perp^{OD} & 0 \\
0 & 0 & [F]_\perp^D & 0 & 0 & [F]_\perp^{OD} \\
\hline
[F]_\parallel^{OD} & 0 & 0 & [F]_\parallel^D & 0 & 0 \\
0 & [F]_\perp^{OD} & 0 & 0 & [F]_\perp^D & 0 \\
0 & 0 & [F]_\perp^{OD} & 0 & 0 & [F]_\perp^D
\end{array}
\end{bmatrix}
\quad (7.10)
$$

Note that the matrices F have only 4 (rather than 21) distinct nonvanishing elements. These are the components diagonal (D) and off-diagonal (OD) in the solute-atom indices, which are parallel to ($\parallel$) and perpendicular to ($\perp$) the bond axis. Also note that the matrices are symmetric in the solute-atom indices, as is expected for homonuclear diatomic solutes.

The matrix structure and the friction kernel $\beta_1(t; x)$ is somewhat more complex. Elements that are off-diagonal in the Cartesian indices are non-vanishing for $t > 0$. Our MD simulations on model solution 2 of Table III, however, showed that off-diagonal Cartesian components are typically 10 to 20 times smaller than the corresponding diagonal components. Thus $\beta_1(t; x)$ and hence $\beta_1(x)$ approximately have the matrix structure of Eq. (7.10). Within the matrix Gaussian model developed in Section VII.C, $\beta_1(t; x)$ rigorously has the matrix structure of Eq. (7.10).

The derived MTGLE matrices $\omega_{ep}^2(x)$, $\omega_{cp}^2(x)$, $\beta_{p+1}(x)$, and $\Omega_p^2(x)$ ($p \geqslant 1$) required to construct the truncated matrix chain models [see Eqs. (7.17) and (8.1) below] have a somewhat more complex structure. The symmetry between solute-atom indices is only approximately maintained. This is because the Choleski prescription[57] for matrix square roots does not rigorously maintain this symmetry.

The matrices $G \equiv \omega_{ep}^2(x), \Omega_p^2(x)$ and $\beta_{p+1}(x)$, $p \geqslant 1$, have the following symmetric form:

$$
G =
\begin{bmatrix}
a & 0 & 0 & e & 0 & 0 \\
0 & b & 0 & 0 & f & 0 \\
0 & 0 & b & 0 & 0 & f \\
\hline
e & 0 & 0 & c & 0 & 0 \\
0 & f & 0 & 0 & d & 0 \\
0 & 0 & f & 0 & 0 & d
\end{bmatrix}
\quad (7.11)
$$

[Note that the matrix structure of Eq. (7.11) differs from that of Eq. (7.10) to the extent that c differs from a and d differs from b. The actual numerical differences are often small. See Table 3 of Ref. 23(b).]

Finally, the matrices $\omega_{cp+1}^2(x)$ have the following lower triangular structure:

$$
\omega_{cp+1}^2(x) =
\begin{bmatrix}
a & 0 & 0 & 0 & 0 & 0 \\
0 & b & 0 & 0 & 0 & 0 \\
0 & 0 & b & 0 & 0 & 0 \\
\hline
e & 0 & 0 & c & 0 & 0 \\
0 & f & 0 & 0 & d & 0 \\
0 & 0 & f & 0 & 0 & d
\end{bmatrix}
\tag{7.12}
$$

3. Molecular-Dynamics and Superposition-Approximation Results for Interacting Reagents

We next present MD results of the nonvanishing components of $\omega_{e0}^2(x)$ and $\omega_{c1}^4(x)$ and for the dominant components of $\beta_1(t; x)$ for model solution 2 (I_2 in CCl_4). The results are calculated, respectively, from Eqs. (6.3a, b) and Eq. (6.6). Details of the MD simulation methods are presented elsewhere.[23(a)] We also present SA results for the nonvanishing components of $\omega_{e0}^2(x)$ and $\omega_{c1}^4(x)$. These are calculated, respectively, from Eqs. (7.4) and (7.5) using the SA equation (7.8). The solute–solvent pair distributions required to implement the SA are determined as solutions to the Percus–Yevick equation.

We begin with results for model solution 1 (argon solute atoms in an argon solvent). The SA results for the matrix components of $\omega_{e0}^2(x)$ and $\omega_{c1}^4(x)$ for this system are plotted, respectively, in Figs. 13 and 14. These results clearly illustrate the general features of the R-dependence of solvent response in Lennard–Jones systems. The main features of the plots are as follows: (i) The R-dependence of the diagonal components is substantial; for example, values of $[\omega_{e0}^2(x)]_{\parallel}^D$ and $[\omega_{e0}^2(x)]_{\perp}^D$ at $x = 1.5$ are, respectively, $\simeq 15\%$ higher and 30% lower than the asymptotic $x = \infty$ values. (ii) The overall shapes of the curves are dominated by excluded-volume structure arising from successive "squeeze-outs" of solvent molecules at $x = 2, 3, \ldots$. (iii) The off-diagonal elements $[\omega_{c1}^4(x)]_{\parallel}^{OD}$ and $[\omega_{c1}^4(x)]_{\perp}^{OD}$ (which, we recall, roughly measure the coupling between solute atom 1 and the cage of solute atom 2) are approximately one order of magnitude smaller than the diagonal elements. [The off-diagonal elements of $\omega_{e0}^2(x)$ rigorously vanish if the solute–solvent forces are pairwise additive; see Eq. (7.4). The physical reason is mentioned in Section III.J.]

The SA results just presented clearly show that solvent response is substantially modified by reagent approach and common-cage formation.

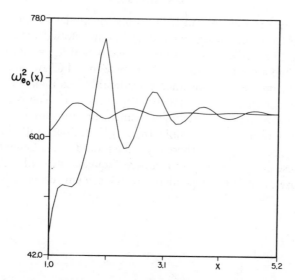

Fig. 13. The matrix parameter $\omega_{e0}^2(x)$ for model solution 1. Superposition-approximation (SA) results for the parallel and perpendicular components of $\omega_{e0}^2(x)$ in psec^{-2} for model solution 1 [Ar$_2$ in Levesque–Verlet Ar] of Table III are plotted vs. reduced internuclear distance $x = R/\sigma$. Parallel component displays excluded-volume maxima at $x \simeq 2, 3, \ldots$.

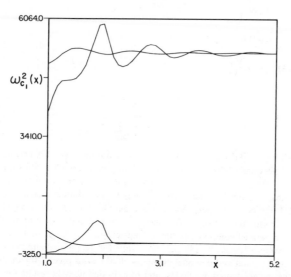

Fig. 14. The matrix parameter $\omega_{c1}^4(x)$ for model solution 1. SA results for the parallel and perpendicular components of the diagonal and off-diagonal elements of $\omega_{c1}^4(x)$ in psec^{-4} are plotted vs. $x = R/\sigma$. The diagonal elements are an order of magnitude larger than the off-diagonal elements. Parallel components display excluded-volume maxima at $x \cong 2, 3, \ldots$.

197

We next turn to the SA and MD results for model solution 2. The non-vanishing components of $\omega_{e0}^2(x)$ and $\omega_{c1}^4(x)$ are plotted vs. $x = R/R_{eq}$ for $x \geqslant 1$ in Fig. 15a and b. Note that the I_2 bond length $R_{eq} = 2.667$ Å is smaller than the iodine-atom Lennard–Jones diameter $\sigma = 4.67$ Å. Thus $x = 1$ corresponds to an internuclear distance considerably less than for contact. The main points concerning the results of Fig. 15 are: (i) The overall shape of the curves is dominated by excluded-volume structure similar to that found for model solution 1. For example, $[\omega_{c1}^4(x)]_{\parallel}^D$ and $[\omega_{c1}^4(x)]_{\perp}^D$ have peaks at $x \simeq \sigma_{12}/R_{eq}, 2\sigma_{12}/R_{eq}, \ldots$, where σ_{12} is the iodine–CCl_4 Lennard–Jones diameter. These peaks are due to successive "squeeze-outs" of CCl_4 molecules as x decreases. (ii) The R-dependence of the solvent response is substantial.

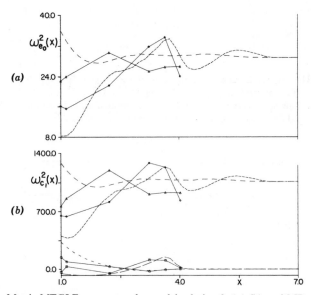

Fig. 15. Matrix MTGLE parameters for model solution 2. (a) SA and MD results for the parallel and perpendicular components of $\omega_{e0}^2(x)$ in $psec^{-2}$ are plotted for model solution 2 of Table III [I_2 in Lennard–Jones CCl_4]. Matrix elements of $\omega_{e0}^2(x)$ are plotted vs. $x = R/R_{eq}$, where $R_{eq} = 2.67$ Å is the I_2 bond length. Triangular and diamond-shaped points denote, respectively, MD results for the perpendicular and parallel components. Long-dashed and short-dashed curves denote, respectively, SA results for the perpendicular and parallel components. Note maxima in parallel components due to "squeeze-out" of CCl_4 molecules during the formation of the common cage. (b) SA and MD results for the parallel and perpendicular components of the off-diagonal and diagonal elements of $\omega_{c1}^4(x)$ in $psec^{-4}$ for model solution 2, vs. $x = R/R_{eq}$. Triangles, diamonds, circles, and squares denote MD results for, respectively, the diagonal perpendicular, diagonal parallel, off-diagonal perpendicular, and off-diagonal parallel components. SA curves are (from top at $x = 1$) diagonal perpendicular, diagonal parallel, off-diagonal perpendicular, and off-diagonal parallel. Note excluded-volume structure. Also note semiquantitive agreement of SA and MD results for $x \geq 2$.

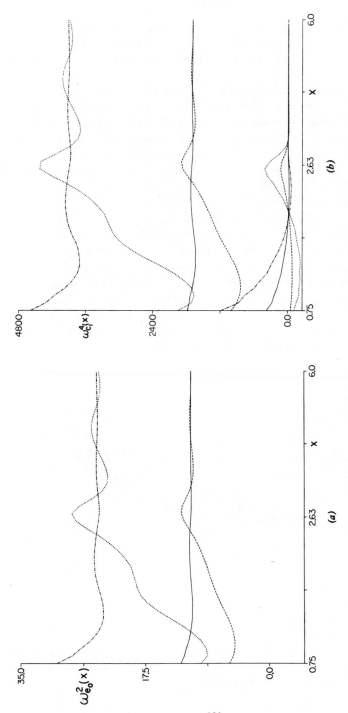

Fig. 16. SA results for the matrix parameters $\omega_{e_0}^2(x)$ and $\omega_{cl}^4(x)$ vs. x for model solutions of I_2 in C_2H_6. Dashed and solid curves denote, respectively, parallel and perpendicular matrix components for model solution 3 of Table III. Dotted and dot-dashed curves denote, respectively, parallel and perpendicular matrix components for model solution 4. (a) $\omega_{e_0}^2(x)$ (psec^{-2}) vs. x; (b) $\omega_{cl}^4(x)$ (psec^{-4}) vs. x. Upper curves are the diagonal components; lower curves are the off-diagonal components.

For example, $[\omega_{e0}^2(x=1)]_{\parallel}^{\mathrm{D}} \simeq 16$ psec^{-1}, while $[\omega_{e0}^2(x=\infty)]_{\parallel}^{\mathrm{D}} \simeq 30$ psec^{-2}.
(iii) Comparison of SA and MD results shows that the SA is usually qualitatively correct for $x \geqslant 1$ and is at least of semiquantitative accuracy for $x \gtrsim \sigma_{12}/R_{\mathrm{eq}}$.

The fact that the SA is reasonably accurate greatly enhances the ease of application of the theory of Section VI. This is because the combination of the SA and the matrix Gaussian modeling methods [see Section VII.C] permits one to bypass MD simulations and to construct chain models for interacting reagents directly from equilibrium pair correlation functions.

In Fig. 16a and b we present SA plots of the nonvanishing components of $\omega_{e0}^2(x)$ and $\omega_{c1}^4(x)$ for model solutions 3 and 4 (I_2 in ethane). These plots are qualitatively similar to those for I_2 in CCl_4.

We next turn to MD results for the diagonal friction-kernel components $[\beta_1(t;x)]_{\parallel}^{\mathrm{D}}$ and $[\beta_1(t;x)]_{\perp}^{\mathrm{D}}$ for model solution 2. These are shown in three-dimensional plots in Figs. 17 and 18. The main features of the plots follow from these facts: (i) By Eq. (6.16a) $\beta_1(0;x) = \omega_{e0}^2(x) - \Omega_0^2(x) \simeq \omega_{e0}^2(x)$ [since numerical calculations show that all components of $\Omega_0^2(x)$ are much smaller than the diagonal components of $\omega_{e0}^2(x)$]. Thus initially the shape of the friction kernel is similar to that of $\omega_{e0}^2(x)$ (see Fig. 15a). (ii) The time dependence of $\beta_1(t;x)$ is well described by the matrix Gaussian model of Eq. (7.16) below. (iii) The friction-kernel matrix $\beta_1(t;x)$ is nearly diagonal

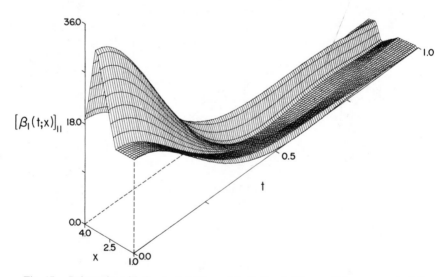

Fig. 17. R-dependent friction kernel for model solution 2. The parallel component of the friction kernel in psec^{-1} is plotted vs. $x = R/R_{\mathrm{eq}}$ and vs. time t in psec. The results are well described by the matrix Gaussian fit of Eq. (7.16).

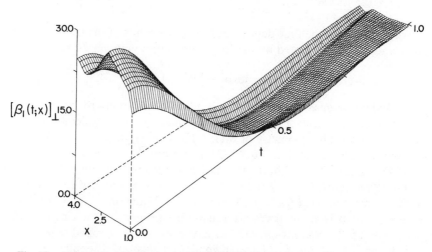

Fig. 18. *R*-dependent friction kernel for model solution 2. Same as Fig. 17 except that the perpendicular component of the friction kernel is plotted.

for most values of x. Thus the diagonal components of $\beta_1(t; x)$ are approximately scalar Gaussian functions of time for most values of x. (iv) The off-diagonal elements of $\beta_1(t; x)$ become appreciable near $x = \sigma_{12}/R_{eq}$, the "squeeze-out" distance of a CCl_4 molecule. This excluded-volume effect gives rise to the "fold" in $\beta_1(t; x)$ near $x = \sigma_{12}/R_{eq}$.

C. Matrix Gaussian Modeling of the Dynamic Solvent Response

We next introduce the matrix Gaussian model for $\beta_1(t; r_0)$. We also present a numerical check of its accuracy for model system 2.

1. Short-Time Expansion of $\beta_1(t; r_0)$

To construct the matrix Gaussian model for $\beta_1(t; r_0)$ we require the short-time expansion of $\beta_1(t; r_0)$. This follows directly from Eqs. (6.3b), (6.6), and (6.16a) as

$$\beta_1(t; r_0) = \left[\beta_1^{1/2}(0; r_0)\right]^{1/2}\left[1 - \tfrac{1}{2}B_0(r_0)t^2 + \cdots\right]\left[\beta_1^{1/2}(0; r_0)\right]^{\mathsf{T}}$$
$$(7.13)$$

where

$$\beta_1^{1/2}(0; r_0) = \left[\omega_{e0}^2(r_0) - \Omega_0^2(r_0)\right]^{1/2} \qquad (7.14)$$

and

$$B_0(r_0) = \left[\beta_1^{-1/2}(0; r_0)\right]\omega_{c1}^4(r_0)\left[\beta_1^{-1/2}(0; r_0)\right]^{\mathsf{T}} \qquad (7.15)$$

2.   Matrix Gaussian Model

We assume that the time dependence of the elements of the friction-kernel matrix may be represented by the following matrix Gaussian form:

$$\beta_1(t; r_0) = \beta_1^{1/2}(0; r_0)\exp\left[-\tfrac{1}{2}B_0(r_0)t^2\right]\left[\beta_1^{1/2}(0; r_0)\right]^{\mathsf{T}} \quad (7.16)$$

This model is a generalization of the familiar Gaussian approximation[61] for conventional (i.e., scalar) time correlation functions. The model has two essential features. The symmetry of $\beta_1(t; r_0)$ [Eq. (6.9a)] is preserved. The matrix power-series expansion [Eq. (7.13)] is preserved to order t^2.

A key point is that within the matrix Gaussian approximation the full dynamic response of the solvent to solute motion is determined by the basic MTGLE matrices $\omega_{e0}^2(r_0)$, $\omega_{c1}^4(r_0)$, and $\Omega_0^2(r_0)$. Equations (7.4)–(7.6), however, show that these matrices are *equilibrium* properties. That is, they depend only on the equilibrium cage structure about the clamped reagents.

Thus within the matrix Gaussian model all quantities required to construct the equation of motion for interacting reagents [Eq. (6.15)] may be calculated from Monte Carlo[5] simulations of solvent structure about the clamped reagents rather than MD simulations of solvent dynamics. Moreover, if one evaluates the matrices $\omega_{e0}^2(r_0)$, $\omega_{c1}^4(r_0)$, and $\Omega_0^2(r_0)$ using the superposition approximations (7.8) and (7.9), then the matrix Gaussian approximation to $\beta_1(t; r_0)$, and hence the equation of motion, (6.15), may be constructed from equilibrium solute–solute and solute–solvent pair correlation functions.

3.   Numerical Test of the Matrix Gaussian Model

We next present a numerical test of the matrix Gaussian approximation for model system 2. The matrices $\omega_{e0}^2(x)$, $\omega_{c1}^4(x)$, and $\Omega_0^2(x)$ required to construct $\beta_1(t; x)$ are determined as follows. The matrices $\omega_{e0}^2(x)$ and $\omega_{c1}^4(x)$ are determined from the MD data plotted in Fig. 15. The matrix $\Omega_0^2(x)$ is calculated from Eq. (7.6) using the Percus–Yevick approximation to the I_2 cavity function $w^{(2)}(x)$. Numerical values of the matrices used in the construction of the matrix Gaussian approximation for model system 2 are tabulated elsewhere.[62]

As mentioned earlier, within the matrix Gaussian approximation off-diagonal Cartesian components of $\beta_1(t; r_0)$ vanish. Thus the only distinct nonvanishing elements of $\beta_1(t; x)$ are the self-friction kernels $[\beta_1(t; x)]_{\parallel \text{ or } \perp}^{\mathrm{D}}$ and the cross-friction kernels $[\beta_1(t; x)]_{\parallel \text{ or } \perp}^{\mathrm{OD}}$. Matrix-Gaussian-approximation results for these nonvanishing components were compared with the corresponding MD results for the full range $1 < x < \infty$. Agreement was excellent except in the region of CCl_4 "squeeze-out" and common-cage formation, $x \simeq 3.2$. Results for the self- and cross-friction kernels are plotted in Fig. 19a for $x = 1.13$ (best agreement) and Fig. 19b for $x = 3.19$ (poorest agreement).

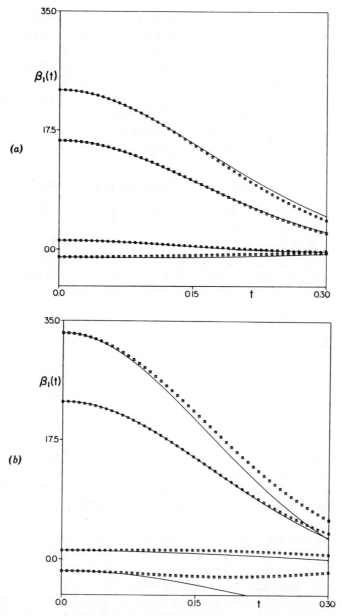

Fig. 19. Comparison of MD and matrix Gaussian approximations for $\beta_1(t) = \beta_1(t; x)$. The friction kernel $\beta_1(t)$ in psec^{-2} is plotted vs. time t in psec for $x = 1.13$ and 3.19 for model solution 2 of Table III. Upper curves are self-friction (D) kernels; lower curves are cross-friction (OD) kernels. Squares and circles denote, respectively, matrix Gaussian components parallel and perpendicular to the I_2 bond axis. Solid lines are MD results. (a) Reduced internuclear separation $x = R/R_{eq} = 1.13$; (b) $x = 3.19$.

Notice that the diagonal (i.e., self-friction) elements are roughly an order of magnitude larger than the off-diagonal (cross-friction) elements. The agreement is poorest for the parallel friction components at the squeeze-out distance $x = 3.19$. Even for this case, however, the Gaussian model provides a qualitatively reasonable description of the solvent response.

We conclude that at least for weakly solvated systems [no strong temporal oscillations in $\beta_1(t; r_0)$], the matrix Gaussian approximation apparently provides a convenient and qualitatively reliable modeling method, which reduces the problem of constructing the matrix equivalent chains to a problem in *equilibrium* statistical mechanics.

D. Heatbath Modeling for Interacting Reagents

In Section III.G we outlined and tested (Fig. 5) a heatbath-modeling method, valid for a single primary-system atom, based on the rigorous equivalent-chain representation (3.60). This modeling method may be extended to the case of interacting reagents in a straightforward manner. The exact matrix chain equations of motion for response functions [Eq. (6.12)], upon modeling, become the following approximate equations of motion [cf. Eq. (3.65)]:

$$\ddot{\rho}_1(t) = - \omega_{e1}^2(r_0)\rho_1(t) + \omega_{c1}^2(r_0)\rho_2(t)$$

$$\ddot{\rho}_2(t) = - \omega_{e2}^2(r_0)\rho_2(t) + \left[\omega_{c2}^2(r_0)\right]^{\mathsf{T}}\rho_1(t) + \omega_{c3}^2(r_0)\rho_3(t)$$

$$\vdots \qquad\qquad\qquad\qquad\qquad\qquad\qquad (7.17)$$

$$\ddot{\rho}_p(t) = \left[\omega_{cp}^2(r_0)\right]^{\mathsf{T}}\rho_{p-1}(t) - \Omega_p^2(r_0)\rho_p(t) - \beta_{p+1}(r_0)\dot{\rho}_p(t)$$

The method for determining the matrix truncation parameters $\Omega_p^2(r_0)$, $\beta_{p+1}(r_0)$ from the dynamic solvent response has been sketched in Section VI.A.6. Within the matrix Gaussian model, the truncation parameters are determined in terms of the basic equilibrium quantities $\omega_{e0}^2(r_0)$, $\omega_{c1}^4(r_0)$, and $\Omega_0^2(r_0)$.

Errors due to heatbath modeling may be assessed by a matrix generalization of the method outlined in Section III.G. The truncation of Eq. (7.17) yields an *approximant* $\theta_1^{(p)}(t; r_0)$ to the exact heatbath response function $\theta_1(t; r_0)$. This approximant may be calculated from Eq. (7.17) via the chain trajectory method and then compared with the corresponding exact result.

Specifically, the matrix components of $\dot{\theta}_1^{(p)}(t; r_0)$ are given by the following prescription [cf. Eqs. (6.13) and (6.14)]:

$$\left[\theta_1^{(p)}(t; r_0)\right]_{ij}^{\alpha\beta} = \left[\rho_1(t)\right]_i^{\alpha} \qquad\qquad (7.18)$$

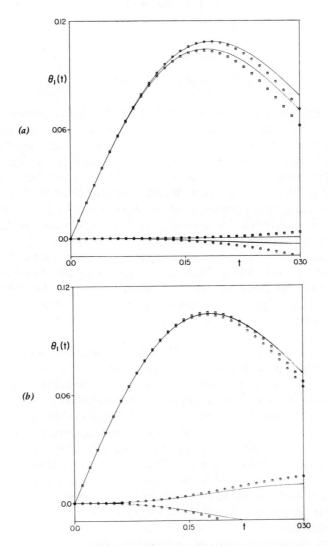

Fig. 20. Test of the truncated matrix chain models. The "exact" matrix Gaussian approximation for $\theta_1(t; x)$ in psec is compared with its two-atom truncated-chain approximant [Eq. (7.17) with $p = 2$] for model solution 2 at $x = 1.13$ and 3.19. The chain parameters are determined via a matrix generalization of the algorithms of Appendix A from the short-time and low-frequency behavior of the matrix Gaussian friction kernels. Input matrices for constructing the matrix Gaussian kernels are determined from the MD results of Fig. 15 [$\omega_{e0}^2(x)$ and $\omega_{c1}^4(x)$] and from the I_2 cavity function [$\Omega_0^2(x)$] determined by solving the Percus–Yevick equation for the Lennard–Jones mixture. Squares and circles respectively denote truncated-chain results for components parallel and perpendicular to the I_2 bond axis. Solid lines are the corresponding exact results. (a) $x = 1.13$; (b) $x = 3.19$.

205

given the initial conditions

$$\left[\dot{\rho}_1(0)\right]_k^\gamma = \delta_{kj}\delta_{\gamma\beta}$$

$$\dot{\rho}_{q+1}(0) = \rho_q(0) = 0, \qquad q = 1, 2, \ldots, p \tag{7.19}$$

A test of the heatbath-modeling method just discussed is presented in Fig. 20a, b for the I_2-in-CCl_4 model solution. "Exact" results for the matrix components of $\theta_1(t; x)$ are constructed from the MD results for $\omega_{cl}^4(x)$ (Fig. 15b) and from the matrix Gaussian approximation to $\beta_1(t; x)$ using Eq. (6.5). The matrix Gaussian approximation is constructed from the MD data as discussed in Section VII.C. Two "atom" truncated-chain models [$p = 2$ in Eq. (7.17)] are then constructed from the MD results for $\omega_{cl}^4(x)$ and the "exact" results for $\theta_1(t; x)$. The approximant $\theta_1^{(2)}(t; x)$ is then calculated from Eq. (7.17) via the chain trajectory method (7.18), (7.19).

The exact results for $\theta_1(t; x)$ are compared with their truncated-chain approximants at $x = 1.13$ (Fig. 20a) and $x = 3.19$ (Fig. 20b). Good agreement is obtained in the chemically important short-time region. For longer times, the response functions of the truncated-chain models are well behaved and decay smoothly to zero with only weak oscillations (i.e., they show no recurrent behavior).

The results of Sections VI and VII may be summarized as follows. Given the clamping approximation, the rigorous formalism of Section V may be made the basis of an approximate framework (Section VI) for liquid-solution reaction dynamics. This framework, when combined with the modeling methods of Sections VII.C and VII.D (implemented either precisely using molecular dynamics or Monte Carlo data, or more approximately using simplified methods like the SA), permit one to deal effectively with the dynamical problems in solution chemistry. The two "atom" heatbath models just discussed, for example, reduce the problem of I_2 photodissociation dynamics in simple solvents to physically realistic effective six-body stochastic classical trajectory calculations (two solvation shells of the common cage treated explicitly).

We next discuss the results of such calculations.

VIII. DYNAMICS OF I_2 RECOMBINATION IN SIMPLE SOLVENTS

We describe in this section the results of MTGLE classical trajectory simulations[23(c)] of an important liquid-state chemical process. This process is the vibrational energy relaxation of a highly excited diatomic molecule, which occurs following its photolysis, and the subsequent recombination of its component atoms. This process is a prototype that is basic to an understanding of the stabilization of vibrationally excited products following a chemical reaction in a liquid.

The chemical systems treated in this section are model solutions 2, 3, and 4 of Table III. These systems, we recall, are designed to simulate the interaction of a pair of iodine atoms dissolved in carbon tetrachloride and ethane. The contents of this section are a preliminary description of a much more detailed study of I_2 photodissociation dynamics in simple solvents, which has been presented elsewhere.[23(c)]

While only a limited set of results are presented here, these results are sufficient to illustrate a number of important general points concerning the physics of liquid-solution reaction dynamics. Most importantly, the simulations illustrate the fundamental role played by solvation-shell motions (i.e., dynamical cage effects) in influencing the course of liquid-solution chemical reactions. We shall see, in particular, that rapid vibrational-energy relaxation of the excited I_2 molecules only occurs when the I_2 vibrational frequency is, roughly speaking, in near resonance with the solvation-shell frequencies.

In order to make the basic role played by dynamical cage effects particularly clear, a comparison of MTGLE and Langevin-model stochastic classical trajectory simulations are made. The MTGLE chain models, since they include the effects of solvation-shell motions [see Fig. 4], give a qualitatively realistic account of dynamical caging. The Langevin model, since it is based on a bulk-solvent rather than a solvation-shell picture of the local solvent environment (cf. Fig. 2c and d), does not realistically describe dynamical caging.

A. Equations of Motion

The MTGLE simulations presented here are based on three-atom truncated-chain models. These correspond to two-atom heatbath models [i.e., $p = 2$ in Eq. (7.17)]. The chain models are obtained by specializing Eq. (6.5) to the equal-mass case and then truncating as in Eq. (7.17). The resulting equations of motion in matrix notation are [cf. Eq. (3.65)]

$$\ddot{r}_0(t) = m^{-1}\langle F \rangle_{R(t)} - \omega_{e0}^2[R(t)]r_0(t) + \omega_{c1}^2[R(t)]r_1(t)$$

$$\ddot{r}_1(t) = -\omega_{e1}^2[R(t)]r_1(t) + [\omega_{c1}^2[R(t)]]^T r_0(t) + \omega_{c2}^2[R(t)]r_2(t) \quad (8.1)$$

$$\ddot{r}_2(t) = [\omega_{c2}^2[R(t)]]^T r_1(t) - \Omega_2^2[R(t)]r_2(t)$$
$$\quad - \beta_3[R(t)]\dot{r}_2(t) + f_3[t; R(t)]$$

These equations require a few words of explanation. The quantity m is the mass of an iodine atom rather than the mass matrix $\mathbf{m}$ of Eq. (6.1), which appears in Eq. (6.15). As before, $R = |\mathbf{r}_{02} - \mathbf{r}_{01}|$ is the I_2 internuclear separation. Also $f_3(t; R) = (\mathbf{f}_{31}(t; R), \mathbf{f}_{32}(t; R))$ is an R-dependent vector white-noise Gaussian random force. It obeys a fluctuation–dissipation theorem that is a matrix generalization of Eq. (3.63), namely

$$\langle f_{3i}^\alpha(\tau; R) f_{3j}^\beta(\tau'; R) \rangle = \frac{2k_B T}{m} [\beta_3(R)]_{ij}^{\alpha\beta} \delta(\tau - \tau'). \qquad (8.2)$$

Equation (8.1) reduces the problem of I_2 recombination dynamics in a liquid to an effective six-particle problem. Two of the particles, coordinates $r_0 = (\mathbf{r}_{01}, \mathbf{r}_{02})$, are the physical iodine atoms, and four of the particles, coordinates $r_1 = (\mathbf{r}_{11}, \mathbf{r}_{12})$ and $r_2 = (\mathbf{r}_{21}, \mathbf{r}_{22})$, are collective variables, which (qualitatively speaking) describe the motions of the first two solvation shells of the common cage.

The Langevin-model simulations are based on a Langevin equation with an R-dependent matrix friction coefficient.[32] The Langevin-model equation of motion in matrix notation is

$$\ddot{r}_0(t) = m^{-1}\langle F \rangle_{R(t)} - \beta_1[R(t)]\dot{r}_0(t) + f_1[t; R(t)] \qquad (8.3)$$

where $f_1(t; R) = (\mathbf{f}_{11}(t; R), \mathbf{f}_{22}(t; R))$ is an R-dependent vector white-noise Gaussian random force. This force satisfies a fluctuation–dissipation theorem analogous to Eq. (8.2), namely,

$$\langle f_{1i}^\alpha(\tau; R) f_{1j}^\beta(\tau'; R) \rangle = \frac{2k_B T}{m} [\beta_1(R)]_{ij}^{\alpha\beta} \delta(\tau - \tau') \qquad (8.4)$$

In the simulations presented below, the mean force $\langle F \rangle_{R(t)}$ is computed from the potential of mean force $W(R)$ via $\langle F \rangle_R = -\partial W(R)/\partial r_0$. The potential of the mean force is determined from the electronic-ground-state I_2 gas-phase potential $U_{uu}(R)$ and the I_2 cavity function $w^{(2)}(R)$ via Eq. (6.27), that is, $W(R) = U_{uu}(R) + w^{(2)}(R)$. The gas-phase I_2 potential was taken to be of the Morse form in Eq. (3.76) [i.e., $U_{uu}(R) = V(R)$] with dissociation energy $D/k_B = 18{,}045$ K, with repulsion parameter $\alpha = 1.87$ Å^{-1}, and with $b = R_{eq} = 2.67$ Å.

The liquid-state quantities required to calculate the parameters in Eqs. (8.1) and (8.3) are the I_2 cavity function $w^{(2)}(R)$; the squared adiabatic frequency matrix of the reagents, $\Omega_0^2(R)$; the friction matrix of the reagents, $\beta_1(R)$; the matrix MTGLE parameters $\{\omega_{ep}^2(R), \omega_{cp+1}^2(R)\}$; and the matrix truncation parameters $\{\Omega_p^2(R), \beta_{p+1}(R)\}$.

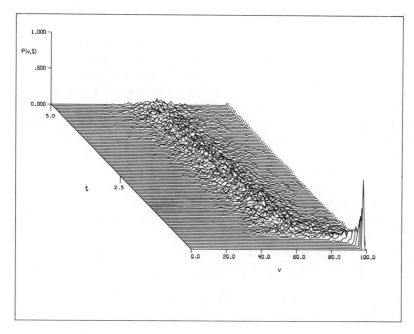

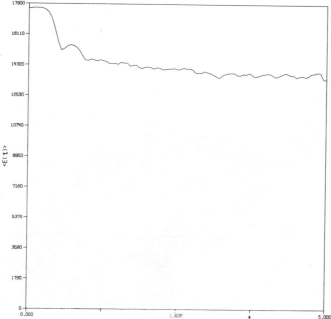

Fig. 21. Vibrational-energy relaxation of I_2 in CCl_4. Top: probability $P[v, t]$ that recombining I_2 molecules are in vibrational state v at time t in psec. Bottom: Average vibrational energy $\langle E(t) \rangle$ of recombining molecules in kelvins vs. time t in psec. Results are derived from MTGLE classical trajectory simulations based on Eq. (8.1) for model solution 2 of Table III.

209

The quantities $w^{(2)}(R)$ and $\Omega_0^2(R)$ are constructed from the Percus–Yevick approximation to the I_2 pair distribution function as described in Section VII. The remaining quantities may be computed from $\omega_{e0}^2(R)$, $\omega_{c1}^4(R)$, and $\beta_1(t; R)$ using Eq. (6.17) and the prescriptions of Appendix A of Ref. 23(b).

Thus to construct the MTGLE chain models we require explicit results for $\omega_{e0}^2(R)$, $\omega_{c1}^4(R)$, and $\beta_1(t; R)$. For model solution two we used the MD results for these quantities plotted in Figs. 15, 17, and 18. For model solutions 3 and 4, we used the superposition approximation results for $\omega_{e0}^2(R)$ and $\omega_{c1}^4(R)$ plotted in Fig. 16. The friction kernel $\beta_1(t; R)$ is constructed using the matrix Gaussian approximation of Eq. (7.16). This approximation is constructed using the results of Fig. 16 and also Fig. 4 of Ref. 23(b).

Finally, the random forces and the initial conditions of the equivalent chain atoms are sampled by matrix generalizations of the Monte Carlo procedures sketched in Section III.H.

B. Simulation Results

In Figs. 21–24, we plot the main results of our stochastic classical trajectory studies of vibrational-energy relaxation of highly excited I_2 molecules.

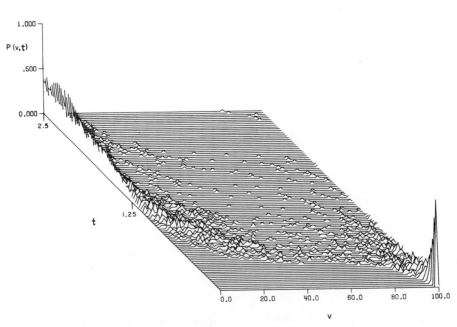

Fig. 22.   Brownian-dynamics model for vibrational-energy relaxation of I_2 in CCl_4. Same as Fig. 21 except calculations are based on the Langevin model of Eq. (8.3).

In each figure we plot two different measures of vibrational-energy relaxation. The first is the PDF $P[v, t]$, which gives the probability that the iodine molecules are in vibrational state v at time t given that they are in $v = 100$ at $t = 0$. Note that these plots are sophisticated analogues of the Langevin-model PDF plot in Fig. 3. The vibrational quantum number v of an I_2 molecule is determined from its vibrational energy $G(v)$ by the standard Morse formula for vibrational energy levels,

$$G(v) = r_1\left[v - \tfrac{1}{2}\right] - r_2\left[v + \tfrac{1}{2}\right]^2 \tag{8.5}$$

where r_1 and r_2 are related to the Morse parameters of Eq. (3.76) and to the reduced mass μ of an iodine atom by

$$r_1 = b\left[\frac{Dh^2}{2\pi^2\mu}\right]^{1/2} \tag{8.6a}$$

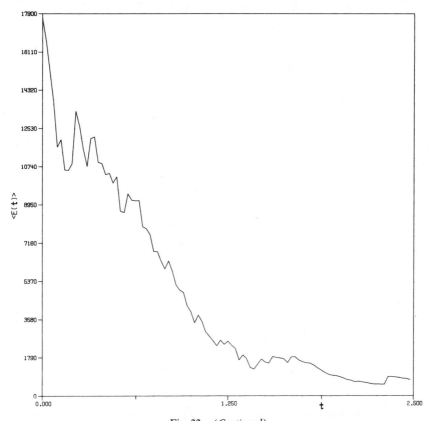

Fig. 22. (*Continued*).

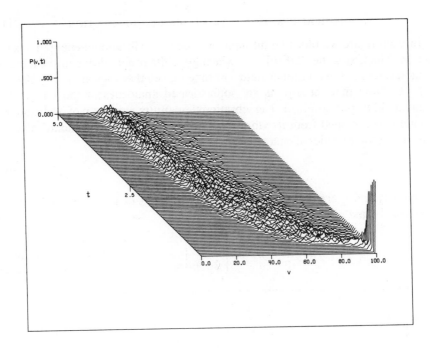

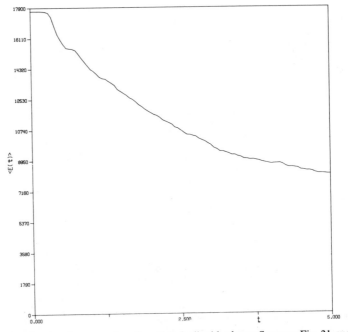

Fig. 23. Vibrational-energy relaxation of I_2 in liquid ethane. Same as Fig. 21 except that MTGLE simulations are performed for model solution 4 of Table III.

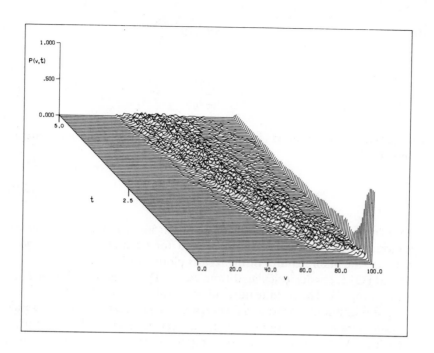

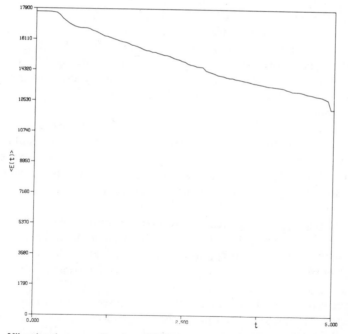

Fig. 24. Vibrational-energy relaxation of I_2 in dense gaseous ethane. Same as Fig. 21 except that MTGLE simulations are performed for model solution 3 of Table III.

and

$$r_2 = \frac{h^2 b^2}{8\pi^2 \mu} \tag{8.6b}$$

The second quantity plotted in Figs. 21–24 is the average vibrational energy at time t,

$$\langle E(t) \rangle \equiv \int G[v] P[v, t] \, dv \tag{8.7}$$

Both $P[v, t]$ and $\langle E(t) \rangle$ are calculated from 50–100 stochastic classical trajectories. Each trajectory was monitored for 60 psec. We next briefly summarize the main features of the results plotted in Figs. 21–24.

The MTGLE results, calculated from Eq. (8.1), for model solution 2 are plotted in Fig. 21. The main features of the results of Fig. 21 are: (i) An initial rapid relaxation of vibrational energy occurs, so that after $t \simeq 2$–3 psec the mean vibrational quantum number has relaxed from $v = 100$ to roughly $v = 60$. (ii) For $t \gtrsim 3$ psec the relaxation rate slows down markedly. The mean vibrational quantum number "stagnates" near $v = 60$. Very little additional relaxation is observed in the time interval $3 \leqslant t \leqslant 60$ psec. Thus two distinct time scales for vibrational relaxation are evident. There is an initial fast relaxation occurring on a picosecond time scale, followed by a much slower relaxation (extrapolation of the energy decay curve in Fig. 21b leads to the expectation that this slower relaxation occurs on nanosecond time scales). Similar behavior has been predicted by Nesbitt and Hynes[63] from a collisional model.

We next compare these MTGLE results with the corresponding Langevin-model results, calculated from Eq. (8.3), for model solution 2. The Langevin-model results are plotted in Fig. 22. These results show unphysically rapid relaxation of vibrational energy. The excess vibrational energy is thermalized on a picosecond time scale. No stagnation effect is observed, and relaxation occurs only on a single time scale.

The initial rapid energy relaxation predicted by the MTGLE model may be qualitatively analyzed[23(c)] as arising from near-resonant energy exchange between the highly excited I_2 molecule and its solvation shells. Below $v = 60$, the effective I_2 vibrational frequency falls out of resonance with the shell frequencies, and the relaxation slows down.

These reagent–shell resonance effects are absent in the Langevin model. Rather, the frequency spectrum of the Langevin-model random force contains all Fourier components with equal weighting. Thus there is significant overlap between this frequency spectrum and the effective vibrational frequency of the I_2 molecule at all vibrational quantum numbers v. Hence the quick monotonic relaxation.

MTGLE results for model solutions 4 (liquid "ethane") and 3 (dense gaseous "ethane") are plotted, respectively, in Figs. 23 and 24. These are qualitatively similar to those of Fig. 21 and may similarily be understood as arising from reagent-shell resonance effects. Notice that for model solution 4 the initial rapid relaxation stagnates near $v = 30$ rather than near $v = 60$ (model solution 2). This result may be understood as being due to stronger coupling between the I_2 and liquid "ethane" solvent than between the I_2 and liquid CCl_4 solvent. The diagonal coupling-constant elements $[\omega_{ci}^2(R)]^D$ are roughly a factor of 2 larger for model solution 4 than for model solution 2 [cf. Figs. 15 and 16]. This difference in coupling constant is mainly due to the fivefold mass difference between CCl_4 and C_2H_6.

Finally, in Fig. 24, we plot results for $P[v, t]$ and $\langle E(t) \rangle$ for dense gaseous "ethane." The vibrational-energy PDF $P[v, t]$ stagnates near $v = 60$, as for model solution 2, but is broader than that for model solution 2. This greater breadth may be qualitatively understood as an Einstein-frequency effect. The diagonal elements of the squared Einstein-frequency matrix, $[\omega_{e0}^2(R)]^D$, for model solution 3, are roughly a factor of 2 smaller than for model solution 2 [cf. Figs. 15 and 16]. Hence the broader distribution.

Acknowledgments

Support of the work described in this article by the National Science Foundation, under Grants CH3 7828043 and CHE 8018285, and by the Purdue NSF-MRL program under Grant DMR 7723798 is gratefully acknowledged. Much of the work described in this article was performed in collaboration with Drs. M. Berkowitz, M. Balk, P. K. Swaminathan, and C. L. Brooks III. We also wish to thank Dr. David Ceperly for supplying us with the CLAMPS molecular-dynamics package used in the MD simulations described in Section VII.

APPENDIX A.   RECURSIVE CONSTRUCTION OF THE MTGLE MODELS

Here we outline recursive algorithms for construction of the MTGLE parameters $\{\omega_{ep}^2, \omega_{c\,p+1}^4\}$ and the truncation parameters $\{\Omega_p^2, \beta_{p+1}\}$ ($p = 0, 1, 2, \ldots$) from the normalized VAF $\dot{\chi}(t)$. The algorithms are based on the rigorous hierarchy for response functions $\dot{\chi}(t), \dot{\theta}_1(t), \ldots$ presented in Section V.B. Given the prescriptions for the parameters developed here, the truncated-chain models for isolated solute atoms [Eq. (3.65)] may be constructed. The corresponding chain models for interacting solute atoms [Eqs. (7.17) and (8.1)] may be constructed in an analogous manner using matrix generalizations of the algorithms presented here. These matrix generalizations are given in Appendix A of Ref. 23(b).

In our development of the algorithms, we employ the following abbreviated notation for initial derivatives of arbitrary functions of time, $f(t)$:

$$f^{(q)} \equiv \left[\frac{d^q f(t)}{dt^q} \right]_{t=0} \tag{A.1}$$

We first present the algorithm for the parameters $\{\omega_{ep}^2, \omega_{cp+1}^4\}$. We begin with the following expressions for the basic parameters ω_{e0}^2 and ω_{c1}^4:

$$\omega_{e0}^2 = -\chi^{(3)} \tag{A.2a}$$

$$\omega_{c1}^4 = \chi^{(5)} - \left[\chi^{(3)} \right]^2. \tag{A.2b}$$

These follow from Eqs. (5.6b) and (5.10). The following expressions for the higher-order parameters $\{\omega_{eq}^2, \omega_{cq+1}^4\}$, $q \geq 1$, analogously follow from Eqs. (5.14b) and (5.19):

$$\omega_{eq}^2 = -\theta_q^{(3)} \tag{A.3a}$$

$$\omega_{cq+1}^4 = \theta_q^{(5)} - \left[\theta_q^{(3)} \right]^2, \qquad q \geq 1 \tag{A.3b}$$

Equations (A.2) provide the required formulae relating the basic parameters ω_{e0}^2 and ω_{c1}^4 to $\dot{\chi}(t)$. Moreover, using the MTGLE hierarchy, the quantities $\theta_q^{(3)}$ and $\theta_q^{(5)}$ may be recursively determined in terms of the odd derivatives ($t = 0$) of $\chi(t)$. We denote these derivatives by $\chi^{(2n+1)}$, where $n = 1, 2, \ldots$. Thus Eqs. (A.2) and (A.3) show that the complete set of MTGLE parameters $\{\omega_{ep}^2, \omega_{cp+1}^4\}$ may be constructed from the derivatives $\chi^{(2n+1)}$ and hence from $\dot{\chi}(t)$.

We next illustrate this construction by determining $\theta_1^{(3)}$ and $\theta_1^{(5)}$ from $\dot{\chi}(t)$. This may be accomplished by differentiating Eq. (5.12b) $2m - 1$ times, setting $t = 0$ in the resulting equation, and using Eqs. (5.5) and (5.13). A brief calculation yields the following result:

$$\omega_{c1}^4 \theta_1^{(2m-3)} = \chi^{(2m+1)} + \omega_{e0}^2 \chi^{(2m-1)} - \omega_{c1}^4 \sum_{\alpha=0}^{m-3} \theta_1^{(2\alpha+1)} \chi^{(2m-3-2\alpha)}, \qquad m \geq 3 \tag{A.4}$$

Eqs. (A.3) and (A.4) permit one to construct ω_{e1}^2 and ω_{c2}^4 from of the initial derivatives $\chi^{(2n+1)}$. For example, setting $m = 3$ in Eq. (A.4) and using the fact that $\dot{\theta}_1(0) = 1$ yields the relation

$$\omega_{c1}^4 \theta_1^{(3)} = \chi^{(7)} + \omega_{e0}^2 \chi^{(5)} - \omega_{c1}^4 \chi^{(3)}$$

Combining the above result with Eq. (A.3a) then gives

$$\omega_{e1}^2 = \omega_{c1}^{-4}\left\{\omega_{c1}^2\chi^{(3)} - \omega_{e0}^2\chi^{(5)} - \chi^{(7)}\right\}$$

which [given Eq. (A.2)] determines ω_{e1}^2 in terms of $\dot{\chi}(t)$.

We next turn to the construction of the higher-order MTGLE parameters. We begin with the following relation connecting the initial derivatives of $\theta_q(t)$ and $\theta_{q-1}(t)$ [cf. Eq. (A.4)]:

$$\omega_{cq}^2\theta_q^{(2m-3)} = \theta_{q-1}^{(2m+1)} + \omega_{eq-1}^2\theta_{q-1}^{(2m-1)} - \omega_{cq}^2\sum_{\alpha=0}^{m-3}\theta_q^{(2\alpha+1)}\theta_{q-1}^{(2m-3-2\alpha)},$$

$$m \geqslant 3 \quad \text{and} \quad q \geqslant 2 \quad (A.5)$$

This equation may be derived by differentiating Eq. (5.20b) $2m-1$ times and evaluating the result at $t = 0$. It determines the initial derivatives of $\theta_q(t)$ in terms of those of $\theta_{q-1}(t)$. The initial derivatives of $\theta_{q-1}(t)$ may in turn be related to those of $\theta_{q-2}(t)$, and so on. Thus Eqs. (A.4) and (A.5) permit one to generate recursively the derivatives $\theta_q^{(3)}$ and $\theta_q^{(5)}$ for $q = 1, 2, \ldots$ from the normalized VAF $\dot{\chi}(t)$. Given this result and Eqs. (A.2) and (A.3), it is clear that the complete set of MTGLE parameters $\{\omega_{ep}^2, \omega_{cp+1}^4\}, p = 0, 1, 2, \ldots$, may be generated from $\dot{\chi}(t)$.

We next turn to the recursive construction of the adiabatic frequencies Ω_p, $p = 0, 1, \ldots$, from $\dot{\chi}(t)$. This construction is based on the following expressions [see Eqs. (5.81)] for $\Omega_0^2, \Omega_1^2, \ldots$:

$$\Omega_0^{-2} = \hat{\chi}(z = 0) \quad (A.6a)$$

and

$$\Omega_p^{-2} = \hat{\theta}_p(z = 0), \quad p = 1, 2, \ldots \quad (A.6b)$$

Equation (A.6a) permits one to determine Ω_0 directly from $\dot{\chi}(t)$. The higher-order frequencies $\Omega_1, \Omega_2, \ldots$ may also be constructed from $\dot{\chi}(t)$ via recursion relations, which we now develop. We begin with the relation

$$\omega_{c1}^4\hat{\theta}_1(z) = z^2 + \omega_{e0}^2 - \hat{\chi}^{-1}(z) \quad (A.7)$$

which is a rearrangement of Eq. (5.12a). Evaluating Eq. (A.7) at $z = 0$ and using Eq. (A.6) yields the relation

$$\Omega_1^2 = \frac{\omega_{c1}^4}{\omega_{e0}^2 - \Omega_0^2} \quad (A.8)$$

which determines Ω_1^2 in terms of ω_{e0}^2, ω_{c1}^4, Ω_0^2, and hence $\dot{\chi}(t)$. Equation (A.8) is the first member of the following infinite set of relations:

$$\Omega_p^2 = \frac{\omega_{cp}^2}{\omega_{e\,p-1}^2 - \Omega_{p-1}^2}, \qquad p = 2, 3, \ldots \tag{A.9}$$

Equation (A.8) and (A.9) recursively determine $\Omega_1^2, \Omega_2^2, \ldots$ in terms of $\dot{\chi}(t)$.

The friction coefficients β_{p+1}, $p = 0, 1, \ldots$, may similarily be constructed from $\dot{\chi}(t)$ via the MTGLE hierarchy. The relation between β_1 and $\dot{\chi}(t)$ may be developed as follows. We begin with Eq. (3.56b) rewritten as

$$\hat{\beta}_1(z) = z^{-1} \left[\hat{\chi}^{-1}(z) - \Omega_0^2 - z^2 \right] \tag{A.10}$$

Thus the friction coefficient $\beta_1 \equiv \lim_{z \to 0} \hat{\beta}_1(z)$ is equal to

$$\beta_1 = \lim_{z \to 0} \left\{ z^{-1} \left[\hat{\chi}^{-1}(z) - \Omega_0^2 \right] \right\} \tag{A.11}$$

Comparing Eqs. (A.6a) and (A.11) yields the following alternative result for β_1:

$$\beta_1 = \lim_{z \to 0} \left[\frac{d\hat{\chi}^{-1}(z)}{dz} \right] \tag{A.12}$$

The $(p+1)$th friction coefficient β_{p+1} may similarily be expressed as

$$\beta_{p+1} = \lim_{z \to 0} \left[\frac{d\theta_p^{-1}(z)}{dz} \right] \tag{A.13}$$

Eq. (A.11) gives a relation for β_1 in terms of $\dot{\chi}(t)$. The higher-order friction coefficients may be expressed in terms of $\dot{\chi}(t)$ using Eq. (A.13) and the MTGLE hierarchy. For example, using Eqs. (A.6b) and (A.13), β_2 may be expressed as

$$\beta_2 = -\Omega_1^4 \lim_{z \to 0} \left[\frac{d\hat{\theta}_1(z)}{dz} \right]$$

However, Eqs. (A.7) and (A.12) give the additional relation

$$\lim_{z \to 0} \frac{d\hat{\theta}_1(z)}{dz} = -\beta_1 \omega_{c1}^{-4}$$

Combining the above two equations gives the following result for β_2:

$$\beta_2 = \beta_1 \left[\frac{\Omega_1^4}{\omega_{c1}^4} \right] \tag{A.14a}$$

Finally, using Eq. (A.8), Eq. (A.14a) may be rewritten as

$$\beta_2 = \beta_1 \frac{\omega_{c1}^4}{\left[\omega_{e0}^2 - \Omega_0^2 \right]^2} \tag{A.14b}$$

Equations (A.14) determine β_2 in terms of $\dot{\chi}(t)$. They are the first members of the following infinite sequence of equations:

$$\beta_{p+1} = \beta_p \left[\frac{\Omega_p^4}{\omega_{cp}^4} \right] = \beta_p \frac{\omega_{cp}^4}{\left[\omega_{ep-1}^2 - \Omega_{p-1}^2 \right]^2} \tag{A.15}$$

Eqs. (A.14) and (A.15) permit one to generate the friction coefficients $\beta_1, \beta_2, \ldots$ recursively from $\dot{\chi}(t)$.

APPENDIX B

We prove here that the Liouville operator L is Hermitian if $\langle \ \rangle$ is interpreted as an average over the canonical-ensemble distribution function $f_{CA}[p, q]$ of Eq. (4.6), that is [see Eq. (4.43)],

$$\langle XLY^* \rangle = \langle YLX^* \rangle^* \tag{B.1}$$

We begin with

$$LY^* = i^{-1} [Y^*, H]$$

$$= i^{-1} \sum_{\lambda=1}^{N_s} \left\{ \frac{\partial Y^*}{\partial \mathbf{q}_{\lambda 0}} \cdot \frac{\partial H}{\partial \mathbf{p}_{\lambda 0}} - \frac{\partial Y^*}{\partial \mathbf{p}_{\lambda 0}} \cdot \frac{\partial H}{\partial \mathbf{q}_{\lambda 0}} \right\} \tag{B.2}$$

which follows from Eqs. (4.36) and (4.37). We additionally require the following relationship:

$$\exp[-\beta H] [Y^*, H] = [Y^* \exp[-\beta H], H] \tag{B.3}$$

which may readily be proven from the definition of the Poisson bracket [see Eq. (B.2)] and the fact that $[\exp[-\beta H], H] = 0$, since $\exp[-\beta H]$ is a constant of the motion.

Using Eqs. (4.6) and (4.42), one may explicitly express $\langle XLY^* \rangle$ as

$$\langle XLY^* \rangle = Z^{-1}(\beta) \int dp\, dq\, X(p,q) \exp[-\beta H(p,q)] LY^*(p,q) \quad \text{(B.4)}$$

Finally combining Eqs. (B.2)–(B.4) yields

$$\langle XLY^* \rangle = [iZ(\beta)]^{-1} \int dp\, dq\, X(p,q)$$
$$\times [Y^*(p,q)\exp[-\beta H(p,q)]], H(p,q)] \quad \text{(B.5)}$$

This may be written as

$$\langle XLY^* \rangle = [-iZ(\beta)]^{-1} \int dp\, dq \exp[-\beta H(p,q)]$$
$$\times Y^*(p,q)[X(p,q), H(p,q)] \quad \text{(B.6)}$$

Equation (B.6) may be derived from Eq. (B.5) in the following manner: (i) Explicitly evaluate the Poisson bracket in Eq. (B.5) using Eq. (B.2). (ii) Integrate the terms $\partial[Y\exp(-\beta H)]/\partial \mathbf{q}_{\lambda 0}$ and $\partial[Y\exp(-\beta H)]/\partial \mathbf{p}_{\lambda 0}$ by parts. (iii) Use the fact that the surface terms vanish for a fluid confined to a finite volume V.

Finally, we may rewrite Eq. (B.6) as

$$\langle XLY^* \rangle = \left\{ [iZ(\beta)]^{-1} \int dp\, dq \exp[-\beta H(p,q)] \right.$$
$$\left. \times Y(p,q)[X^*(p,q), H(p,q)] \right\}^*$$
$$= \langle YLX^* \rangle^*$$

This completes the proof that L is Hermitian if $\langle\ \rangle$ is a canonical-ensemble average.

References

1. See, for example, H. F. Schaefer III, Ed., *Modern Theoretical Chemistry, Methods of Electronic Structure Theory*, Plenum, New York, 1977.

2. See, for example, W. H. Miller, Ed., *Modern Theoretical Chemistry, Dynamics of Molecular Collisions*, Plenum, New York, 1976.

3. Some key developments in gas-phase chemical-reaction theory include the development of natural collision coordinates by Marcus [see for example R. A. Marcus, *J. Chem. Phys.* **49**, 2610 (1968)]; the formulation of classical S-matrix theory by Marcus and Miller [see,

e.g., R. A. Marcus, *J. Chem. Phys.* **54**, 3965 (1971), and W. H. Miller, *Adv. Chem. Phys.* **25**, 69 (1974)]; and the development of general formulations of transition-state theory by Miller, Pechukas, and others [see, e.g., W. H. Miller, *J. Chem. Phys.* **61**, 1823 (1974), and P. Pechukas and F. J. McLafferty, *J. Chem. Phys.* **58**, 1622 (1973)].

4. See for example the classical trajectory study of the hydrogen exchange reaction by M. Karplus, R. N. Porter, and R. D. Sharma, *J. Chem. Phys.* **43**, 3259 (1965).

5. N. Metropolis, A. W. Rosenbluth, M. N. Rosenbluth, A. H. Teller, and E. Teller, *J. Chem. Phys.* **21**, 1087 (1953).

6. B. J. Alder and T. E. Wainwright, *J. Chem. Phys.* **31**, 459 (1959).

7. A. Rahman, *Phys. Rev.* **136**, A405 (1964).

8. D. Levesque and L. Verlet, *Phys. Rev. A* **2**, 2514 (1970).

9. See for example, A. Rahman and F. H. Stillinger, *J. Chem. Phys.* **55**, 3336 (1971); F. H. Stillinger, *Adv. Chem. Phys.* **31**, 1 (1975).

10. (a) Linear-response theory is developed in general quantum-mechanical form by R. Kubo, *J. Phys. Soc. Japan* **12**, 570 (1957). (b) The construction of generalized Langevin representations is discussed by R. Kubo, *Rep. Prog. Theor. Phys.* **29**, 235 (1966). This paper presents, in implicit form, the response-function approach to generalized Langevin representations that is developed in more detail here.

11. (a) H. Mori, *Prog. Theor. Phys.* **33**, 423 (1965). (b) H. Mori, *Prog. Theor. Phys.* **34**, 399 (1965).

12. The basic principles of the projection-operator approach to generalized master equations are presented in R. Zwanzig, *J. Chem. Phys.* **33**, 1338 (1960).

13. See Eq. (3.10) of Ref. 11(a).

14. The general phenomenological theory of linear irreversible processes is presented in L. Onsager and S. Machlup, *Phys. Rev.* **91**, 1505 (1953); S. Machlup and L. Onsager, *Phys. Rev.* **91**, 1512 (1953).

15. See for example *Water: A Comprehensive Treatise*, Felix Franks, Ed., Plenum, New York, 1972.

16. See, for example, C. Tanford, *The Hydrophobic Effect*, Wiley, New York, 1973.

17. See, for example, R. W. Gurney, *Ionic Processes in Solution*, McGraw-Hill, New York, 1953.

18. See, for example, J. D. Weeks, D. Chandler, and H. C. Andersen, *J. Chem. Phys.* **54**, 5237 (1971).

19. See, for example, S. A. Adelman, *J. Chem. Phys.* **64**, 724 (1976); S. A. Adelman and J. H. Chen, *J. Chem. Phys.* **70**, 4291 (1979).

20. See, for example, B. Bigot and W. L. Jorgensen, *J. Chem. Phys.* **75**, 1944 (1981).

21. This framework is the molecular-time-scale generalized-Langevin-equation (MTGLE) theory. (a) The basic principles of the MTGLE theory are presented in S. A. Adelman, *Adv. Chem. Phys.* **44**, 143 (1980). (b) The mathematical structure of the MTGLE theory is given in S. A. Adelman, *J. Chem. Phys.* **74**, 4646 (1981). (c) The formal framework for application of the MTGLE theory to liquid-solution reactions is presented in S. A. Adelman, *J. Chem. Phys.* **73**, 3145 (1980). There are unfortunately a number of misprints in this paper. These are corrected here and in Refs. 23. The MTGLE theory emerged as a generalization of the work of S. A. Adelman and J. D. Doll, *J. Chem. Phys.* **64**, 2375 (1976) on gas–solid collisions.

22. For simple applications of the MTGLE theory (a) to correlation-function modeling see M. Berkowitz, C. L. Brooks III, and S. A. Adelman, *J. Chem. Phys.* **72**, 3889 (1980); (b) to dynamics see C. L. Brooks III, M. Berkowitz, and S. A. Adelman, *J. Chem. Phys.* **73**, 4353 (1980). These papers are based on the numerical methods presented in S. A. Adelman, *J. Chem. Phys.* **71**, 4471 (1979).

23. For a sophisticated application of the MTGLE theory (a) to the calculation of reagent-configuration-dependent correlation functions see C. L. Brooks III and S. A. Adelman, *J. Chem. Phys.* **76**, 1007 (1982); (b) to modeling of reagent-configuration-dependent correlation functions see C. L. Brooks III and S. A. Adelman, *J. Chem. Phys.* **77**, 424, (1982); (c) to liquid-solution reaction dynamics see C. L. Brooks III, M. Balk, and S. A. Adelman, *J. Chem. Phys.* **77**, (1983).

24. First steps in the development of a quantum generalization of the classical MTGLE theory are presented in S. A. Adelman, *J. Chem. Phys.* **75**, 5837 (1981).

25. See, for example, M. Eigen and L. De Maeyer, *Proc. Roy. Soc. London A* **247**, 505 (1958).

26. See, for example, R. P. Bell, *The Proton in Chemistry*, Cornell University Press, Ithaca, 1973.

27. See, for example, D. A. McQuarrie, *Statistical Mechanics*, Harper and Row, New York, 1975.

28. See Appendix 4 of Ref. 21(a).

29. The basic references on phenomenological Brownian-motion theory are S. Chandrasekhar, *Rev. Mod. Phys.* **15**, 1 (1943); M. Wang and G. Uhlenbeck, *Rev. Mod. Phys.* **17**, 323 (1945).

30. P. Langevin, *Comptes Rendus* **146**, 530 (1908).

31. A. Einstein, *Investigations on the Theory of Brownian Movement*, Dover, New York, 1924.

32. See, for example, H. Yamakawa, *Modern Theory of Polymer Solutions*, Harper and Row, New York, 1971.

33. W. Edgell and M. Balk, private communication.

34. M. Balk, C. L. Brooks III, and S. A. Adelman, unpublished calculations.

35. See Tables 1, 2, 4, and 5 of Ref. 22a.

36. R. Zwanzig, *J. Chem. Phys.* **32**, 1173 (1960).

37. For algorithms for sampling random forces, see, for example, J. C. Tully, G. H. Gilmer, and M. Shugard, *J. Chem. Phys.* **71**, 163 (1979).

38. See for example R. Zwanzig and M. Bixon, *Phys. Rev. A* **2**, 2005 (1970).

39. For a rather complete coverage of topics in the equilibrium theory of simple liquids see *Classical Fluids*, H. L. Frisch and J. L. Lebowitz, Eds., Benjamin, New York, 1964.

40. See, for example, J. K. Percus, in *Classical Fluids*, H. L. Frisch and J. L. Lebowitz, Eds., Benjamin, New York, 1964.

41. For a discussion of the continuum dielectric solvent approximation see Ref. 19.

42. M. S. Wertheim, *J. Math. Phys.* **5**, 643 (1964). Wertheim's solution of the Percus–Yevick equation is based on a remarkable application of analytic-function theory.

43. J. L. Lebowitz, *Phys. Rev.* **133**, A895 (1964).

44. G. J. Throop and R. J. Bearman, J. Chem. Phys. **42**, 2408 (1965).

45. See, for example, Chapter 8 of H. Goldstein, *Classical Mechanics*, Addison-Wesley, Reading, Mass., 1950.

46. For a discussion of this point see W. H. Miller, *Adv. Chem. Phys.* **25**, 69 (1974).

47. See for example A. Messiah, *Quantum Mechanics*, Wiley, New York, 1966.

48. See, for example, H. Feshbach, *Ann. Phys.* **19**, 287 (1962).

49. See, for example, A. L. Fetter and J. D. Walecka, *Quantum Theory of Many-Particle Systems*, McGraw-Hill, New York, 1971.

50. See, for example, Chapter 3 of K. Huang, *Statistical Mechanics*, Wiley, New York, 1963.

51. R. W. Zwanzig, *Ann. Rev. Phys. Chem.* **16**, 67 (1965).

52. The distinction between internal and external transport coefficients has been discussed by Kubo in Ref. 10(b).

53. W. Kohn and J. M. Luttinger, *Phys. Rev.* **108**, 590 (1957).

54. S. Nakajima, *Prog. Theor. Phys.* **20**, 948 (1958).

55. See, for example, L. Landau and E. M. Lifshitz, *Fluid Mechanics*, Pergamon, London, 1959.

56. The normalization factor is discussed in Ref. 21(b).

57. For a description of the Choleski prescription for matrix square roots see, for example, E. Bodewig, *Matrix Calculus*, Part I, North-Holland, Amsterdam, 1956.

58. See Appendix A of Ref. 23(b).

59. See the discussion in Ref. 21(c).

60. See Fig. 5 of Ref. 23(a).

61. Use of the Gaussian approximation and its extensions to model conventional scalar friction kernels has been discussed, for example, by Levesque and Verlet in Ref. 8. For more recent applications and further justification see R. F. Grote and J. T. Hynes, *J. Chem. Phys.* **75**, 2191 (1981).

62. See Table 2 of Ref. 23(b).

63. D. J. Nesbitt and J. T. Hynes, *J. Chem. Phys.* **77**, 2130 (1982).

SOLITONS IN CHEMICAL PHYSICS

MICHAEL A. COLLINS

Research School of Chemistry
Australian National University
Canberra, ACT, 2600 Australia

CONTENTS

I. INTRODUCTION

It is probably fair to say that most practitioners in the field of chemical physics are not intimately familiar with the concept of a soliton. This review is intended to provide an introduction to this phenomenon while still providing information about the state of the art in several areas. Essentially, I use the recent work in five fields of soliton research to engender an understanding of the nature of soliton dynamics and a knowledge of current achievements.

What is a soliton? It is almost traditional in introducing solitons to avoid a rigorous definition until after presenting John Scott-Russell's first report[1] of a soliton observation:

> *I was observing the motion of a boat which was rapidly drawn along a narrow channel by a pair of horses, when the boat suddenly stopped — not so the mass of water in the channel which it had put in motion; it accumulated round the prow of the vessel in a state of violent agitation, then suddenly leaving it behind, rolled forward with great velocity, assuming the form of a large solitary elevation, a rounded, smooth and well defined heap of water, which continued its course along the channel apparently without change of form or diminution of speed. I followed it on horseback, and overtook it still rolling on at a rate of some eight or nine miles an hour, preserving its original figure some thirty feet long and a foot to a foot and a half in height. Its height gradually diminished, and after a chase of one or two miles I lost it in the windings of the channel. Such, in the month of August 1834, was my first chance interview with that singular and beautiful phenomenon. . . .*

Solitons are *solitary* or localized waves, rather than wave trains, of *large amplitude*. They propagate at constant velocity without change of shape or loss of energy. They are stable to small perturbations. Moreover, they have the remarkable property of *surviving collisions with other solitons*. Had Scott-Russell been lucky enough to observe another solitary wave coming down the channel towards the first, he might have witnessed the violent collision and even more stunning separation of the waves as they continued on their separate courses. Anyone who has seen the backwash–wave collision at a beach has witnessed something like this.

Needless to say, this review is not concerned with the fluid dynamics of British waterways, but deals with the dynamics of matter at the atomic and molecular level, where solitons also abound. How deeply must we treat the dynamics of matter to encounter solitons? You may recall the story of the mathematical physicist who, when challenged to calculate the surface area of a cow, said, "First assume a spherical cow. . . ." No doubt that same scholar, if asked to describe the locomotion of a cow, would have begun by assuming that the cow's muscles could be approximated by harmonic oscillators. So it is that much of our understanding of the dynamics of molecules and crystals is based on a harmonic picture of the interatomic forces. To encounter solitons we must go beyond this linear mechanics to consider the anharmonic nature of atomic interactions. This does not mean that we shall be mired in tedious calculations of anharmonic corrections to a normal-mode description of molecular motion. Rather, we look for strikingly new phenomena that lie outside the range of linear physics.

Solitons arise in systems that are both nonlinear and dispersive. A dispersive medium is one in which the velocity of low-amplitude waves is dependent on their wavelength: some examples are spin waves in magnetic solids, excitons in excited molecular crystals, and phonons in vibrating crystals. Large-amplitude solitary waves can exist in such systems by balancing the dispersion effect, which tends to spread the pulse out, against the nonlinear forces that operate in large-amplitude motion. The review explores this process in more detail by considering particular systems.

I have not attempted to cover every area of soliton research of relevance to chemical physics. Perhaps the most notable subject omitted is that of self-induced transparency. In this process a short (10^{-9}–10^{-12} sec), intense pulse of light can pass through an absorbing medium as if it were transparent. The leading edge of the pulse populates the upper level while the trailing edge depopulates the level by stimulated emission. Pulses of the "right" shape pass through at speeds well below the velocity of light without change in form or loss of intensity. These pulses are solitons (see Refs. 2, 3, 4, and references therein).

For the sake of coherence, the examples considered here are all applications to dynamics in the solid state. This is not accidental, since coherent phenomena are natural in an ordered phase. However, the equations employed are much more general, and additional applications can be found in the references or will spring naturally into the reader's mind. While Sections II–V form a distinct whole, each of the other sections can be understood if read in isolation, so that any one subject of particular interest may be examined separately. Section X is specifically recommended to students of intramolecular dynamics.

Section II shows how the spin dynamics in some quasi-one-dimensional transition-metal compounds may be described classically in terms of the sine–Gordon equation. This is one of the best-studied soliton-bearing equations, and much recent effort has been directed towards establishing the importance of solitons in the low-temperature spin dynamics of compounds like $CsNiF_3$ and $(CH_3)_4NMnCl_3$. Section III explores the dynamics determined by the sine–Gordon equation, with particular emphasis on the soliton motion as well as other nonlinear excitations. Section IV shows how the statistical mechanics of a sine–Gordon system can be understood in terms of the solitons and low-amplitude motions by evaluating the free energy associated with the soliton degrees of freedom at low temperature. Section V explains how the solitons can have significant effect on the static and dynamic correlation functions of a sine–Gordon system. The soliton-model predictions are compared with experimental data. Section VI contains a brief description, with references, of other magnetic systems that display soliton and related dynamics.

Section VII considers the Brownian motion of solitons and related nonlinear waves that results from their interaction with thermal fluctuations. The example dealt with in most detail is the ϕ^4 model of ferrodistortive materials. Section VIII investigates the solitons that arise from resonance interactions in an electronically excited molecular crystal. Section IX presents a brief exposition of the nature of solitonlike states in the π electron configuration of long polyenes, derived from a simple Hückel model. Section X considers at some length the appearance and nature of solitons in the longitudinal vibrations of atomic lattices. Shock waves are considered in terms of solitons. The importance of solitons in the low-amplitude (thermal) dynamics is indicated, and some applications of these ideas to molecular vibrations are suggested.

Each of these topics is concerned with motion in one spatial dimension. This preoccupation with one dimension is in good measure due to the nature of the solitons themselves. It is not obvious how a localized wave can preserve its shape and energy density if its constituent parts are moving along more than one spatial axis at the same time. However, the frequent restriction herein to one degree of freedom per particle is more directly due to a desire for mathematical simplicity. If we are unwilling to simplify the dynamics of a cow by invoking an harmonic approximation, the mechanics is almost certainly intractable analytically unless we consider a single-coordinate description of each muscle. Of course, with high-speed computers and molecular-dynamics techniques now commonplace, most of the problems studied here could be attacked without such restrictions. In fact molecular dynamics has been employed on these systems. However, it is very difficult to understand such numerical experiments without a qualitative knowledge of the elementary excitations of these nonlinear systems. The purpose of these simple models is to develop a clearer picture of the nature of the anharmonic dynamics, particularly the unexpected role of solitons.

The rapid growth in soliton research over the last decade has accelerated in recent years in terms of fundamental mathematical development and applications to plasma physics, commensurate–incommensurate phase transitions, fluid dynamics in liquid helium, and many other areas also not considered here. A comprehensive review of the soliton literature prior to 1973, with emphasis on the mathematical aspects, is contained in the article by Scott, Chu, and McLaughlin.[4] A very readable account of soliton applications in modern physics has been provided by Bullough.[5] Recently, a number of books and contributed volumes have appeared. Mathematical methods for dealing with soliton-bearing equations of motion are presented by Eilenberger[6] and in the contributed volume edited by Bullough and Caudrey.[7] Applications of soliton theory to condensed-matter physics are contained in the volume edited by Bishop and Schneider,[8] while the powerful mathematical technique called the inverse-scattering transformation is presented in detail by Eckhaus and Van Harten.[9]

Here we concentrate on the physical mechanisms that are most likely to operate in systems of interest to physical chemists. The mathematics employed is generally straightforward and is intended to aid the physical interpretation of soliton phenomena rather than represent an end in itself. A more detailed knowledge of the mathematical techniques must be obtained from the references provided.

II. SPIN SYSTEMS AND THE SINE–GORDON EQUATION

One of the most active fields of current soliton research is that of quasi-one-dimensional magnets such as $CsNiF_3$, $(CH_3)_4NMnCl_3$ (TMMC), and $CsCoCl_3$. The recent theoretical models for the spin dynamics in the first two of these crystals make use of the sine–Gordon equation, which is among the most studied and best-understood "soliton-bearing" systems. This provides us with an excellent context in which to meet the sine–Gordon (SG) equation and a whole arsenal of useful results.

The crystal structure of the compounds listed above[10, 11] can generally be described as follows. The transition-metal ion lies at the center of an octahedron of halide ions. The octahedra, stacked face to face, form a chain. Each chain is well separated from other parallel chains by the large cations, Cs^+ or $(CH_3)_4N^+$, as depicted in Figure 1.

The paramagnetic transition-metal ions interact more strongly with their neighbors along the chains than with ions in the adjacent chains. In general, these crystals are paramagnetic at high temperature; below some temperature, the spins become partially ordered along the chains, but remain uncorrelated with the adjacent chains. Finally, at very low temperature, the weak interchain coupling is sufficient to produce complete three-dimen-

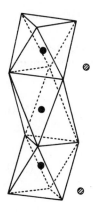

Fig. 1. The crystal structure of $CsNiF_3$ and $(CH_3)_4NM_nCl_3$ is shown schematically. Each transition-metal ion (dark circles) lies at the center of an octahedron of halide ions. These octahedra share faces to form a filament. Each filament is well separated from its neighbors by the large cations (shaded circles), C_s^+ or $(CH_3)_4N^+$.

sional order. In the intermediate temperature regime from about 2.6 to 50 K for $CsNiF_3$, it is reasonable to describe the separate chains as quasi-one-dimensional magnets: $CsNiF_3$ is a ferromagnet, with adjacent spins predominantly parallel; $CsCoCl_3$ and $(CH_3)_4NMnCl_3$ are antiferromagnets, with adjacent spins antiparallel.

Adopting the Heisenberg model for these chains, we write the Hamiltonian for a single chain as

$$H = -2J\sum_l \mathbf{S}_l \cdot \mathbf{S}_{l+1} \qquad (2.1)$$

where J is the exchange interaction between the spins on adjacent lattice sites $l, l+1$. J is positive (negative) for a ferromagnet (antiferromagnet).

In addition to the spin coupling, there may be effects, such as spin–orbit coupling in the transition-metal ions, that create an anisotropy energy at each site. This tends to align the spin with some crystallographic direction, producing a plane or axis of "easy magnetization." All three crystals considered here have an easy plane perpendicular to the chain axis. Denoting the chain axis by z, we amend (2.1) to include this effect for ferromagnets such as $CsNiF_3$:[12]

$$H_F = \sum_l -2J\mathbf{S}_l \cdot \mathbf{S}_{l+1} + A\sum_l (S_l^z)^2 \qquad (2.2)$$

If the anisotropy energy A (> 0) is sufficiently large, we can assume that the spin vector is basically confined to the xy plane. For the antiferromagnet TMMC, the spin–orbit coupling is weak and the anisotropy is attributed[13] to a dipole–dipole interaction between sites:

$$H_{AF} = \sum_l 2|J|\mathbf{S}_l \cdot \mathbf{S}_{l+1} - 2DS_l^z S_{l+1}^z \qquad (2.3)$$

If D is sufficiently large ($D > 0$, $S_l^z \simeq -S_{l+1}^z$), these spins are also confined to the xy plane.

We complete the Hamiltonian to be studied by introducing an external magnetic field along, say, the x-axis to break the symmetry in the xy plane. We have finally[14]

$$H_F = \sum_l -2J\mathbf{S}_l \cdot \mathbf{S}_{l+1} + A(S_l^z)^2 - g\mu_B B^x S_l^x \qquad (2.4)$$

and[13]

$$H_{AF} = \sum_l 2|J|\mathbf{S}_l \cdot \mathbf{S}_{l+1} - 2DS_l^z S_{l+1}^z - g\mu_B B^x S_l^x \qquad (2.5)$$

where μ_B is the Bohr magneton and g is the gyromagnetic ratio.

The ferromagnetic chain is somewhat easier to study, so we begin with (2.4) and return to (2.5) a little later.

Mikeska and Patzak[15] have shown that quantum effects in the spin dynamics of (2.4) can be neglected (except for a rescaling of the anisotropy energy) so long as

$$\frac{A}{JS(S+1)} \ll 4\pi^2 \tag{2.6}$$

In this case we can consider the spin $\mathbf{S}$ as a classical vector:[16]

$$\mathbf{S}_l = (S_l^x, S_l^y, S_l^z) = S(\sin\theta_l\cos\phi_l, \sin\theta_l\sin\phi_l, \cos\theta_l) \tag{2.7}$$

with the angles θ_l, ϕ_l defined in Fig. 2. If the anisotropy dominates thermal fluctuations, $\theta_l(t) \simeq \frac{1}{2}\pi$, while the ground state configuration has all spins parallel to the x axis ($\theta_l = \frac{1}{2}\pi$, $\phi_l = 0$, as in Figure 2a).

Using the angular variables in (2.7), the Hamiltonian of (2.4) becomes

$$H_F = \sum_l -2JS^2 \left[\sin\theta_l\sin\theta_{l+1}\cos(\phi_{l+1} - \phi_l) + \cos\theta_l\cos\theta_{l+1} \right]$$

$$- g\mu_B SB^x \sum_l \sin\theta_l\cos\phi_l + AS^2 \sum_l \cos^2\theta_l \tag{2.8}$$

Beginning with Mikeska,[16] several papers[17-20] have considered a continuum approximation to (2.8) and the associated equations of motion for the spins.

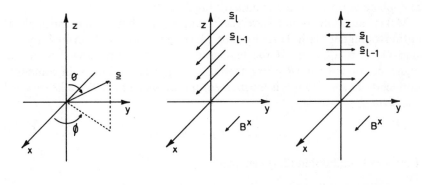

(a) ferromagnet.   (b) antiferromagnet.

Fig. 2.   The angles θ and ϕ defining the orientation of the spin vector $\mathbf{S}$ are shown. The spin configuration in the ground state of the 1-D magnets is represented for (a) the xy-easy-plane ferromagnet and (b) the xy-easy-plane antiferromagnet, in a magnetic field, B^x, along the x-axis.

Essentially, one assumes that $\theta_l \simeq \frac{1}{2}\pi$ and that the rate of change of the planar angles ϕ_l is sufficiently small that

$$\phi_{l+1} - \phi_l = a\frac{d\phi}{dz}(z), \qquad \text{etc.} \tag{2.9}$$

where a is the lattice spacing along the z (chain) axis. Starting from the Heisenberg equations of motion for the spin variables S^x, S^y, and S^z, we obtain in the continuum approximation

$$\frac{\partial^2\phi}{\partial z^2}(z,t) - \frac{1}{c^2}\frac{\partial^2\phi}{\partial t^2}(z,t) = m^2\sin\phi(z,t) \tag{2.10}$$

$$\cos\theta = \frac{1}{2AS}\frac{\partial\phi}{\partial t} + O\left(\frac{\mu_B g B^x}{2AS}\right) \tag{2.11}$$

where $c = aS(4AJ)^{1/2}$ and $m = (g\mu_B B^x/2JSa^2)^{1/2}$.

Since the derivation of (2.10) and (2.11) is, unfortunately, not given explicitly in the literature, we include it in Appendix A. There, we also show that the Hamiltonian (2.8) becomes, in the continuum limit,

$$H_F = 2JS^2a^2\int\frac{dz}{a}\left\{\frac{1}{2c^2}\left(\frac{\partial\phi}{\partial t}\right)^2 + \frac{1}{2}\left(\frac{\partial\phi}{\partial z}\right)^2 + m^2(1-\cos\phi)\right\} \tag{2.12}$$

The equation of motion (2.10) is the well-studied sine–Gordon (SG) equation, which describes the motion of a one-dimensional array of coupled pendula, equivalent in this continuum limit to the 1-D ferromagnet with easy-plane anisotropy in a symmetry-breaking field.

Maki[21] and Villain and Loveluck[22] have shown that the dynamics of the antiferromagnet with Hamiltonian (2.5) can also be described by the sine–Gordon equation. If the temperature is sufficiently low that out-of-plane motion is small ($\theta \simeq \frac{1}{2}\pi$), then $\cos\theta_l \equiv p_l \ll S$ and the spin commutation relations show that p_l is the momentum conjugate to ϕ_l. We take $\sin\theta_l = 1$ and write[21]

$$S_l = (S\cos\phi_l, S\sin\phi_l, p_l). \tag{2.13}$$

Then the Hamiltonian (2.5) becomes

$$H = \sum_l\left\{-[D-2|J|]p_l p_{l+1} + 2S^2|J|\cos(\phi_l - \phi_{l+1}) - g\mu_B B^x S\cos\phi_l\right\}$$

$$\tag{2.14}$$

There are two degenerate ground states for the system, one with

$$\cos\phi_{2l} = \cos\phi_{2l+1} = \tfrac{1}{4}h, \qquad \sin\phi_{2l} = -\sin\phi_{2l+1} = \left(1 - \tfrac{1}{16}h^2\right)^{1/2}$$

$$(2.15)$$

where $h = (g\mu_B B^x)/2JS$, and the other given by interchanging ϕ_{2l} and ϕ_{2l+1} in (2.15). If the magnetic field is small, so that $h \ll 1$ (this is generally true for the recent experiments on TMMC), then

$$\phi_l = (-1)^l \tfrac{1}{2}\pi, (-1)^{l+1} \tfrac{1}{2}\pi \qquad (2.16)$$

for these degenerate equilibria. This configuration, shown in Fig. 2b, has the spins lined up in antiferromagnetic fashion, with the spins *perpendicular* to the small magnetic field in the x-direction.

Clearly, the spin vector changes rapidly between sites l and $l+1$. However, a continuum theory may describe the relative motion of alternate spin vectors, which are equal at equilibrium.

To study the dynamics of every second spin only, a "decimation" procedure is employed.[21,22] We define a new Hamiltonian H_{DAF}, given by[21]

$$\exp\left(-\int_0^\beta d\tau\, H_{DAF}\right) = \int \prod_l D(\phi_{2l+1}(\tau)) \exp\left(-\int_0^\tau d\tau\, H_{AF}\right) \quad (2.17)$$

where $\beta = (k_B T)^{-1}$ and (2.17) involves a functional integral over dynamical variables $\phi_{2l+1}(\tau)$ at odd sites. In the low-temperature regime $(2\beta S^2|J| \gg 1)$, (2.17) can be evaluated by the saddle-point method with the momentum contribution to H_{AF} neglected $(p \ll S)$.[21] Maki obtains[21]

$$H_{DAF} = \sum_l \tfrac{1}{8}[2|J| - D]\frac{Y(\phi_{2l}, \phi_{2l+2})}{X(\phi_{2l}, \phi_{2l+2})}(p_{2l} + p_{2l+2})^2$$
$$- 2|J|S^2\left[X^{1/2}(\phi_{2l}, \phi_{2l+2}) + \tfrac{1}{2}h(\cos\phi_{2l} + \cos\phi_{2l+2})\right] \quad (2.18)$$

where

$$X(\phi_{2l}, \phi_{2l+2}) = 2 + 2\cos(\phi_{2l} - \phi_{2l+2}) - h(\cos\phi_{2l} + \cos\phi_{2l+2}) + h^2$$
$$Y(\phi_{2l}, \phi_{2l+2}) = 2 + 2\cos(\phi_{2l} - \phi_{2l+2}) - \tfrac{1}{2}h(\cos\phi_{2l} + \cos\phi_{2l+2}) \quad (2.19)$$

Now, we can assume that ϕ_{2l} changes slowly with l and proceed to a continuum limit, with

$$\phi_{2l+2} - \phi_{2l} = 2a\frac{\partial\phi}{\partial z}, \qquad \text{etc.} \qquad (2.20)$$

giving

$$H_{\text{DAF}} = \int \frac{dz}{2a} \left\{ \frac{1}{2} [2|J| - D] p^2(z) \right.$$
$$\left. + 2a^2 |J| S^2 \left[2X_0^{-1/2} \left(1 - \frac{h}{2} \cos\phi \right) + \frac{1}{2} h \cos\phi \right] \left(\frac{\partial\phi}{\partial z} \right)^2 \right.$$
$$\left. - 2|J| S^2 \left[X_0^{1/2} + h \cos\phi \right] \right\} \tag{2.21}$$

with $X_0 = 4(1 - h\cos\phi + \frac{1}{4}h^2)$

Finally, with small magnetic field, $h \ll 1$,

$$H_{\text{DAF}} = \int \frac{dz}{2a} \left\{ \frac{1}{2} (2|J| - D) p^2(z) + 2a^2 |J| S^2 \left(\frac{\partial\phi}{\partial z} \right)^2 \right.$$
$$\left. + \frac{1}{4} \frac{(g\mu_B B^x)^2}{2|J|} \cos^2\phi \right\} \tag{2.22}$$

That is,

$$H_{\text{DAF}} = 2|J| S^2 a \int dz \left\{ \frac{1}{2C_A^2} \left(\frac{\partial\phi}{\partial t} \right)^2 + \frac{1}{2} \left(\frac{\partial\phi}{\partial z} \right)^2 + \frac{m_A^2}{2} (1 + \cos 2\phi) \right\} \tag{2.23}$$

with

$$C_A = [(2|J| - D)2|J|]^{1/2} \sqrt{2} \, Sa, \qquad m_A = \frac{g\mu_B B^x}{4\sqrt{2} \, JSa}$$

This "decimated" continuum-limit Hamiltonian for the antiferromagnet is just that for a SG system like (2.12) except that the potential term involves $\cos 2\phi$, which has half the period of the corresponding term in the ferromagnet Hamiltonian. Newton's equation of motion for $\phi(z, t)$ is

$$\frac{\partial^2\phi}{\partial z^2} - \frac{1}{C_A^2} \frac{\partial^2\phi}{\partial t^2} = -m_A^2 \sin 2\phi \tag{2.24}$$

Having derived a continuum description of the 1-D magnets as SG systems, we consider in some detail the dynamics given by the SG equation.

III. DYNAMICS OF THE SINE–GORDON SYSTEM

Both (2.10) and (2.24) can be written in the form

$$\frac{\partial^2 u}{\partial z^2} - \frac{1}{c^2}\frac{\partial^2 u}{\partial t^2} = m^2 \sin u \tag{3.1}$$

where $u = 2\phi - \pi$, $m^2 = 2m_A^2$ corresponds to (2.24). We can associate a continuum Hamiltonian with (3.1), given by

$$H = A\int_{-\infty}^{\infty} dz \left\{ \frac{1}{2}\left(\frac{du}{dt}\right)^2 + \frac{1}{2}c^2\left(\frac{du}{dz}\right)^2 + c^2 m^2 (1 - \cos u) \right\} \tag{3.2}$$

where we have set the rest energy equal to zero ($u = 0$). Equation (3.1) is clearly a wave equation with a nonlinear force term. It seems sensible, then, to look for traveling-wave solutions to (3.1) in which

$$u(z, t) = u(\omega t - qz)$$
$$= u(x) \tag{3.3}$$

Then, we have the ordinary differential equation

$$\left(q^2 - \frac{\omega^2}{c^2}\right)\frac{d^2 u}{dx^2} = m^2 \sin u \tag{3.4}$$

Now it seems appropriate to indulge ourselves in a little explicit algebra to add a bit of flavor to the stew. Multiplying (3.4) by du/dz and integrating gives

$$\frac{\omega^2/c^2 - q^2}{2m^2}\left(\frac{du}{dx}\right)^2 + (1 - \cos u) = 1 - \cos u_0 \tag{3.5}$$

where $(du/dx)_{u=u_0} = 0$. This is a "Hamiltonian" for a particle with mass $(\omega^2/c^2 - q^2)/m^2$ in a potential $V(u) = 1 - \cos u$, with energy $E = 1 - \cos u_0$. This periodic potential is shown in Figure 3. Below some critical energy, $E = 2$, there are periodic waves, which correspond to oscillations about one minimum of $V(u)$. With $0 < u_0 < \pi$, we put $y = (\sin\frac{1}{2}u)/(\sin\frac{1}{2}u_0)$; then (3.5) gives

$$\pm\left[\frac{\omega^2/c^2 - q^2}{m^2}\right]^{1/2}\frac{dy}{[(1 - y^2)(1 - \sin^2(\frac{1}{2}u_0)y^2)]^{1/2}} = dx \tag{3.6}$$

Integrating from $y' = 0$ to y gives[23]

$$\pm \left[\frac{\omega^2/c^2 - q^2}{m^2} \right]^{1/2} F\left[\arcsin y | \sin\tfrac{1}{2}u_0 \right] = x(y) - x(0) \qquad (3.7)$$

where $F(\phi|k)$ is an elliptic integral of the first kind. Working in the soliton area often involves encounters with these functions of the first, second, and even third kinds[23]. We can unravel this implicit solution for $u(x)$ to write

$$\sin\tfrac{1}{2}u(x) = \sin\tfrac{1}{2}u_0 \operatorname{sn}\left(\frac{m}{(\omega^2/c^2 - q^2)^{1/2}} \left[x - x(0) \right] | \sin\tfrac{1}{2}u_0 \right) \qquad (3.8)$$

where sn is a Jacobi elliptic sine function.[23] In the limit of small-amplitude oscillations ($u_0 \to 0$), $\operatorname{sn}(x) \sim \sin x$, and we obtain the familiar sine or cosine solutions to the linearized version of (3.1) where $\sin u \simeq u$. These trigonometric waves have a dispersion relation

$$\omega^2(q) = c^2 \left[m^2 + q^2 \right] \qquad (3.9)$$

In the 1-D magnets these are the classical versions of the "magnons" or spin waves.

In addition to these periodic waves there is a solitary-wave solution to (3.1). Setting $u_0 = 2\pi$, $q^2 = -1$, and $\omega^2/q^2 = v^2$, (3.5) becomes

$$\frac{1 - v^2/c^2}{2m^2} \left(\frac{du}{dx} \right)^2 = 1 - \cos u$$

$$= 2\sin^2\tfrac{1}{2}u \qquad (3.10)$$

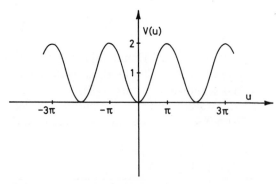

Fig. 3.   The periodic potential $V(u) = 1 - \cos u$ of Eq. (3.5).

so that

$$\pm \frac{\left(1 - v^2/c^2\right)^{1/2}}{2m} \frac{du}{dx} = \sin\tfrac{1}{2}u \qquad (3.11)$$

where $du/dx = 0$ when $u = 0, 2\pi$. Integrating (3.11), we find

$$\tan\tfrac{1}{4}u = \exp\{\pm m\gamma(z - z_0 - vt)\} \qquad (3.12)$$

where $\gamma = (1 - v^2/c^2)^{-1/2}$. This is a solitary wave traveling with velocity v, and z_0 is an arbitrary initial position. For the 1-D ferromagnet, the solitary wave describes a spin rotation of $\pm 2\pi$ as z increases, while the rotation for the antiferromagnet is $\pm \pi$ ($u = 2\phi - \pi$). Figure 4 presents the wave profile given by (3.12) together with a schematic representation of the spin profiles.

The sharp jump in u, shown in Figure 4, is often referred to as a "kink" or an "antikink" when the sign in (3.12) is $+$ or $-$, respectively. We note that the kinks and antikinks display a Lorentz invariance arising from the form of (3.1), which sets an upper limit c on the velocity v.

The energy of this solitary wave is found by substituting the waveform (3.12) in the Hamiltonian (3.2). A little algebra gives the solitary-wave energy E_s as

$$E_s(v) = \gamma E_s(0) = \left[E_s^2(0) + p_s^2 c^2 \right]^{1/2} \qquad (3.13)$$

where

$$E_s(0) = 8A\, mc^2 \qquad (3.14)$$

and p_s is defined to be the soliton momentum. We see that the "kink" energy diverges in relativistic fashion as $v \to c$. We might consider the localized waveform (3.12) as an "extended particle" with rest mass

$$m_s = \frac{E_s(0)}{c^2} = 8Am \qquad (3.15)$$

Now, there are no prizes for guessing that the wave of (3.12) is in fact a soliton. Firstly, it is easy to demonstrate that the wave is stable to small perturbations.[24] Suppose that

$$u(z, t) = u_s(x) + \delta u(z, t) \qquad (3.16)$$

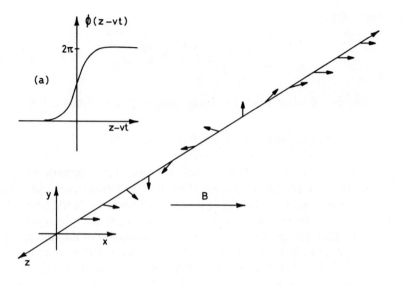

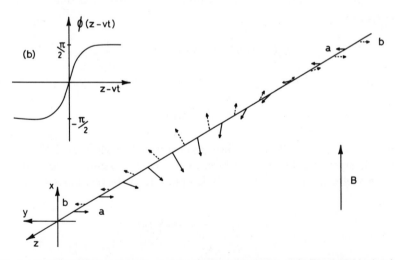

Fig. 4. (a) The solitary wave $\phi(z - vt)$ for the ferromagnet [$\phi = u$ in Eq. (3.12)] is sketched together with a schematic representation of the spin-vector profile. The spin vectors are initially parallel to the x-axis, then perform a twist of 2π to return to the original configuration. Each spin performs a complete loop as the soliton passes by. (b) The solitary wave $\phi(z - vt)$ for the antiferromagnet [$\phi = \frac{1}{2}u - \frac{1}{2}\pi$ in Eq. (3.12)] is sketched together with a schematic representation of the spin profile. The even- and odd-numbered spins simultaneously perform a twist of π radians so as to reverse the direction of the a and b sublattices.

238

where $u_s(x)$ is the solitary-wave solution and $\delta u(z, t)$ is a small perturbation. Then, linearizing (3.1) about $u_s(x)$ gives

$$\frac{\partial^2 \delta u}{\partial z^2} - \frac{1}{c^2} \frac{\partial^2 \delta u}{\partial z^2} = m^2 \cos u_s(x) \, \delta u(z, t) \qquad (3.17)$$

Equation (3.17) is solved in Appendix B. There it is shown that any small perturbation δu can be expressed as a linear combination of perturbed periodic waves with the dispersion (3.9). The perturbation does not grow in magnitude, so that $u_s(x)$ is a stable wave. However, the low-amplitude periodic waves undergo "elastic collisions" with the large-amplitude solitary wave, in which they suffer some change of form in the vicinity of the solitary wave and an asymptotic phase shift. This has consequences for the statistical mechanics, which we shall consider a little later.

It has been demonstrated[25, 26] that these kinks also undergo elastic collisions with other kinks or antikinks, and so are solitons in our strict meaning. Solutions of (3.1) that represent collisions between two kinks or one kink and one antikink are

$$\tan\tfrac{1}{4}u = \frac{v}{c} \frac{\sinh(m\gamma z)}{\cosh(m\gamma vt)} \qquad (3.18)$$

and

$$\tan\tfrac{1}{4}u = \frac{c}{v} \frac{\sinh(m\gamma vt)}{\cosh(m\gamma z)} \qquad (3.19)$$

respectively.[4] As $t \to \pm\infty$, before and after the collision, these solutions reduce to separated ($x \to \pm\infty$) kinks and antikinks. These collisions are shown in Fig. 5. As the solitons approach one another, they accelerate before combining and finally separating. This acceleration produces a shift in position of the solitons compared to their unperturbed motion, as expected for the elastic collision of two interacting "particles."

There are two common and important techniques for obtaining solutions like (3.18) and (3.19) which involve multiple solitons. These are the Bäcklund transformation[25, 27] and the inverse scattering transform (IST).[4, 28, 29] The IST is an important technique for solving the initial-value problem for a nonlinear, completely integrable system like the SG equation.[28] A detailed survey of the mathematical methods used in analyzing soliton-bearing nonlinear equations is outside the spirit of this review. However, those so inclined can find in Appendix C a description of the Bäcklund transformation as used to derive Eqs. (3.18) and (3.19). Readable accounts of the IST can be found in the references cited above.

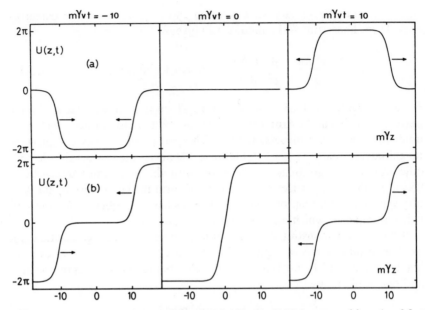

Fig. 5.   (a) The kink–antikink collision described by Eq. (3.19) is presented for $v/c = 0.5$ at three times, $m\gamma vt = -10, 0, 10$. The variation of $u(z, t)$ with $m\gamma z$ shows the solitons approaching, colliding [$u(z,0) = 0$ but $\frac{du}{dt}(z,0) \neq 0$], and emerging in a different 2π range. (b) The kink–kink collision described by Eq. (3.18) is shown under the same conditions as (a).

The SG equation allows still one more type of solution, known as the "breather": a propagating, oscillating pulse. These have the form (Ref. 30, Appendix C)

$$\tan\tfrac{1}{4}u_B = \left(\frac{\omega_0^2}{\omega_B} - 1\right)^{1/2} \frac{\sin\left[\omega_B\left(1 - v^2/c^2\right)^{-1/2}\left(t - t_0 - vz/c^2\right)\right]}{\cosh\left[m\left(1 - v^2/c^2\right)^{-1/2}\left(z - z_0 - vt\right)\left(1 - \omega_B^2/\omega_0^2\right)^{1/2}\right]}$$

$$(3.20)$$

where $\omega_0 = mc$. This is an envelope that propagates with velocity $v < c$ and oscillates with frequency $\omega_B(1 - v^2/c^2)^{-1/2}$, where $0 < \omega_B < \omega_0$. The "breather" solution is sketched in Fig. 6. The breather has energy $E_B(v, \omega_B)$, given by[30]

$$E_B(v, \omega_B) = 2E_s(v)\left(1 - \frac{\omega_B^2}{\omega_0^2}\right)^{1/2}$$

$$(3.21)$$

This solution can be considered as a bound kink-and-antikink pair: it reduces to (3.19) as $\omega_B \to 0$. On the other hand, as $\omega_B \to \omega_0$ the breather becomes more like an extended mode, an anharmonic periodic wave. In a quantum treatment of spin chains it appears as a multimagnon bound state.[31]

The breathers are perhaps the least understood of the "natural" excitations in a SG system. They pose considerable difficulties for a complete understanding of the temperature dependence of thermodynamic properties of these systems.[30, 32]. However, since they appear to involve multiple occupancy of soliton or magnon "states," they may have little bearing on the low-temperature properties of SG systems. We shall proceed, as with few exceptions[33] the literature has previously done, to avoid breathers. Effort is currently directed towards clarifying the position here.[34]

We have seen that the high-energy nonlinear kink excitations of the SG system have the remarkable stability properties of strict solitons. In order to understand the importance of the solitons in determining the thermodynamic properties of these systems, we consider the equilibrium statistical mechanics of a 1-D SG chain.

IV. STATISTICAL MECHANICS OF SINE–GORDON CHAINS

The classical partition function of a 1-D SG system can be evaluated exactly using the transfer-integral method developed by Scalapino, Sears, and

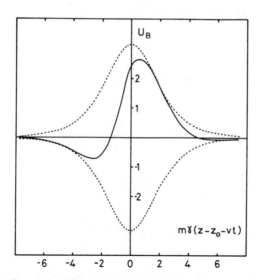

Fig. 6. The breather mode of Eq. (3.20) is presented for $\omega_B/\omega_0 = v/c = 1/\sqrt{2}$ as a function of $m\gamma(z - vt)$. Setting $x = m\gamma(z - vt)$, the breather solution is $\tan \frac{1}{4}u_B = (\omega_0^2/\omega_B^2 - 1)^{1/2}(\sin[(\omega_B v/\omega_0 c)x + \gamma t])/\cosh[(1 - \omega_B^2/\omega_0^2)^{1/2}x]$. The time t is fixed at $\gamma t = 1/\sqrt{2}$ to give the solid curve for u_B vs. $m\gamma(z - vt)$ shown. The dashed line represents the envelope within which the breather oscillates and is given by setting $\sin[(\omega_B v/\omega_0 c)x + \gamma t] = \pm 1$.

Ferrell.[35] While numerical analysis is normally required for this method, analytic results can be obtained in the low-temperature limit.[35-37] We will follow the approach used by Krumhansl and Schrieffer[38] in studying a related soliton-bearing system, by using the exact transfer-integral results to measure the accuracy of a phenomenological model. Since Krumhansl and Schrieffer, many authors have followed this path in order to highlight the contribution of solitons to the thermodynamic properties: an insight that is obscure in the formal transfer-integral method. The phenomenological approach views the SG chain as containing a gas of quasiparticles: the soliton kinks and antikinks together with the "magnons" or low-amplitude periodic modes. As we shall see below, this picture yields an exact form for the free energy at low temperature.[39]

The partition function for the SG system has been evaluated by Gupta and Sutherland[36] and by Currie, Fogel, and Palmer.[37] In the continuum approximation, the canonical partition function is written as a functional integral,[36]

$$Z = \int [\delta(u) \, \delta(\dot{u})] e^{-\beta H} \tag{4.1}$$

where $\beta = 1/k_B T$, H is the Hamiltonian of Eq. (3.2), and the functional integral denotes a sum over all possible paths in the phase space of u and $\dot{u}$.

The form of H leads to the factorization of Z:

$$Z = Z_{\dot{u}} Z_u \tag{4.2}$$

where $Z_{\dot{u}}$ is the usual contribution from the momenta of N particles in a chain,[30]

$$Z_{\dot{u}} = \left(\frac{2\pi A a k_B T}{h^2} \right)^{N/2} \tag{4.3}$$

in which h, Planck's constant, is a normalizing factor. This leaves the problem of evaluating the configuration contribution Z_u:

$$Z_u = \int \delta(u) \exp\left\{ -A \int dz \left[\frac{1}{2} c^2 \left(\frac{du}{dz} \right)^2 + \omega_0^2 (1 - \cos u) \right] \right\} \tag{4.4}$$

where $\omega_0 = cm$. For 1-D systems the configurational contribution can be evaluated by the transfer-integral method. This very useful technique has been well described elsewhere,[30,40] so that we need only quote the results we require.

In the thermodynamic limit, the length of the chain, L, and the number of lattice sites, N, become infinite, while $L/N = a$ is finite. Then, the free-energy density F/L is[30]

$$\frac{F}{L} = -\frac{k_B T}{L} \ln Z = -\frac{k_B T}{L} [\ln Z_{\ddot{u}} + \ln Z_u] \tag{4.5}$$

$$\underset{L \to \infty}{\to} -\frac{k_B T}{L} \ln Z_{\ddot{u}} + A\omega_0^2 [E_0 - E_1] \tag{4.6}$$

where in the low-temperature limit $(T \to 0)$

$$E_0 = \left[\frac{2A\omega_0 c}{k_B T}\right]^{-1} + \left[\frac{2Aa\omega_0^2}{k_B T}\right]^{-1} \ln\left[\frac{Ac^2}{2\pi a k_B T}\right] + O(T^2) \tag{4.7}$$

$$E_1 = \frac{32}{\sqrt{2\pi}} \left[\frac{8Amc^2}{k_B T}\right]^{-1/2} \exp\left\{-\frac{8Amc^2}{k_B T}\right\} \tag{4.8}$$

The numerical prefactor in (4.8) is the best *approximation* currently available.[30] We note that the second term, E_1, in (4.6) appears in the transfer-integral result because the potential, $1 - \cos u$, in H has more than one degenerate minimum. Moreover, recalling the calculated soliton energy $E_s(v)$ of Eqs. (3.13) and (3.14), we see that

$$\frac{F_1}{L} = -A\omega_0^2 E_1 = -\frac{32}{\sqrt{2\pi}} A\omega_0^2 [\beta E_s(0)]^{-1/2} \exp\{-\beta E_s(0)\}$$

$$= -\frac{4m}{\sqrt{2\pi}} E_s(0) \left[\frac{E_s(0)}{k_B T}\right]^{-1/2} \exp\left\{-\frac{E_s(0)}{k_B T}\right\} \tag{4.9}$$

which suggests that this contribution to the free energy is associated with the creation of soliton kinks. To understand the form of the free-energy density in (4.6), we pursue a phenomenological approach.

Let us begin simply by assuming that at low temperature the solitons (kinks and antikinks) form an ideal gas of "quasiparticles," which move freely along the chain. This seems reasonable in view of the soliton property by which the kinks and antikinks survive mutual collisions. The chain also contains a gas of "magnons" or "phonons"—the low-amplitude, periodic, harmonic modes. These harmonic modes, labeled by the wave number q, have frequencies $\omega(q)$ given by (3.9). Now the free energy of a harmonic oscillator with frequency ω is $k_B T \ln(\hbar\omega/k_B T)$.[40] Hence, the free-energy density arising from all the modes is given by integrating over the first Brillouin zone:[40]

$$\frac{F_0}{L} = k_B T \frac{1}{2\pi} \int_{-\pi/a}^{\pi/a} dq \ln\left(\frac{\hbar\omega(q)}{k_B T}\right) \tag{4.10}$$

$$= \frac{k_B T}{a} \left(\ln\left(\frac{\hbar\omega_0}{k_B T}\right) + \ln\left\{ \frac{1}{2}\left[1 + \left(1 + \frac{4}{a^2 m^2}\right)^{1/2} \right] \right\} \right) \tag{4.11}$$

$$\xrightarrow[a \to 0]{} k_B T \left\{ \frac{1}{a} \ln\left[\frac{\hbar\omega_0}{mak_B T} \right] + \frac{m}{2} \right\}$$

$$= -\frac{k_B T}{L} \ln Z_{\ddot{u}} + A\omega_0^2 E_0 \tag{4.12}$$

Hence, the low-amplitude periodic modes account exactly for the first two terms in the free energy at low temperature, Eq. (4.6).

Now, to describe a gas of soliton kinks and antikinks whose number may be temperature dependent, we use the grand canonical partition function $\Theta(T, L, \mu_s, \mu_{\bar{s}})$ where μ_s and $\mu_{\bar{s}}$ are the chemical potentials for kinks and antikinks, respectively. Assuming the two species are noninteracting:[40]

$$\Theta(T, L, \mu_s, \mu_{\bar{s}}) = \Theta(T, L, \mu_s)\Theta(T, L, \mu_{\bar{s}}) \tag{4.13}$$

where

$$\Theta(T, L, \mu_s) = \sum_{N_s = 0}^{\infty} e^{\beta\mu_s N_s} Z_s(N_s) \tag{4.14}$$

in which N_s is the number of solitons, and $\Theta(T, L, \mu_{\bar{s}})$ is of the same form. Here $Z_s(N_s)$ is the canonical partition function for N_s solitons, given by[40]

$$Z_s(N_s) = \frac{1}{N_s!} \left\{ \frac{1}{h} \int_0^L dq_s \int_{-\infty}^{\infty} dp_s \exp\{-\beta E_s(p_s)\} \right\}^{N_s}$$

where q_s and p_s are the soliton position and momentum. Using the soliton kink energy given by (3.13), we can evaluate $Z_s(N_s)$ and hence $\Theta(T, L, N_s)$. Pursuing this standard equilibrium statistical mechanics, we can obtain the free energy and the average soliton density, $n_s = N_s/L$. Since there are no external forces on the system, $\mu_s = \mu_{\bar{s}} = 0$ and

$$F_s = -k_B T \ln \Theta(T, L, \mu_s, \mu_{\bar{s}})|_{\mu_s = \mu_{\bar{s}} = 0} \tag{4.15}$$

$$n_s^{tot} = n_s + n_{\bar{s}} = \frac{2}{\beta L} \left[\frac{\partial}{\partial \mu_s} \ln \Theta(T, L, \mu_s) \right]_{\mu_s = 0} \tag{4.16}$$

where F_s is the soliton contribution to the free energy, and n_s^{tot} is the total average density of solitons and antisolitons. This process can be carried through, and in the low-temperature limit one obtains

$$
\begin{aligned}
n_s^{tot} &= \frac{2}{hL} \int dq_s \int_{-\infty}^{\infty} dp_s \exp\{-\beta E_s(p_s)\} \\
&= \frac{2}{h} \int_{-\infty}^{\infty} dp_s \exp\{-\beta(E_s^2(0) + p_s^2 c^2)^{1/2}\} \\
&\simeq \frac{2}{h} \int_{-\infty}^{\infty} dp_s \exp\left\{-\beta E_s(0)\left[1 + \frac{1}{2}\frac{p_s^2 c^2}{E_s^2(0)}\right]\right\} \\
&= \frac{2E_s(0)}{hc}(2\pi)^{1/2}\left[\frac{E_s(0)}{k_B T}\right]^{-1/2} \exp\left\{-\frac{E_s(0)}{k_B T}\right\}
\end{aligned}
\tag{4.17}
$$

The expansion of $E_s(p_s)$ is valid at low temperature, when we expect p_s to be small.

Similarly we find

$$
\frac{F_s}{L} = -k_B T \frac{2E_s(0)}{hc}(2\pi)^{1/2}\left[\frac{E_s(0)}{k_B T}\right]^{-1/2}\exp\left\{-\frac{E_s(0)}{k_B T}\right\}
\tag{4.18}
$$

which implies

$$
\frac{F_s}{L} = -k_B T n_s^{tot}
\tag{4.19}
$$

Now, the exponential temperature dependence in (4.18) agrees with the exact low-temperature result (4.9), but the prefactor is wrong in both T dependence and numerical factors. So the ideal-gas model is not quite right. This is the level of phenomenology introduced by Krumhansl and Schrieffer.[38]

Fortunately, Currie, Krumhansl, Bishop, and Trullinger[30] have seen how the model can be modified to reach exact results. We have used an ideal-gas model. However, the solitons do suffer elastic collisions with the "phonons" and other solitons. At low temperature the soliton density is low because of the large activation energy $E_s(0)$ appearing in the exponential of (4.17). Hence, soliton–soliton collisions may be neglected at low T. However, the "phonon" density is significant even at very low T, so that soliton–phonon collisions may be important. As shown in Appendix B, the colliding low-amplitude waves suffer a phase shift $\Delta(q, v)$ that is a function of the soliton velocity v and the wave number q. This affects the phonon density of states, $\rho(q)$.[30] For a chain of length L with periodic boundary conditions, the phonon wave numbers are given by q_n:

$$Lq_n + \Delta(q_n, v) = 2\pi n, \qquad n = 0, \pm 1, \pm 2, \ldots, \frac{L}{2a} \qquad (4.20)$$

The phonon density of states is given by

$$\rho(q) = \frac{1}{2\pi} \frac{dn}{dq} \qquad (4.21)$$

Thus in the presence of a single soliton, the density of states is changed by $\Delta\rho(q)$:[30]

$$\Delta\rho(q, v) = \frac{1}{2\pi} \frac{d\Delta}{dq}(q, v) \qquad (4.22)$$

Using the phase shift for the SG system given in Appendix B, one can evaluate the total change in the number of phonon states N_{ph}:[40]

$$\Delta N_{ph} = P \int dq \, \Delta\rho(q, v)$$

$$\dot{=} -1 \qquad (4.23)$$

where P denotes the Cauchy principal value. This is just a statement of the obvious fact that if we create a soliton we must eliminate a phonon mode in order to conserve the total number of degrees of freedom. Our statistical mechanics above did not, then, take proper account of the number of degrees of freedom. The soliton must always have at least two coordinates to describe its position and momentum. In other soliton-bearing systems[30, 38] there are additional vibrational modes localized to the soliton, so that the soliton "wobbles" as it propagates. The number of extended phonon modes is decreased by one for each local mode.

It is fairly straightforward to evaluate the effect of this change in density of phonon states on the free energy. Recalling our calculation for the phonon free energy (4.10), we amend this to read

$$F_{ph} = F_0 + k_B T P \int_{-\pi/a}^{\pi/a} dq \, \Delta\rho(q) \ln\left(\frac{\hbar\omega(q)}{k_B T}\right)$$

$$\equiv F_0 + \Delta F \qquad (4.24)$$

This gives,[40] as $a \to 0$,

$$\Delta F = -k_B T \ln\left[\frac{\hbar\omega_0\left\{1 + (1 - v^2/c^2)^{-1/2}\right\}}{k_B T}\right] \qquad (4.25)$$

This is the change in the phonon free energy caused by the presence of a single kink. Therefore, one can regard ΔF as a soliton self (free) energy[30,41] and write the effective soliton energy as $E_s^*(v)$:

$$E_s^*(v) = E_s(v) + \Delta F(v) \tag{4.26}$$

Now the calculation for the soliton free energy and average number density, Eqs. (4.13)–(4.18), can be redone using $E_s^*(v)$ in place of $E_s(v)$. One finds[30,40]

$$n_s^{\text{tot}} = \frac{4m}{(2\pi)^{1/2}} \left(\frac{E_s(0)}{k_B T} \right) \left[1 + \frac{5}{8} \left(\frac{E_s(0)}{k_B T} \right)^{-1} \right.$$

$$\left. + O\left(\left(\frac{E_s(0)}{k_B T} \right)^{-2} \right) \right] \exp\left\{ -\frac{E_s(0)}{k_B T} \right\} \tag{4.27}$$

so that to lowest order in $(E_s(0)/k_B T)^{-1}$,

$$\frac{F_s}{L} = -k_B T n_s^{\text{tot}}$$

$$= -\frac{4m}{(2\pi)^{1/2}} E_s(0) \left[\frac{E_s(0)}{k_B T} \right]^{-1/2} \exp\left\{ -\frac{E_s(0)}{k_B T} \right\} \tag{4.28}$$

in precise agreement with F_1/L of (4.9).

Thus, at least at low temperature, we can consider the free energy of a SG system as arising from a gas of solitons and phonons.

The gas is not ideal. The quasiparticles undergo elastic collisions, which produce phase shifts, that is, shifts in their spatial positions. These interaction effects account exactly at low temperature for the redistribution of the degrees of freedom between the extended modes and the localized soliton modes of the system. The phenomenological approach to the high-temperature thermodynamics is complicated by the problem of accounting correctly for the "breather" modes.[30,32]

We can see from the form of (4.28) that the soliton contribution to the free energy is limited by the activation energy for soliton formation, ensuring a low soliton density at low temperature. This does not necessarily imply that solitons have a negligible effect on observed properties. The form of static and dynamic correlation functions can be strongly affected by the solitons. We can best illustrate these effects by considering some experiments performed on the spin systems $CsNiF_3$ and TMMC.

V. OBSERVATION OF SOLITONS IN SPIN SYSTEMS: CORRELATION FUNCTIONS

We introduced the SG equation as an approximate description of the spin dynamics in crystals like $CsNiF_3$ and TMMC. While this review is not devoted to the understanding of these particular substances, it seems reasonable to spend some time to describe the experimental investigations of these crystals and discuss how the soliton theories account for the observations. This serves two purposes. Firstly, you may be a little less skeptical about the whole soliton concept if some foundation in experiment is demonstrated. Secondly, it is important to understand how solitons can have significant effects on some properties and not on others.

Of course, many difficulties arise when one attempts to use a simple equation like the SG to describe a range of observations. There were many assumptions invoked in Section II that are not perfectly valid for either $CsNiF_3$ or TMMC. In particular, we assumed that the spin dynamics were well confined to the easy plane. Molecular dynamics has shown that this breaks down for $CsNiF_3$ in the temperature range of interest.[19] Moreover, our use of a continuum approximation is dangerous in at least two ways. The trigonometric dependence of terms in the Hamiltonians (2.4) and (2.5) is partially lost, which alters the topology of the phase space.[42] Luckily, exact transfer-integral calculations using the discrete Hamiltonian (2.4) have been performed,[42] which show that the SG equation gives reasonably accurate approximations to thermodynamic properties. The continuum approximation also oversimplifies the dynamics by allowing a free propagation of solitons. Discrete-lattice effects can give rise to a small barrier to propagation, which immobilizes the solitons at low enough energy.[43,44] Putting all such quibbles aside, let's see how well the SG soliton description performs.

There are a number of experimental probes of spin dynamics, including measurements of magnetic neutron scattering,[45] nuclear spin–lattice relaxation,[46] average magnetization,[47] and optical properties.[48, 49] These techniques are all concerned in whole or in part with the measurement of the spin correlation functions $\langle S^\alpha(z,t)S^\alpha(0,0)\rangle$, where $\alpha = x, y$ and $\langle \rangle$ denotes an equilibrium average. The most direct probe is magnetic neutron scattering. This technique is well described by Marshall and Lovesey.[50] With the external magnetic field in the x-direction, we define the longitudinal and transverse correlations, respectively, as

$$s_{\parallel}(z,t) = \langle S^x(z,t)S^x(0,0)\rangle \tag{5.1}$$

and

$$s_{\perp}(z,t) = \langle S^y(z,t)S^y(0,0)\rangle \tag{5.2}$$

The scattered intensity of magnetically aligned neutrons as a function of scattering angle and energy can be analyzed to measure the time and space transforms of these correlation functions:

$$S_{\parallel}(q,\omega) = \frac{1}{(2\pi)^2} \int_{-\infty}^{\infty} dz \int_{-\infty}^{\infty} dt\, e^{-i(\omega t - qz)} \langle S^x(z,t) S^x(0,0) \rangle \quad (5.3)$$

and

$$S_{\perp}(q,\omega) = \frac{1}{(2\pi)^2} \int_{-\infty}^{\infty} dz \int_{-\infty}^{\infty} dt\, e^{-i(\omega t - qz)} \langle S^y(z,t) S^y(0,0) \rangle \quad (5.4)$$

Under normal resolution restrictions, it is very difficult to separate these two functions and some combination only is observed, $S(q,\omega)$:

$$S(q,\omega) = A S_{\parallel}(q,\omega) + B S_{\perp}(q,\omega) \quad (5.5)$$

where A and B are constants that depend on the scattering geometry.

There is, at present, no exact method for calculating the dynamic correlation functions of (5.1) and (5.2) for a SG chain. However, the static correlations[32]

$$F_{\parallel}(q) = \int_{-\infty}^{\infty} S_{\parallel}(q,\omega)\, d\omega = \frac{1}{2\pi} \int_{-\infty}^{\infty} dz \left[e^{iqz} \langle S^x(z,0) S^x(0,0) \rangle \right] \quad (5.6)$$

and

$$F_{\perp}(q) = \int_{-\infty}^{\infty} S_{\perp}(q,\omega)\, d\omega = \frac{1}{2\pi} \int_{-\infty}^{\infty} dz \left[e^{iqz} \langle S^y(z,0) S^y(0,0) \rangle \right] \quad (5.7)$$

can be evaluated exactly by the transfer-integral method. Thus, one can use the phenomenological model of a soliton–phonon gas to calculate $S_{\parallel}(q,\omega)$ and $S_{\perp}(q,\omega)$ while having exact expressions for $F_{\parallel}(q)$ and $F_{\perp}(q)$ for reference purposes. The soliton contributions to $F_{\parallel}(q)$ and $F_{\perp}(q)$ can then be identified as they were for the free energy.

If we continue to neglect spin fluctuations out of the easy plane,

$$\langle S^x(z,t) S^x(0,0) \rangle = S^2 \langle \cos\phi(z,t) \cos\phi(0,0) \rangle \quad (5.8)$$

and

$$\langle S^y(z,t) S^y(0,0) \rangle = S^2 \langle \sin\phi(z,t) \sin\phi(0,0) \rangle \quad (5.9)$$

What effect will soliton kinks and antikinks have on these functions? Let's suppose that at time $t = 0$ there is a kink (or antikink) at position $z = z_1$ on the chain. For the ferromagnet, Eq. (3.12) with $u = \phi$ gives

$$\cos\phi(z,t)\cos\phi(0,0) = \{1 - 2\operatorname{sech}^2[m\gamma(vt - z + z_1)]\}$$
$$\times \{1 - 2\operatorname{sech}^2[m\gamma(z_1)]\} \qquad (5.10)$$

Note that this product is approximately one, as it is for the chain at rest, unless

$$|z_1| < (m\gamma)^{-1} \quad \text{or} \quad |z - (vt + z_1)| < (m\gamma)^{-1} \qquad (5.11)$$

That is, the soliton only has an effect on $\cos\phi(z,t)\cos\phi(0,0)$ if the soliton was near 0 or z at time 0 or t respectively. Only those lattice sites within the soliton "width," $2(m\gamma)^{-1}$, are affected. The same holds for $\langle\sin\phi(z,t)\sin\phi(0,0)\rangle$:

$$\sin\phi(z,t)\sin\phi(0,0)$$
$$= 4\cos\phi(z,t)\cos\phi(0,0)\operatorname{sech}^2[m\gamma(vt - z + z_1)]\operatorname{sech}^2[m\gamma z_1]$$
$$(5.12)$$

which vanishes unless position 0 or z lies within the soliton width. If only those particles very near a kink are affected, then the total effect on the correlation functions will be proportional to the soliton number density. Since this density is low at low temperature, soliton effects will be hard to find in a ferromagnet like $CsNiF_3$. This point has only recently been appreciated.[32, 51] However, the situation for antiferromagnets is much more interesting. From (3.12) with $u = 2\phi - \pi$,

$$\sin\phi(z,t)\sin\phi(0,0) = \tanh[m\gamma(vt - z + z_1)]\tanh(m\gamma z_1) \quad (5.13)$$
$$\cos\phi(z,t)\cos\phi(0,0) = \operatorname{sech}[m\gamma(vt - z + z_1)]\operatorname{sech}(m\gamma z_1) \quad (5.14)$$

for a single kink. The form of $\cos\phi(z,t)\cos\phi(0,0)$ for the antiferromagnet leads to the same dependence on soliton density as for the ferromagnet. Important effects arise for the transverse correlation. If the kink lies anywhere between 0 and z at time t,

$$\sin\phi(z,t)\sin\phi(0,0) \simeq -1 \qquad (5.15)$$

—the opposite of the value for the chain at rest. Now at low temperature most of the solitons will be at rest or moving very slowly. Then the product in (5.13) is -1 whenever there is one kink or one antikink (but not both) between 0 and z. This will be the case so long as z is no larger than the average distance between kinks and antikinks. This distance is roughly *inversely proportional* to the soliton number density. Hence as temperature decreases and the soliton number density decreases, the effect of solitons on $S_\perp(q, \omega)$ increases. We can understand this quite simply. The kinks and antikinks in antiferromagnets like TMMC are π rotations of the spins. As shown in Fig. 4, the spins on each side of the kink are perfectly ordered antiferromagnetically. Each side is one state of a doubly degenerate ground state. The states are energetically equivalent but distinguishable by the sign of $\sin \phi$, that is, the direction of (say) even-numbered spins. Now a 1-D system can only be perfectly ordered at $T = 0$. At $T = 0$ we have perfect antiferromagnetic order, and generally a new Bragg peak will appear in the magnetic neutron diffraction pattern. Above zero temperature the expected loss of correlation is provided by the soliton kinks and antikinks. One soliton, for example, halves the average distance over which the spins are ordered or correlated. As we shall see in more detail below, this loss of order is reflected in a weakening of the Bragg peak intensity and the appearance of a more diffuse scattering pattern around the Bragg peak position, a so called "central peak," whose *width* is related to the soliton density.

The phenomenological derivation of $S(q, \omega)$ has been performed by Mikeska[16] for the ferromagnet and by Mikeska[52] and Maki[21] for the antiferromagnet. It was assumed therein that the solitons and magnons make noninteracting contributions to the dynamic structure factor, though the correct soliton density, renormalized by the self-free-energy effect, has been used.[21]

For the ferromagnet we follow Mikeska's results. With the magnetic field in the x-direction, the magnon contribution to the longitudinal fluctuations will be suppressed and soliton effects may dominate. Mikeska then considers only soliton effects on $S_\parallel(q, \omega)$ and sets

$$\cos(z, t) = 1 + \sum_{n=1}^{N_s^{\text{tot}}} (\cos \phi_n^s - 1)$$

$$= 1 - 2 \sum_{n=1}^{N_s^{\text{tot}}} \text{sech}^2 m\gamma(z - v_n t - z_{0n}) \tag{5.16}$$

which represents a nonoverlapping (low-density) collection of soliton kinks and antikinks. One can then determine $\langle \cos \phi(z, t) \cos \phi(0, 0) \rangle$, where the average is over the arbitrary soliton positions z_{0n} and velocities v_n with statistical weight $\gamma \exp\{-E_s(v)/k_B T\}$. At low temperature, the v_n are small, so that $\gamma \simeq 1 + v^2/2c^2$. Mikeska has performed the average and computes $S_\parallel$.

$(q, \omega):$[17]

$$S_{\|}(q, \omega) = \frac{8}{\pi c m q} \frac{2n_s^{\text{tot}}}{m} \sqrt{\frac{bm}{\pi}} \exp\left\{-\frac{4m\omega^2 b}{c^2 q^2}\right\}\left[\frac{\pi q/2m}{\sinh(\pi q/2m)}\right]^2$$

$$+ \left(1 - \frac{4n_s^{\text{tot}}}{m}\right)^2 \delta(q)\delta(\omega) \qquad (5.17)$$

where $b = 2JS^2 a/k_B T$, m, and c are defined in (2.11) and where $\delta(q), \delta(\omega)$ are Dirac delta functions corresponding to the Bragg peak. The transverse component $S_{\perp}(q, \omega)$ can be constructed in the same fashion. Independent magnon contributions are also significant here, so that Mikeska finds[17]

$$S_{\perp}(q, \omega) = S_{\perp}^{\text{magnon}} + S_{\perp}^{\text{soliton}} \qquad (5.18)$$

where

$$S_{\perp}^{\text{magnon}} = \frac{1}{4\pi b}\left[m^2 + q^2\right]^{-1}\{\delta(\omega - \omega(q)) + \delta(\omega + \omega(q))\} \qquad (5.19)$$

and

$$S_{\perp}^{\text{soliton}} = \frac{8}{\pi c m q} \frac{2n_s^{\text{tot}}}{m} \sqrt{\frac{bm}{\pi}} \exp\left\{-\frac{4m\omega^2 b}{c^2 q^2}\right\}\left[\frac{\pi q/2m}{\cosh(\pi q/2m)}\right]^2 \qquad (5.20)$$

It is possible to perform these calculations retaining the relativistic velocity effects approximated above; however, the model is invalid at the high temperatures for which relativistic corrections are important.[18] At low q-values the soliton contribution to $S_{\|}(q, \omega)$ is larger than that to $S_{\perp}(q, \omega)$ because of the form factor $(\pi q/2m)/\sinh(\pi q/2m)$. Concentrating on this, we see that the solitons reduce the intensity of the Bragg peak only in proportion to their number density, as expected from our previous arguments. This intensity appears as a Gaussian "quasielastic" (peaking at $\omega = 0$) central peak (peaking at $q = 0$). The scattering of magnetically aligned neutrons from a "gas" of solitons peaks near $q = 0$ because of their translational invariance; the solitons can be anywhere along the chain. The frequency response peaks at $\omega = 0$ because the solitons are freely translating, not vibrational, modes of motion.

Needless to say (or we would never have mentioned the compound), just such a quasielastic central peak has been observed in the neutron scattering from $CsNiF_3$ by Steiner and co-workers.[53-56] Early comparisons of the experiment with the dominant soliton contribution (5.17) were impressively good, as shown in Figs. 7 and 8. The q- and T-dependence of the frequency

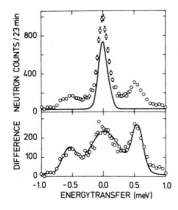

Fig. 7. A reproduction of Fig. 1 from the article by J. K. Kjems and M. Steiner, *Phys. Rev. Lett.* **41**, 1137 (1978) (with permission). Upper: Observed inelastic neutron scattering spectrum of CsNiF$_3$ at reciprocal-lattice vector $(0,0,1.9)$ for $T = 9.3$ K and magnetic field H (here B^x) $= 5$ kG (circles). The full line is the observed profile at $T = 3.1$ K and 30 kG, which is assumed to be background. Lower: Difference between the two spectra in the upper part of the figure. The full line is the result of a least-squares fit assuming a Gaussian line shape for the central quasielastic peak and Lorentzian profiles for the spin-wave features at ± 0.55 meV (see Ref. 53).

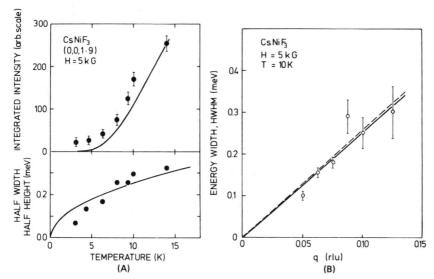

Fig. 8. A reproduction of Fig. 1 from the article by M. Steiner, J. Kjems, K. Kakurai, and H. Dachs, *J. Magn. Magn. Mater.* **15–18**, 1057 (1980) (with permission). (*a*) Temperature dependence of the integrated intensity (upper) and energy half width at half maximum (lower) of the quasielastic component of the neutron scattering from CsNiF$_3$, seen in Fig. 7, for H (here B^x) $= 5$ kG at the reciprocal-lattice vector $(0,0,1.9)$. (*b*) Energy width ΔE (HWHM) of the central (quasielastic) component vs. wave number q (in reciprocal-lattice units). The points are the experimental results; the solid lines are the theoretical predictions of Mikeska's soliton model.[54, 16]

width and the T-dependence of the intensity of the central peak were well described by Mikeska's theory. However, the q-dependence of the intensity was *quantitatively* wrong.[51] The simple description of these experimental features in terms of SG solitons has lost some conviction as the importance of the magnon terms in $S(q, \omega)$ and especially multimagnon contributions has been appreciated. As we noted above, there are few solitons but many magnons at low T, where these experiments are performed. Loveluck et al.[19] have used molecular dynamics to demonstrate both the inadequacy of the simple SG model (2.12) as opposed to the original Hamiltonian (2.4), and the fact that multimagnon processes or perhaps pulse solitons[52] can contribute to $S(q, \omega)$ near $q = \omega = 0$. Recently, Reiter[57] has shown that a theory

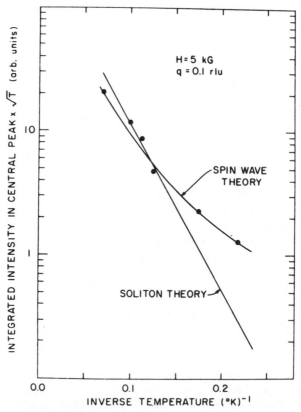

Fig. 9. A reproduction of Fig. 2 from the article by G. Reiter, *Phys. Rev. Lett.* **46**, 202, 518 (1981) (with permission). The temperature dependence of the integrated intensity of the central (quasielastic) peak observed in $CsNiF_3$. The experimental points are the data of Kjems and Steiner.[53] Here H and q represent the applied magnetic field and the scattering wave number, respectively.

based on spin-wave density fluctuations describes the observed central peak as well as or better than the soliton theory (see Figs. 9 and 10).

More accurate soliton theories for $CsNiF_3$, which retain discrete-lattice effects along the lines of Eq. (2.4), have appeared.[58,59] Indeed, they show that the SG theory of Mikeska is reasonably accurate in describing the soliton contribution to $S(q, \omega)$. No doubt the solitons are significant in $CsNiF_3$, but the inherent dependence on a low N_s^{tot} makes it essential for very detailed and accurate scattering experiments to separate these effects from the equally significant magnon processes. Recent experiments are directed towards this.[51,60]

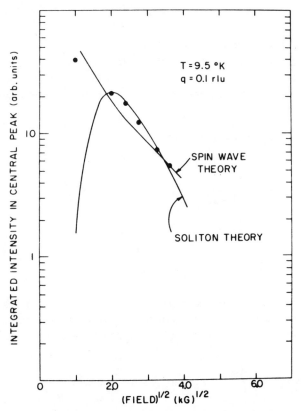

Fig. 10. A reproduction of Fig. 3 from the article by G. Reiter (see Figure 9) (with permission). Variation of the intensity in the central peak, observed in $CsNiF_3$, with field. From Ref. 57, the effect of the variation of the soliton shape with field is responsible for the downturn of the soliton prediction, omitted in Ref. 53, and accounts for the difference between this figure and Fig. 3 of Ref. 53.

The static and dynamical correlation functions for the antiferromagnet have been presented by Mikeska[52] and Maki.[21] We consider Maki's results below. Following Krumhansl and Schrieffer[38] once more, one assumes a gas of noninteracting solitons distributed along the chain, to obtain

$$\langle \sin\phi(z,t)\sin\phi(0,0)\rangle = \langle \sin^2\phi\rangle\exp\left[i\pi\frac{z}{a} - 4\int_{-\infty}^{\infty}\frac{dp}{2\pi}n_s(v)|z - vt|\right] \quad (5.21)$$

where $n_s(v)$ is the density of solitons with velocity v, p is a momentum (see Ref. 21), and $\langle \sin^2\phi\rangle$ gives the average magnetic order. Maki finds the approximate result

$$\langle \sin\phi(z,t)\sin\phi(0,0)\rangle = \langle \sin^2\phi\rangle\exp\left[i\pi\frac{z}{a} - K_\perp\left(z^2 + v_0^2t^2\right)^{1/2}\right] \quad (5.22)$$

where $v_0 = (2k_BT/\pi B_0)^{1/2}C_A$ is the average thermal velocity of the solitons, $B_0 = g\mu_BSB^x$, and $K_\perp = a\xi_\perp^{-1}$ is the *inverse spin-correlation length* for S^y along the z-axis. Both $\langle \sin^2\phi\rangle$ and $K_\perp$ can be evaluated exactly via the transfer-integral method. For $(\frac{1}{4}B_0/k_BT)^2 \gg 1$ (the experimentally relevant case) Maki finds

$$K_\perp = 4n_s^{\text{tot}} \quad [\text{see Eq. (4.27)}] \quad (5.23)$$

$$\langle \sin^2\phi\rangle = 1 - \langle \cos^2\phi\rangle$$

$$= 1 - \left(\frac{B_0}{k_BT}\right)^{-1}\left[1 + \left(\frac{B_0}{k_BT}\right)^{-2} + 3\left(\frac{B_0}{k_BT}\right)^{-3} + \cdots\right] \quad (5.24)$$

Then the structure factor $S_\perp(q,\omega)$ is written as

$$S_\perp(q,\omega) = S^2\frac{\xi_\perp^{-1}}{2\pi v_0}\langle \sin^2\phi\rangle\left[q^{*2} + \left(\frac{\omega}{v_0}\right)^2 + \xi_\perp^{-2}\right]^{-3/2} \quad (5.25)$$

where $q^* = q - \pi/a$ vanishes at the magnetic Bragg peak. While Mikeska[52] derives a somewhat different result for $S_\perp(q,\omega)$, both expressions give a frequency-integrated result, $F_\perp(q)$, in agreement with the transfer-integral method:

$$F_\perp(q) = \int_{-\infty}^{\infty}S_\perp(q,\omega)\,d\omega = S^2\frac{\xi_\perp^{-1}}{\pi}\frac{\langle \sin^2\phi\rangle}{\xi_\perp^2 + q^{*2}}$$

$$= S^2\frac{\langle \sin^2\phi\rangle}{\pi}\frac{4n_s^{\text{tot}}/a}{\left(4n_s^{\text{tot}}/a\right)^2 + q^{*2}} \quad [\text{using (5.23)}] \quad (5.26)$$

$$\xrightarrow[q^*\to 0]{} S^2\langle \sin^2\phi\rangle\frac{a}{4\pi}\left(n_s^{\text{tot}}\right)^{-1} \quad (5.27)$$

Thus the integrated intensity of the central peak is inversely proportional to n_s^{tot} at the Bragg peak position, as expected from our previous arguments. Here the external magnetic field suppresses the magnon contributions to $S_\perp(q, \omega)$.

The derivation of $S_\parallel(q, \omega)$ follows along the same lines as for the ferromagnet:[21]

$$S_\parallel(q, \omega) = S_\parallel^{magnon} + S_\parallel^{soliton} \tag{5.28}$$

where

$$S_\parallel^{magnon}(q, \omega) = S^2 \langle \sin^2\phi \rangle \left\{ 1 + \left[\exp\left[\frac{\hbar\omega(q^*)}{k_B T} \right] - 1 \right]^{-1} \right\}$$
$$\times \{ \delta(\omega - \omega(q^*)) + \delta(\omega + \omega(q^*)) \} \tag{5.29}$$

with

$$\omega^2(q^*) = c^2 q^{*2} + \Delta^2, \qquad \Delta = \frac{\mu_B g B^x (1 - D/2|J|)^{1/2}}{\sqrt{2}} \tag{5.30}$$

and

$$S_\parallel^{soliton}(q, \omega) = \frac{2 B_0 S^2}{C_A m q^* k_B T} \exp\left\{ -\frac{B_0}{k_B T} \left[1 + \frac{1}{2}\left(\frac{\omega}{c q^*} \right)^2 \right] \right\} \operatorname{sech}^2\left(\frac{\pi q^*}{2m} \right) \tag{5.31}$$

where $m = \sqrt{2}\, m_A$ [see (2.23)]. The soliton contributions to both $S_\perp$ and $S_\parallel$ predict a "central peak." However, the magnitude of $S_\perp$ is now large, due to the diverging correlation length, and dominates $S(q, \omega)$ for $\omega \simeq q \simeq 0$.[61]

The magnetic neutron scattering from TMMC has been observed and described in terms of these expressions for $S(q, \omega)$ by Boucher et al.[61] who write (5.25) as

$$S_\perp(q, \omega) = S^2 \frac{\langle \sin^2\phi \rangle}{2\pi} \frac{\Gamma_\omega^2/\Gamma_q}{[\Gamma_\omega^2(1 + q^{*2}/\Gamma_q^2) + \omega^2]^{3/2}} \tag{5.32}$$

where

$$\Gamma_q = \xi^{-1} = \frac{4 n_s^{tot}}{a}$$

$$= \frac{16 m}{a(2\pi)^{1/2}} \left(\frac{E_s(0)}{k_B T} \right)^{1/2} \exp\left\{ -\frac{E_s(0)}{k_B T} \right\}$$

$$= A B^x \left[\frac{B^x}{T} \right]^{1/2} \exp\left\{ -\frac{\alpha B^x}{T} \right\} \tag{5.33}$$

where $\alpha = g\mu_B S/k_B$ from the value of $E_s(0)$ in (3.14) and the Hamiltonian (2.23). Similarly

$$\Gamma_\omega = BB^x \exp\left\{ -\frac{\alpha B^x}{T} \right\}. \tag{5.34}$$

Boucher et al.[61] treat A, B, and α as adjustable parameters to fit the observed neutron scattering intensity, as a function of ω, q, B^x, and T, to $S_\perp(q, \omega)$ in (5.32). As shown in Figure 11, the data are well described by

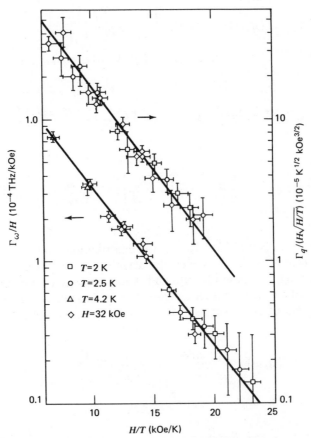

Fig. 11. A reproduction of Fig. 2 from the article by Boucher, Regnault, Rossat-Mignod, Renard, Bouillot, and Stirling, *J. Appl. Phys.* **52**, 1956 (1981) (with permission). The central peak in the inelastic neutron scattering spectrum of TMMC as a function of ω and q has been fitted with Eq. (5.32), yielding the scaled values of Γ_ω and Γ_q shown above vs. H/T (here $H = B^x$). The soliton theory predicts a linear variation of these scaled quantities with H/T. The fitted straight lines yield the values for Γ_ω and Γ_q of (5.35) and (5.36).

(5.32) with Γ_q and Γ_ω from (5.33) and (5.34) as

$$\Gamma_q = (2.5 \pm 0.5) \times 10^3 B^x \left[\frac{B^x}{T} \right]^{1/2} \exp\left\{ -\frac{(0.26 \pm 0.02) B^x}{T} \right\} \quad (5.35)$$

$$\Gamma_\omega = (2.1 \pm 0.3) \times 10^1 B^x \exp\left\{ -\frac{(0.26 \pm 0.03) B^x}{T} \right\} \quad (5.36)$$

The qualitative form of the structure factor is clearly well described by the soliton theory. However, the theoretical predictions for α, A, and B are all larger than the experimental values ($\alpha = 0.336$ K/kOe as against 0.26 K/kOe above). The situation is somewhat confused here, because Maki's derivation of the SG Hamiltonian given in Section II differs from that of Mikeska[52] and gives a different value for C_A, leading to a different prediction for the parameter B. However, as Boucher et al. point out, if m in the SG Hamiltonian is rescaled to give the correct value for α (0.26 K/kOe), then Mikeska's predictions for A and B agree very well with experiment. Thus a rescaling of the "soliton mass" parameter by about 23% downwards is sufficient to give a quantitative description of the ω, q, B^x, and T dependence of the scattering data. Figure 12 shows how Γ_q and Γ_ω given by (5.35) and (5.36) accurately describe the frequency width and maximum intensity of the central peak.

Another probe of the spin dynamics is provided by measuring the nuclear spin–lattice relaxation (NSLR). The importance of solitons to NSLR in TMMC has been demonstrated by Boucher and Renard.[46] They measure the relaxation time T_1 of ^{15}N in an enriched crystal of TMMC under the same experiment conditions as the neutron scattering. T_1 characterizes the exponential decay of the nuclear magnetization following a $\pi - \frac{1}{2}\pi$ magnetic-field pulse sequence.[46] For magnetic solids the NSLR rate can be written as[46]

$$T_1^{-1} = \frac{2\pi}{N} \sum_q A_q S_{\parallel}(q, \omega_N) + B_q S_{\perp}(q, \omega_N) \quad (5.37)$$

where A_q, B_q are hyperfine coupling constants between the ^{15}N and electron spins, and ω_N is the Larmor frequency for ^{15}N. Again the spin fluctuations described by $S_{\perp}(q, \omega_N)$ will dominate. Clearly, as a soliton passes each point in the chain, a large (π) spin flip occurs and the previous spin configuration is relaxed. T_1^{-1} is large at low T and large B^x. Under reasonable assumptions, the contribution of perpendicular spin fluctuations is given by[46]

$$\left(T_1^{-1} \right)_{\perp} = \frac{B_\pi S^2}{\pi \Gamma_\omega} \quad (5.38)$$

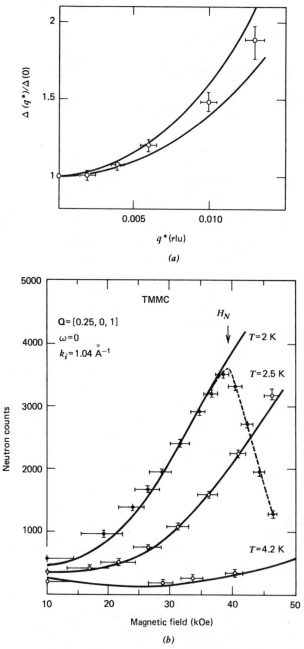

Fig. 12. (a) Reproduced from Fig. 3 of Boucher et al. (see Fig. 11) (with permission). The experimental width (HWHM) of the central peak, for H (here B^x) = 36.2 kOe and T = 2.5 K, as a function of q^* (in reciprocal-lattice units), derived from (5.32) using (5.35). (b) Reproduced from Fig. 4 of Boucher et al.[61] (with permission). Field dependence of the maximum intensity of the central peak in TMMC. The solid curves are derived from (5.32) using (5.35) and (5.36). H_N displays the 3-D ordering. Q is the scattering wave vector and k_i is the incident neutron wave number.

which dominates T_1^{-1} except at low B^x/T. The observed T_1^{-1} were fitted to the soliton theory prediction for Γ_ω as shown in Fig. 13. As shown in Fig. 14, T_1 and Γ_ω (obtained from the neutron-scattering data) have the same slope vs. B^x/T. Thus these two independent measurements of $S_\perp(q, \omega)$ are both well described by the concept of freely moving soliton domain walls dominating the low-temperature spin dynamics.

Boucher[47] has obtained further evidence for these soliton domain walls from the magnetic-field dependence of the 3-D ordering temperature, which is determined by $\xi_\perp$ and hence n_s^{tot} for a range of B^x-values.

The presence of soliton domain walls in a 1-D antiferromagnet should have an effect on the crystal optical properties. This raises the possibility of soliton effects in a traditional field in chemical physics, the study of electronic excitation of crystals by either absorption or Raman spectroscopy.

The absorption[49] and Raman[48] spectra of the ferromagnet $CsNiF_3$ have already been interpreted in terms of two spin correlation functions in the

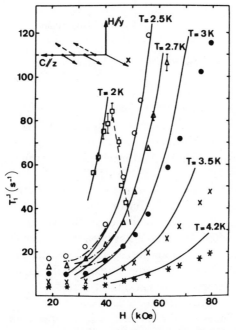

Fig. 13. Reproduced from Fig. 1 of Boucher and Renard, *Phys. Rev. Lett.* **45**, 486 (1980) (with permission). The NSLR rate T_1^{-1} of ^{15}N in TMMC vs. magnetic field H at different temperatures. The full lines represent the soliton-model predictions for the contribution of spin fluctuations perpendicular to the field [see (5.38)]. The dot–dash lines include the fluctuations parallel to the field [see (5.37)]. The dashed line is an aid to the eye that displays the Γ_1^{-1} behavior at the 3-D magnetic ordering.

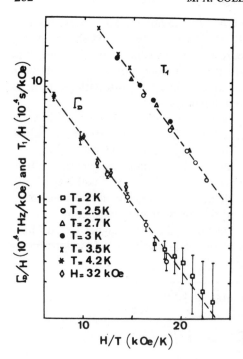

Fig. 14. Reproduced from Fig. 2 of Boucher and Renard (see Fig. 13) (with permission). Experimental values of T_1/H and Γ_D/H (on a logarithmic scale), as a function of H/T. Here Γ_D is the width of the central peak observed by neutron measurements and reported in Ref. 45. (Essentially Γ_D corresponds to Γ_ω of Ref. 61 and Fig. 13.) The slopes of the two lines are the same [with $\alpha \simeq 0.26$ in (5.34)], as predicted by the soliton theory.

ground state. Since visible–UV radiation is essentially a long-wavelength probe, the intensity, energy, and line shape of electronic transitions will depend in part on the long-range order of the ground-state lattice. We know that solitons provide the dominant mechanism determining the spin correlation length in antiferromagnets like TMMC. Thus, it is reasonable to expect soliton effects to be observable in the electronic spectra of this and similar antiferromagnets, while related ferromagnets will be relatively unaffected. The intensity of the magnon sidebands in the optical absorption spectrum of TMMC and related compounds has been the subject of considerable study in physics and chemistry laboratories. The dominant model, due to Tanabe and co-workers[62, 63] explains the unexpectedly high intensity of these spin-forbidden electronic transitions in terms of a pairwise interaction between neighboring transition-metal ions. It may be interesting to look for soliton influences on the intensities in the presence of the symmetry-breaking field B^x.

VI. MORE SPIN SOLITONS

Those brave readers who have come this far without a love of magnetic systems to sustain them may wish to pass on to the next section. The intention herein is merely to mention, with useful references, a number of spin systems that display solitons and related phenomena, and is not essential reading for the general progression of the argument.

We begin with the simplest-looking spin Hamiltonian, that for the 1-D *isotropic* Heisenberg chain:

$$H = -2J \sum_l \mathbf{S}_l \cdot \mathbf{S}_{l+1} \tag{6.1}$$

In the *classical continuum* limit, the ferromagnet $(J > 0)$ displays transverse spin waves[64] and large-amplitude strict solitons,[65-67] first derived by Nakamura and Sasada. These solitons are like the breather modes of the SG chain in having both translational and vibrational motion. Here the spin vector is asymptotically in the z-direction perpendicular to the chain direction (x): the soliton is a localized decrease in the magnitude of S^z, while the phase of S^z is a sum of terms constant in the frame of reference moving with the soliton amplitude plus a term increasing with *any* frequency.

The isotropic ferromagnet in a magnetic field,

$$H = \sum_l -2J \mathbf{S}_l \cdot \mathbf{S}_{l+1} - \mu_B g B^x \sum_l S_l^x \tag{6.2}$$

also displays solitons of this nature, often called "pulse" solitons.[20]

Nakamura and Sasada[68] have also considered the *uniaxial ferromagnet*. If we begin with

$$H = \sum_l -2J \mathbf{S}_l \cdot \mathbf{S}_{l+1} - A(S_l^z)^2 \tag{6.3}$$

then $A > 0$ implies that z is an easy axis of magnetization. It is easy to derive a Hamiltonian density in the continuum limit:

$$H = \frac{1}{N} \left\{ \sum_{l=-N/2}^{N/2} -2J \mathbf{S}_l \cdot \mathbf{S}_{l+1} - A(S_l^z)^2 \right\}$$

$$\simeq \frac{1}{N} \left\{ \int_{-N/2}^{N/2} \frac{dx}{a} -2J \mathbf{S}(x) \cdot \left[\mathbf{S}(x) + a \frac{\partial \mathbf{S}}{\partial x} + \frac{a^2}{2!} \frac{\partial^2 \mathbf{S}}{\partial x^2} \right] - A[S^z(x)]^2 \right\}$$

$$= \frac{1}{N} \int_{-N/2}^{N/2} -2J \left[S^2 - \frac{a^2}{2} \left(\frac{\partial S}{\partial x} \right)^2 \right] - A[S^z(x)]^2 \frac{dx}{a} \tag{6.4}$$

in the limit of large N, where a is the lattice spacing along the chain direction x. This uniaxial ferromagnet also displays pulse solitons[20] similar to those from (6.1) and (6.2). In addition, the Hamiltonian density (6.4) allows another type of nonlinear excitation. This is a *static* domain wall given by

$$S^z = S \tanh\left[\frac{(x - x_0)(J/A)^{1/2}}{a}\right]$$

$$S_\pm = S^x \pm iS^y$$

$$= S \exp\{i\phi_0\}\text{sech}\left[\frac{(x - x_0)(J/A)^{1/2}}{a}\right]$$

where x_0 and ϕ_0 are arbitrary. This solution represents two domains, $S^z \to \pm 1$ as $x \to \pm \infty$, with a wall or transition region between them. Since S^z changes sign across this wall, such excitations will have large effects on the spin correlation functions, just as for the easy-plane antiferromagnet of Section II. Moreover, since the strict solitons are pulses like the SG breathers, the low-temperature statistical mechanics and correlation functions will depend more on the static domain walls than on the solitons. Nakamura and Sasada,[69] and later Bishop, Nakamura, and Sasada,[70] have followed a similar route to that in Section IV in studying the domain-wall contribution to the free energy. The problem here is more general because the *two* spin degrees of freedom are involved, rather than just one as in Section IV.[71,72]

We note that the isotropic ferromagnet in a field, (6.2), differs from that of (2.4) only in the anisotropy term, which produces the *xy* easy plane. Bishop[20] has argued that the vibrating pulse solitons associated with (6.2) deform continuously into the SG breather modes as the anisotropy coefficient A in (2.4) increases from zero to the easy-plane limit.

All the spin Hamiltonians discussed above are one-dimensional. Some analytic and numerical evidence can be found to suggest that nonlinear excitations also play a significant role in the statistical mechanics of the *two-dimensional*, classical, isotropic Heisenberg model.[73]

Analytic soliton or solitary-wave excitations have been found for the easy-plane ferromagnet in low and high fields along the axis by Ivanov, Kosevich, and Manzhos.[74] Also, Bar'yakhtar and Ivanov[75] have investigated a number of domain-wall structures and solitons in the uniaxial antiferromagnet.

Solitons have also been found in the magnetic and elastic dynamics of easy-plane antiferromagnets. The magnetoelastic coupling of the spins with the nuclear position coordinates can produce large anharmonic effects in the lattice dynamics.[76] Ozhogin, Lebedev, and co-workers have shown that this gives rise to soliton propagation[77] and overtone generation in the acoustic modes.[76,78]

VII. SOLITON DIFFUSION AND THE ϕ^4 MODEL

A. The Conservative ϕ^4 Model

In this section we show how the phonon modes can induce a stochastic character in the motion of solitons and solitary waves. While we could pursue this study using the SG solitons, it is more informative to consider a related system, the so-called ϕ^4 model.

This model has been used to describe systems that undergo structural phase transitions.[38,79] This structural change is associated with an instability in the lattice-displacement pattern as the crystal is cooled below its critical temperature T_c. Below this temperature, the lattice has more than one possible stable configuration; in the simplest systems there are just two degenerate structures. In ferroelectric compounds the transition is a displacement of a group of atoms in each unit cell, which gives a net dipole moment to the cell. Each of the possible (two) displacements yields a different direction for the polarization of the crystal. Depending on preparation, the crystal may be entirely in a state of uniform polarization or in a state with coexisting domains separated by walls. When the forces between neighboring unit cells are large, these walls are broad and the systems are called "displacive" ferrodistortive materials. The application of these compounds in circuitry components, including switches, as well as the general interest in this type of phase transition, has contributed to much recent work in the area (Ref. 80, and Refs. 6–11 of Ref. 81). The ϕ^4 model is intended to describe the behavior of uniaxial, essentially one-dimensional, ferrodistortive crystals such as $Pb_5Ge_3O_{11}$ (lead germanate).[38,80,81]

Let us begin by describing the model as expounded by Krumhansl and Schrieffer.[38] We consider a 1-D array of particles (or groups of particles) each of which is characterized by a displacement u. The model Hamiltonian is given by

$$H = \sum_l \tfrac{1}{2}m\left(\frac{du_l}{dt}\right)^2 + \tfrac{1}{2}C\left(u_l - u_{l-1}\right)^2 + \frac{A}{2}u_l^2 + \frac{B}{4}u_l^4 \qquad (7.1)$$

where each lattice site is denoted by l. Here m is the mass of each particle, C is the coefficient of a nearest-neighbor elastic coupling potential, and

$$V(u) = \frac{A}{2}u^2 + \frac{B}{4}u^4 \qquad (7.2)$$

is a potential at each site produced by other groups in the crystal, which are not considered explicitly in the chain. As shown in Fig. 15, if $B > 0$ and $A > 0$, then $V(u)$ has a simple minimum at $u = 0$, but if $A < 0$, then $V(u)$ has two

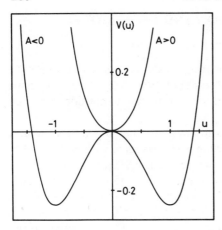

Fig. 15. The on-site potential $V(u) = (A/2)u^2 + (B/4)u^4$ for $B = 1$; $A = -1$ and $A = 1$.

degenerate minima at $u = \pm[|A|/B]^{1/2}$. This potential is intended to be the microscopic analogue of the free-energy-vs.-order-parameter function, which forms the basis of the Landau–Ginzberg theory of second-order phase transitions. Leaving the thermodynamics aside, we consider the dynamics of (7.1) in the interesting case of $A < 0$.

If the coupling constant C is large[38] then the displacement u_l will not change rapidly between neighboring sites and the continuum limit is appropriate. The Hamiltonian becomes

$$H = \int \frac{dx}{a} \left\{ \frac{1}{2m} p^2(x) + \frac{mc^2}{2} \left(\frac{du(x)}{dx} \right)^2 - \frac{|A|}{2} u^2(x) + \frac{B}{4} u^4(x) \right\} \quad (7.3)$$

where $p(x) = m\, du(x)/dt$ is the momentum at each site, $c = Ca^2/m$ is the velocity of low-amplitude sound waves, and a is the lattice spacing along the chain direction x. Newton's equation of motion for the displacement $u(x, t)$ is easily found to be

$$m\frac{\partial^2 u}{\partial t^2}(x, t) = mc^2 \frac{\partial^2 u}{\partial x^2}(x, t) + |A|u - Bu^3 \quad (7.4)$$

The degenerate equilibria are

$$\pm u_0 = \pm \left[\frac{|A|}{B} \right]^{1/2} \quad (7.5)$$

As with the SG chain, we look for traveling-wave solutions. Again there are extended anharmonic phonon modes expressed in terms of elliptic sine functions.[38] In the low-amplitude limit when $u(x, t)$ is always close to one equilibrium, we put

$$u(x, t) = \pm u_0 + \delta u(x, t) \tag{7.6}$$

Then, to first order in δu, (7.4) becomes

$$m\frac{\partial^2 \delta u}{\partial t^2} = mc^2\frac{\partial^2 \delta u}{\partial x^2} - 2|A|\,\delta u \tag{7.7}$$

which has the fundamental solutions

$$\delta u(x, t) = e^{i(\omega t - qx)} \tag{7.8}$$

with

$$\begin{aligned}
\omega^2 &= \frac{2|A|}{m} + c^2 q^2 \\
&= \omega_0^2 + c^2 q^2
\end{aligned} \tag{7.9}$$

In addition, (7.4) has a large-amplitude solitary wave. This can be derived in exactly the same way as the SG soliton in Section III. We find the traveling wave

$$u_s(x, t) = u_0 \tanh\left[\frac{\omega_0 \gamma}{c}(x - vt)\right] \tag{7.10}$$

where $\gamma = (1 - v^2/c^2)^{-1/2}$ (see Fig. 16).

This is very like the SG soliton in that it describes a transition from one minimum of $V(u)$ ($-u_0$ at $x = -\infty$) to the other ($+u_0$ at $x = +\infty$). How-

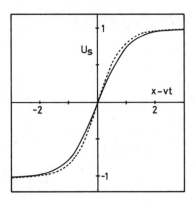

Fig. 16. The solitary-wave or kink solution of the ϕ^4 model, $u_s = u_0\tanh[\omega_0(1 - v^2/c^2)^{-1/2} c^{-1}(x - vt)$ [see Eq. (7.10)], as a function of $x - vt$ for $u_0 = 1$, $\omega_0/c = 1$, and $v/c = 0$ (full line), $v/c = 0.5$ (dashed line). As the velocity increases towards c, the kinks approach a step function.

ever, since there are only two minima in this case, the kinks (u_s) and anti-kinks ($-u_s$) are not strict solitons. There are no additional equivalent minima to allow the waves to pass through one another as in Fig. 5. Nevertheless the solitary waves are stable with respect to small perturbations,[80] and to *first order* in the phonon amplitude, a phonon–kink collision is elastic with a shift in relative positions, just as in Appendix B.

At low kink densities, one can neglect interactions between kinks, and the thermodynamics of Section IV carries through as before.[30] In fact, Krumhansl and Schrieffer initiated the approach presented in Section IV with their study of this ϕ^4 model [named for the ϕ^4 dependence of $V(\phi)$ in (7.2)]. It is worth noting that the displacement variable u undergoes a large change across the kink. The kink represents two domains in which $u = u_0$ and $u = -u_0$, and the domain wall between them. The large change in $u(x, t)$ implies that these kinks will have an effect on certain correlation functions comparable to that seen for the SG antiferromagnet.

Since Krumhansl and Schrieffer, papers have appeared that extend their work on the ϕ^4 model in a number of directions. These include a study of quantum corrections by Bishop, Domany, and Krumhansl, [82] a study of long range interaction effects by Sarker and Krumhansl,[83] an exact solution for solitons in the discrete system by Schmidt,[84] and an investigation of impurity effects on kink propagation by Sasada.[85] We do not discuss these subjects here, but concentrate on the question of Brownian motion of the kinks.

A mechanism for the Brownian motion of a domain-wall kink has been proposed by Wada and Schrieffer.[86] This mechanism relies on the nonlinear interaction of the wall with the *finite*-amplitude phonons. For a chain with one stationary kink, we write

$$u(x, t) = u_s(x) + \delta u(x, t) \tag{7.11}$$

Then, using (7.10), the equation of motion (7.4) becomes

$$m\frac{\partial^2 \delta u}{\partial t^2} = mc^2 \frac{\partial^2 \delta u}{\partial x^2} + |A|\, \delta u - B\,\delta u^3 - 3Bu_s(x)\,\delta u^2 - 3Bu_s^2\,\delta u$$

$$= mc^2 \frac{\partial^2 \delta u}{\partial x^2} - 2|A|\left\{1 - \tfrac{3}{2}\operatorname{sech}^2\left[\frac{\omega_0}{c}x\right]\right\}\delta u$$

$$- 3[|A||B|]^{1/2}\tanh\left[\frac{\omega_0}{c}x\right]\delta u^2 - B\,\delta u^3 \tag{7.12}$$

If the nonlinear terms involving δu^2 and δu^3 are neglected, then the procedure of Appendix B gives the fundamental solutions of (7.12) as

$$\delta u = e^{-i\omega t}f(x) \tag{7.13}$$

with

$$\omega_1 = 0, \qquad f_1(x) = \text{sech}^2\left(\frac{\omega_0}{c} x\right) \tag{7.14a}$$

$$\omega_2^2 = \tfrac{3}{4}\omega_0^2, \qquad f_2(x) = \sinh\left(\frac{\omega_0}{c} x\right)\text{sech}^2\left(\frac{\omega_0}{c} x\right) \tag{7.14b}$$

$$\omega^2(q) = \omega_0^2 + c^2 q^2,$$

$$f_q(x) = e^{iqx}\left[3\tanh^2\left(\frac{\omega_0}{c} x\right) - \frac{6c}{\omega_0} iq\tanh\left(\frac{\omega_0}{c} x\right) - 1 - 2\left(\frac{cq}{\omega_0}\right)^2\right] \tag{7.14c}$$

The major difference between this result and that for the perturbations of an SG soliton is the appearance of a local kink oscillation (ω_2, f_2) in addition to the translational mode (ω_1, f_1) and the perturbed extended phonon modes $(\omega(q), f_q)$. Wada and Schrieffer[86] study the collision between a phonon wave packet and a kink by solving (7.12) to *second* order in the phonon amplitude with initial conditions corresponding to a separated phonon wave packet and kink. The set of functions $\{f_1(x), f_2(x), f_q(x)\}$ forms a complete orthonormal set spanning the space of functions of x and is the natural basis to use in solving the nonlinear problem. Nevertheless, a long calculation is required[86] to show that the phonon wave packet produces a finite shift in the kink position; this shift is proportional to the squared amplitude of the packet and is in the direction of the incoming packet, as the interaction is attractive. In the limit that the wave packet is characterized by a single wave number q and frequency $\omega(q)$, they show[86] that the wave transmitted through the kink consists of the original phonon plus a higher harmonic. The higher harmonic is also reflected by the kink. It may be possible to detect this generation of higher harmonics by the domain walls in molecular-dynamics simulations or in experiments, though the intensity is weak in comparison with the original wave packet.

More importantly, the finite displacement of the wall will lead to a diffusive motion. The collisions between the kink and phonon wave packets incident from the left or right will occur in a random fashion, leading to a random walk of the kink along the chain. Making use of the fluctuation–dissipation theorem to quantify the phonon population, the temperature dependence of the diffusion constant D can be obtained. D is defined by

$$D = \frac{\langle \delta^2(t) \rangle}{2t} \tag{7.15}$$

where $\delta(t)$ is the position of the wall and t is a time long in comparison with the mean free time between collisions. They find[86]

$$D = 0.516\omega_0 a^2 \left(\frac{k_B T}{mu_0^2 \omega_0^2}\right)^2 \qquad (7.16)$$

Since the mechanism for diffusion is second order in the phonon amplitude, it is not surprising that $D \propto T^2$. Moreover, the second-order nature of the process ensures that $D \propto u_0^4$ and is *independent*[86] of the wall width c/ω_0 [see (7.10) with $v = 0$].

There is, however, a *lower-order* mechanism for kink diffusion derived by Collins, Blumen, Currie, and Ross.[80] Since this theory has been applied to the interpretation of experiments by Currie, Blumen, Collins, and Ross,[81] we shall review this approach.

B.   A Langevin Approach and Applications

In writing down the Hamiltonian (7.1) we invoked an on-site potential $V(u)$ produced by other particles in the crystal, which are not included in the chain description. We have projected out of the crystal Hamiltonian all the degrees of freedom except those associated with the motion of the displacement u_l. It is reasonable to expect that the reduced system containing only this subset of the total number of degrees of freedom will not be conservative: energy will be exchanged between the chain degrees of freedom and the other crystal modes. To account for this energy sharing we employ a Langevin-equation approach.[80] The positions and momenta of the other crystal particles give rise to a random fluctuating force on each coordinate u_n as well as two systematic effects: the on-site potential $V(u)$ and a "viscous" damping on the velocities $\{du_n/d_t\}$. Thus, the equation of motion for each coordinate u_n becomes

$$m\frac{d^2 u_n}{dt^2} + m\lambda\frac{du_n}{dt} + Au_n + Bu_n^3 - C(u_{n+1} + u_{n-1} - 2u_n) = R(t) \quad (7.17)$$

where $R(t)$ is a random fluctuating force and λ is the damping coefficient. In the continuum limit, (7.4) is amended to

$$m\frac{\partial^2 u(x,t)}{\partial t^2} + m\lambda\frac{\partial u}{\partial t} - mc^2\frac{\partial^2 u}{\partial x^2} - |A|u + Bu^3 = R(x,t) \quad (7.18)$$

The properties of the Markovian force $R(x, t)$ are described by

$$\langle R(x, t) \rangle = 0 \tag{7.19}$$

$$\langle R(x, t) R(x', t') \rangle = R_0 \delta(x - x') \delta(t - t') \tag{7.20}$$

where $\delta(x - x')$ and $\delta(t - t')$ are Dirac delta functions and $\langle \ \rangle$ denotes an equilibrium average. Assuming that the fluctuation–dissipation theorem[87] holds, the parameters λ and R_0 are related by

$$R_0 = 2k_B Tm\lambda a \tag{7.21}$$

We have simply described the coupling of the chain with its environment in terms of an interaction with a heatbath in addition to the on-site potential. This interaction has significant effects on the chain motion, particularly the solitary-wave motion. Similar effects to those described below can be obtained for the motion of SG solitons if this coupling to a heatbath is included.[88]

Now, since the equilibrium average of $R(x, t)$ vanishes, (7.18) gives

$$\left[m\frac{\partial^2}{\partial t^2} + m\frac{\partial}{\partial t} - mc^2\frac{\partial^2}{\partial x^2} - |A| \right]\langle u \rangle + B\langle u^3 \rangle = 0 \tag{7.22}$$

The obvious lowest-order assumption is that $\langle u^3 \rangle = \langle u \rangle^3$, which should hold at low temperature. With this assumption (7.22) becomes very like (7.4) but has the additional damping term. This term is zero in the stationary case ($\partial u/\partial t = 0$), so that the stationary kink, $v = 0$ in (7.10), is also a solution of

$$\left[m\frac{\partial^2}{\partial t^2} - m\lambda\frac{\partial}{\partial t} - mc^2\frac{\partial^2}{\partial x^2} - |A| \right]\langle u \rangle + B\langle u \rangle^3 = 0 \tag{7.23}$$

Since $\langle u \rangle$ has a stationary-kink solution $u_s(x)$, it seems reasonable to investigate the fluctuations in $u(x, t)$ produced by $R(x, t)$ when $u(x, t) \simeq u_s(x)$. Setting

$$u(x, t) = \langle u \rangle(x, t) + \delta u(x, t)$$
$$= u_s(x) + \delta u(x, t) \tag{7.24}$$

we write the equation of motion (7.18) to *first order* in δu:

$$\left\{ m\frac{\partial^2}{\partial t^2} + m\lambda\frac{\partial}{\partial t} - mc^2\frac{\partial^2}{\partial x^2} + 2|A|\left[1 - \tfrac{3}{2}\operatorname{sech}^2\left(\frac{\omega_0 x}{c}\right) \right] \right\} \delta u = R(x, t)$$

$$\tag{7.25}$$

The fluctuations $\delta u(x,t)$ are expressed in terms of the natural basis functions $\{F_\alpha, \alpha = 1, 2, q\}$ of (7.13) and (7.14):

$$\delta u(x,t) = C_1(t) F_1(x,t) + C_2(t) F_2(x,t) + \int_{-\infty}^{\infty} dq\, C_q(t) F_q(x,t)$$

$$(7.26)$$

Substituting (7.26) in (7.25) and using the orthonormality of the $\{F_\alpha\}$,[80] we obtain a separate equation of motion for each of the coefficients $C_\alpha(t)$:

$$\frac{d^2 C_\alpha}{dt^2} + (\lambda + 2i\omega_\alpha)\frac{dC_\alpha}{dt} + i\omega_\alpha \lambda C_\alpha(t)$$

$$= \frac{1}{m}\int_{-\infty}^{\infty} R(x,t) F_\alpha^*(x,t)\, dx \equiv Q_\alpha(t) \qquad (7.27)$$

where F_α^* is the complex conjugate of F_α. The functions F_2 and F_q describe the local oscillation and extended phonon modes respectively; F_1 is associated with the translation of the wall. If we consider only the F_1-component of the fluctuations,

$$u(x,t) = u_s(x) + C_1(t) F_1(x,t)$$

$$= u_s(x) + C_1(t)\operatorname{sech}^2\left(\frac{\omega_0 x}{c}\right)$$

$$= u_s(x) + C_1(t)\Omega\frac{du_s(x)}{dx}$$

$$= u_s[x + \Omega C_1(t)] + O(C_1^2) \qquad (7.28)$$

where

$$\Omega = \left(\frac{4}{3}\frac{u_0^2 \omega_0}{c}\right)^{-1/2} \qquad (7.29)$$

If we define the position of the kink, $x_s(t)$, as the value of the coordinate corresponding to the kink center $u_s = 0$, then

$$x_s(t) = -\Omega C_1(t) \qquad (7.30)$$

We can see that the random fluctuating force $R(x,t)$ acting on the displacements u_l induces a fluctuating coefficient $C_1(t)$ in (7.27), which produces a random walk of the domain wall via (7.30). As the chain particle positions u_l

fluctuate up and down, the domain wall wobbles back and forth. The instantaneous velocity of the wall is

$$v(t) = \frac{dx_s(t)}{dt} = -\Omega \frac{dC_1(t)}{dt} \tag{7.31}$$

This velocity is described from (7.27) as obeying

$$\frac{dv(t)}{dt} + \lambda v(t) = -\Omega Q_1(t) \tag{7.32}$$

This gives

$$v(t) = -e^{-\lambda t} \int_{-\infty}^{t} e^{\lambda \tau} \Omega Q_1(\tau) \, d\tau \tag{7.33}$$

The Brownian motion of the wall is described by a diffusion coefficient D defined by the Einstein relation[87]

$$D = \int_{0}^{\infty} \langle v(t)v(0) \rangle \, dt \tag{7.34}$$

The velocity autocorrelation function $\langle v(t)v(0) \rangle$ can be obtained from (7.33) as

$$\langle v(t)v(0) \rangle = \Omega^2 e^{-\lambda t} \int_{-\infty}^{0} d\tau_1 \int_{-\infty}^{t} d\tau_2 \, e^{\lambda(\tau_1 + \tau_2)} \langle Q_1(\tau_2)Q_1(\tau_1) \rangle \tag{7.35}$$

where

$$\Omega^2 \langle Q_1(\tau_2)Q_1(\tau_1) \rangle = \frac{2k_B T \lambda}{m^*} \delta(\tau_2 - \tau_1) \tag{7.36}$$

$$m^* = m \left[\frac{4u_0^2 \omega_0}{3ca} \right] \tag{7.37}$$

follows from the properties of $R(x, t)$. Thus,

$$\langle v(t)v(0) \rangle = \frac{k_B T}{m^*} e^{-\lambda t} \tag{7.38}$$

and

$$D = \frac{k_B T}{m^* \lambda} \tag{7.39}$$

The kink-position autocorrelation function can easily be obtained from

$\langle v(t)v(0)\rangle$ as[80]

$$\langle [x_s(t) - x_s(0)]^2 \rangle = 2D\left[t + \frac{e^{-\lambda t} - 1}{\lambda}\right] \qquad (7.40)$$

It is clear from (7.40) that our definition of D is equivalent to that used by Wada and Schrieffer, (7.15).

The Brownian motion of the kink or domain wall is a little unusual in that the random fluctuating force does not act directly on the position co-ordinate $x_s(t)$ of the "Brownian particle" but rather on the particle displacements u_l. The damping of the coordinates u_l is reflected in the same damping on the wall translation. The effective wall mass is proportional to the square displacement u_0^2 and to the wall width ω_0/c. This arises because small fluctuations in $u(x, t)$ near the wall center lead to large fluctuations in the wall position if the wall is broad $[\delta x \simeq \delta u(du/dx)^{-1}]$. This first-order mechanism for solitary-wave diffusion leads to a diffusion coefficient that is linear in T and linearly dependent on the wave width and squared amplitude.

We have considered the diffusive motion of a kink whose average velocity $\langle v(t)\rangle$ is zero. In fact, the stationary kink (or antikink) is the *only* kink solution. Moving kinks are only possible if the kink is driven by an external field. For the ferroelectric materials an applied uniform electric field E couples to an effective charge e^* at each site to give

$$\left(m\frac{\partial^2}{\partial t^2} - m\lambda\frac{\partial}{\partial t} - mc^2\frac{\partial^2}{\partial x^2} - |A|\right)\langle u\rangle + B\langle u\rangle^3 = e^*E \qquad (7.41)$$

The displacement at each site gives rise to a dipole $-u_l e^*$ that has a component parallel or antiparallel to the field if, say, $u_l = u_0$ or $u_l = -u_0$, respectively. In the presence of the field, parallel domains grow in spatial extent by consuming a neighboring antiparallel domain: the domain wall between propagates into the antiparallel domain as a traveling wave. The traveling-wave solutions are given by $\langle u\rangle(x - vt)$. Defining $s = \sqrt{2}(\omega_0\gamma/c)(x - vt)$ and $\eta = \langle u\rangle(s)/u_0$, (7.41) becomes

$$\frac{d^2\eta}{ds^2} + v'\frac{d\eta}{ds} + \eta - \eta^3 - E' = 0 \qquad (7.42)$$

where $v' = (\lambda v/\sqrt{2})(c/\omega_0\gamma)$ and $E' = e^*E/u_0|A|$. Equation (7.42) is well known in the fields of population genetics[89] and nonequilibrium chemical systems.[90] There is a *unique* solitonlike solution

$$\eta(s) = a + (b - a)[1 + \exp(\pm\beta s)]^{-1} \qquad (7.43)$$

where $\beta = (b-a)/\sqrt{2}$, and $a < c < b$ are the roots of

$$F(\eta) = \eta - \eta^3 + E' = -(\eta - a)(\eta - b)(\eta - c) \qquad (7.44)$$

The wave has velocity v such that[90]

$$v' = \pm 3\sqrt{2}\,c \qquad (7.45)$$

Note that there is a unique wave with a unique velocity for each value for the field. At zero field only the static domain wall is a solution. The addition of the damping term eliminates all the moving kinks in the absence of a field. When a field is applied, stationary kinks accelerate until they reach a terminal velocity uniquely determined by the field. Generally the wall velocity will be a nonlinear function of the applied field. However, at low field strength, it is easy to show that $c \simeq -E'$ in (7.45), so that

$$v = \mu e^* E \qquad (7.46)$$

where

$$\mu \equiv \frac{3c}{\lambda |A|} \left\{ \frac{B}{2m} \right\}^{1/2} \qquad (7.47)$$

is defined as the wall *mobility*. The idea of driving a domain wall by a field is not a fanciful notion: it has been observed in ferroelectric materials, and the mobility has been measured.[91] The motion of these domain walls is principally responsible for determining the use of ferroelectric materials as switches.

This model of driven and diffusing domain walls has been applied to the uniaxial displacive ferroelectrics $Pb_5Ge_3O_{11}$ (lead germanate) and SbSI (antimony sulphoiodide) by Currie, Blumen, Collins, and Ross.[81] These systems are well-known examples of the "soft mode" phenomenon. As the temperature of these crystals is raised from well below their critical temperature, the barrier in the ϕ^4 potential decreases in magnitude until there is a single minimum and no lattice instability. The temperature dependence of this barrier height is accounted for by making A in (7.2) temperature dependent:

$$\frac{A(T)}{A(0)} = \frac{T - T_A}{T_A}, \qquad T < T_c < T_A. \qquad (7.48)$$

From (7.9), we see that the phonon frequency ω_0 is then temperature dependent, decreasing with increasing temperature as this vibration goes soft. This effect has been observed experimentally.[92,93] The form of $A(T)$ is reflected in the T-dependence of a central peak in the neutron scattering from $Pb_5Ge_3O_{11}$.[94] This experiment provides a measure of the dynamical structure factor, which is the time and space Fourier transform of $\langle u(x,t)u(0,0)\rangle$.

The procedures of Section V cannot be applied to the calculation of $S(q,\omega)$, since the ferroelectric crystals are patently far from equilibrium.[81]

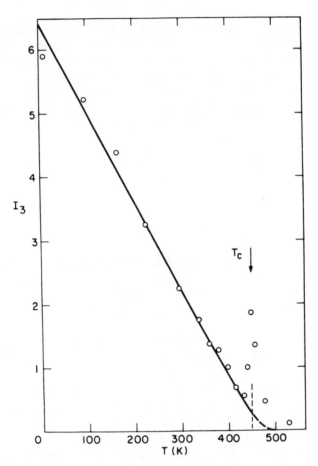

Fig. 17.   Reproduced from Fig. 3 of the article by Currie, Blumen, Collins, and Ross, *Phys. Rev. B* **19** 3645 (1979). The integrated intensity (in arbitrary units) of the central peak in the neutron scattering from $Pb_5Ge_3O_{11}$ at fixed wave number ($k = 2.6 \times 10^6$ cm^{-1}) vs. the crystal temperature T. The open circles are the experimental values represented in Fig. 2 of Ref. 94. The theoretical curve (solid line) was calculated from the soliton model (see Eq. (4.9) of Currie et al. [81]) and normalized to the arbitrary experimental intensity scale at $T = 350$ K. The anomalous central peak, not considered in this model, is evident in the experimental results near T_c.

However, using very simple assumptions about the low domain-wall density, Currie et al.[81] obtained quite good agreement between the central peak predicted by the solitonlike wall dynamics and the experimental data. Figure 17 presents the predicted and observed integrated intensity of the central (quasielastic) peak vs. temperature. The agreement is remarkably good considering the simplicity of the model. In essence, the central peak is due to the scattering from the Brownian domain walls. This quasielastic feature is a Lorentzian peak about $\omega = 0$, similar to the usual Rayleigh peak seen in scattering from fluids. The line shape has a width proportional to the wall diffusion coefficient of (7.39). A more detailed comparison between the "soliton" theory and experiments can be found in Ref. 81.

The model of diffusing solitons[80] has also been applied by Skinner and Wolynes[95, 96] to the dielectric behavior of amorphous polymers, particularly polyethylene. Mansfield and Boyd[97] proposed a mechanism for the dielectric α-relaxation of polyethylene that envisages the destruction of orientational correlation of the CH_2 groups by moving twists in the polymer chain. The twists are interpreted as SG solitons by Mansfield[98] and by Skinner and Wolynes.[96] The latter authors find that the correlation decay produced by Brownian solitons is in general agreement with the empirical forms adopted for amorphous polymers.

As noted at the beginning of this section, the concept of soliton diffusion could easily have been discussed in the context of SG chains. A careful study of this subject has been pursued by Büttiker and Landauer and co-workers, who consider a damped, sometimes driven, SG chain subject to thermal fluctuations.[99-102] They have considered the short- and long-time behavior of the equilibrium mean squared displacement of particles in the SG chain as it is determined by soliton diffusion.[101] They find that the mean squared displacement at long times is critically dependent on the manner in which solitons are created or nucleated in the chain.[101, 103] The Langevin approach is presumably more appropriate to the evaluation of dynamical correlation functions for realistic systems than that in Section V, though Fesser[104] found very little change in the central peak intensity for $CsNiF_3$ using a type of Langevin approach.

And now, for something completely different, the next section considers a soliton that does not involve a multiminimum potential-energy function.

VIII. MOLECULAR CRYSTALS, RESONANT ENERGY TRANSFER, AND THE CUBIC SCHRÖDINGER EQUATION

The UV–visible spectroscopy and photochemistry of molecular crystals has received a great deal of attention in the chemical-physics literature. Ideally, a molecular crystal, as the name suggests, consists of a lattice of clearly identifiable molecules that are only weakly interacting. From a theoretical viewpoint, the interest in molecular crystals arises in part from this notion of weak interactions: the electronic and vibrational properties of the individual molecules ought to be very similar to those of an isolated molecule, and the crystal properties might easily be obtained in a low-order perturbation theory. Naturally there are many practical reasons why the electronically excited states of molecular crystals have warranted so much study, not the least being the very selective nature of the photochemistry.[105] Here we are concerned simply with a demonstration of how solitons arise in the study of this subject and with using this case as an example of a system of identical particles with resonance interactions.

Let us consider the simplest possible molecular lattice, a 1-D lattice of neutral atoms—say, one of the heavier rare-gas atoms. The interatomic potential energy in the ground electronic state is that of the Van der Waals dimer,[106] with a large equilibrium separation and small binding energy (about 1 kJ/mole for Ar_2, for example). The common practice in dealing with molecular crystals is to express the total potential energy as a sum over pairwise atom–atom interactions. Hence the ground-state Hamiltonian is written as

$$H_0 = \sum_n \tfrac{1}{2} m \dot{u}_n^2 + V(r_n) \tag{8.1}$$

where each lattice site is labeled by an integer n, m is the atomic mass, u_n is the displacement of the nth atom along the chain axis from some fixed origin, $\dot{u}_n$ is the associated velocity, and

$$r_n \equiv u_n - u_{n-1} \tag{8.2}$$

is the nth bond length. Only nearest-neighbor interactions have been included in (8.1), for the pair potential $V(r_n)$. Since all the nearest-neighbor interactions are functions of the bond lengths $\{r_n\}$, it is convenient to choose these as the coordinates and write[107]

$$H_0 = \sum_n \frac{1}{2m} (p_n - p_{n+1})^2 + V(r_n) \tag{8.3}$$

where p_n is the momentum conjugate to r_n. At equilibrium the lattice has equidistantly spaced atoms separated by the equilibrium bond length, r_{eq}, of $V(r)$.

The ground-state electronic wavefunction is taken to be a simple product of the atomic (molecular) ground-state functions, as if the atoms were noninteracting. The lattice ground state is denoted by $|0\rangle$, and so

$$|0\rangle = \prod_n |0_n\rangle \tag{8.4}$$

where $|0_n\rangle$ is the ground electronic state of the nth atom. Clearly, with this choice of $|0\rangle$, the total potential energy appearing in (8.1) should be, in first-order perturbation theory, the electrostatic interaction of the atomic charge distributions.[108] Nevertheless, the pairwise form of (8.1) is used, and $V(r)$ is taken to be a typical atom–atom potential. So far all this is standard practice within the field.

The description of the electronically excited states of the lattice also proceeds within a standard "tight-binding" theory as presented by Davydov.[108] The ground-state wavefunction and Hamiltonian are taken as the zeroth-order terms in a perturbation expansion. The first-order wavefunction describing the excited electronic state is

$$|\psi\rangle = \sum_n a_n B_n^\dagger |0\rangle \qquad (8.5)$$

where $B_n^\dagger$ is an operator which destroys the ground state $|0_n\rangle$ on the nth site and creates an excited state $|1_n\rangle$ there. For convenience we assume that each atom has only one relevant excited state. Equation (8.5) represents a system of identical particles only one of which is electronically excited. As the particles are identical, it is not known which is excited, so a sum is taken over all the atoms, with a weighting coefficient a_n for each. The probability that the nth atom is excited is given by $|a_n|^2$, and the fact that only one excitation is present implies that

$$\sum_n |a_n|^2 = 1 \qquad (8.6)$$

When an atom in the lattice is excited, it interacts with its neighbors via a new potential in place of $V(r)$. Moreover, since all the atoms are identical, the "internal" electronic excitation of an atom can be *resonantly* transferred to a neighboring atom. We write the Hamiltonian for the excited lattice as

$$H = H_0 + \sum_n \{ [\Delta\varepsilon + D(r_n) + D(r_{n+1})] B_n^\dagger B_n$$
$$+ M(r_n) B_{n-1}^\dagger B_n + M(r_{n+1}) B_{n+1}^\dagger B_n \} \qquad (8.7)$$

where B_n is the operator that destroys an excited state on atom n and creates a ground state; with $B_n^\dagger$ it obeys the commutation relations[108]

$$[B_n, B_m^\dagger] = \delta_{mn}, \qquad [B_n, B_m] = 0 \qquad (8.8)$$

The new terms in (8.7) are easily understood. If the nth atom is excited, the lattice energy rises by the energy difference, $\Delta\varepsilon$, between the excited and ground states of the isolated atom. In addition the interaction with its near neighbors changes: $D(r)$ is the change in the atom–atom interaction potential upon excitation of one atom. Finally, and most importantly, the resonance sharing of the excitation between the nth atom and its neighbors (deexcitation at n followed by excitation at $n+1$ or $n-1$) changes the energy. The resonance energy $M(r)$ is a function of the interatomic distance.

The Hamiltonian of (8.7), interpreted more generally, describes an array of weakly interacting particles with an internal degree of freedom whose state affects the interparticle coupling and hence is in turn affected by that coupling.

Davydov and Kisluhka[109,110] were the first to consider soliton phenomena in this system. However, that treatment generates the nonlinear behavior from the interaction term $D(r)$ while treating the resonant coupling as a constant. Here we follow the approach of Collins and Craig,[111] which deals more accurately with the resonance effect. In fact it is not essential to consider changes in the interaction potential associated with $D(r)$ in (8.7) to observe soliton phenomena.

The evolution of the electronic state of the lattice is obtained from the time-dependent Schrödinger equation,

$$ i\hbar \frac{\partial}{\partial t}|\psi\rangle = H|\psi\rangle \tag{8.9} $$

with $|\psi\rangle$ of (8.5) and H of (8.7). Assuming the orthogonality of the atomic states, the Schrödinger equation reduces to an equation for each of the amplitudes a_n:[110,111]

$$ i\hbar \frac{da_n}{dt} = \left[H_0 + \Delta\varepsilon + D(r_n) + D(r_{n+1}) \right] a_n $$
$$ + M(r_n) a_{n-1} + M(r_{n+1}) a_{n+1} \tag{8.10} $$

The atomic coordinates evolve on an energy surface that is given by

$$ E = \langle \psi | H | \psi \rangle $$
$$ = H_0 + \sum_n \left\{ \left[\Delta\varepsilon + D(r_n) + D(r_{n+1}) \right] |a_n|^2 + M(r_n) \left[a_n^* a_{n-1} + a_n a_{n-1}^* \right] \right\} \tag{8.11} $$

where a_n^* is the complex conjugate of a_n. To follow the dynamics of the atomic coordinates we employ classical mechanics. So long as the atoms (or molecules) are sufficiently massive, this should be a reasonably accurate approximation, though we must keep in mind when discussing stationary and solitonlike states that zero-point motion has been neglected.[111] From (8.11)

Hamilton's equations of motion are

$$\frac{dr_n}{dt} = \frac{1}{m}[2p_n - p_{n-1} - p_{n+1}] \tag{8.12a}$$

$$\frac{dp_n}{dt} = -V'(r_n) - D'(r_n)[|a_n|^2 + |a_{n-1}|^2] - M'(r_n)[a_n^* a_{n-1} + a_n a_{n-1}^*] \tag{8.12b}$$

where $V'(r) = dV(r)/dr$ and so on.

The dynamics of this model system can be examined numerically in a straightforward manner for an array of finite length, using reasonable forms for the potential functions.[111] Let us concentrate here on simplifying the problem in order to gain an understanding of the fundamentals from an analytic solution. We look for a solitonlike traveling-wave solution for which

$$r_n(t) = r(\phi), \qquad p_n(t) = p(\phi), \qquad a_n(t) = e^{-i\omega t} a(\phi) \tag{8.13}$$

where

$$\phi = n - \nu t \tag{8.14}$$

and $r(\phi) - r_{eq}$, $p(\phi)$, and $a(\phi)$ vanish outside a small range of ϕ. For the moment, we discard the terms involving $D(r)$ in (8.10) and (8.12), since they are not essential to the problem. The solutions of the form (8.13) must now obey

$$\hbar \omega a(\phi) - i\hbar \nu \frac{da}{d\phi} = (H_0 + \Delta\varepsilon) a(\phi) + M[r(\phi)] a(\phi - 1)$$
$$+ M[r(\phi + 1)] a(\phi + 1) \tag{8.15}$$

and using the Newtonian form,

$$m\nu^2 \frac{d^2 r}{d\phi^2} = \Delta^2 \{V'[r(\phi)] + M'[r(\phi)][a^*(\phi) a(\phi - 1) + a(\phi) a^*(\phi - 1)]\} \tag{8.16}$$

where $\Delta^2 f(\phi) \equiv f(\phi + 1) + f(\phi - 1) - 2f(\phi)$.

In molecular crystals the weak resonance interaction is taken to be a long-range electric-dipole–electric-dipole interaction between excited states of neighboring molecules. Thus, we have the form

$$M(r) = -M\left(\frac{r}{r_{eq}}\right)^3 \tag{8.17}$$

where $M > 0$ for dipoles parallel to the axis. The orientation of the dipoles leading to degrees of bonding or antibonding resonance interactions is accounted for by the phase factor in $a(\phi)$.[111]

Let us consider the crudest treatment of (8.15) and (8.16) that preserves the essence of the problem. Set

$$M(r) = -M + \frac{3M}{r_{eq}}(r - r_{eq}) \tag{8.18a}$$

$$V(r) = V(r_{eq}) + \tfrac{1}{2}V''(r_{eq})(r - r_{eq})^2 \tag{8.18b}$$

and let

$$\Delta^2 f(\phi) = \frac{d^2 f}{d\phi^2} \tag{8.19}$$

—the last step being the almost inevitable continuum approximation. Then (8.16) can be integrated to give

$$mv^2(r - r_{eq}) = V''(r_{eq})(r - r_{eq})$$
$$+ \frac{3M}{r_{eq}}[a^*(\phi)a(\phi - 1) + a(\phi)a^*(\phi - 1)] \tag{8.20}$$

Hence—a crucial step—we can express $r(\phi)$ in terms of $a(\phi)$:

$$r(\phi) = r_{eq} - \frac{3M[a^*(\phi)a(\phi - 1) + a(\phi)a^*(\phi - 1)]}{r_{eq}[V''(r_{eq}) - mv^2]} \tag{8.21}$$

When (8.21) is substituted into the r-dependent terms of (8.15), we obtain a nonlinear equation for $a(\phi)$. In the continuum limit, this gives

$$\hbar\omega a(\phi) - i\hbar v \frac{da}{d\phi} = (H_0 + \Delta\varepsilon - 2M)a(\phi) - M\frac{d^2 a(\phi)}{d\phi^2}$$
$$- \left[\frac{3M}{r_{eq}}\right]^2 [V''(r_{eq}) - mv^2]^{-1}$$
$$\times \left\{ 4|a(\phi)|^2 a(\phi) + a^*(\phi)\frac{d^2 a^2(\phi)}{d\phi^2} + a(\phi)\frac{d^2 |a(\phi)|^2}{d\phi^2} \right\} \tag{8.22}$$

To simplify matters further, since $a(\phi)$ is assumed to be both small and slowly varying, the terms in (8.22) involving second derivatives of powers of $a(\phi)$ and $|a(\phi)|$ are neglected. Finally,

$$i\hbar v \frac{da}{d\phi} = M\frac{d^2a}{d\phi^2} + (\hbar\omega - H_0 - \Delta\varepsilon + 2M)a(\phi)$$

$$+ \frac{36M^2}{m(c^2 - v^2)}|a(\phi)|^2 a(\phi) \qquad (8.23)$$

where $c = [r_{eq}^2 V''(r_{eq})/m]^{1/2}$ is the velocity of sound in the lattice and $v = vr_{eq}$ is the velocity of the traveling wave.

Equation (8.23) is a form of the nonlinear, or cubic, Schrödinger equation, which has well-known soliton solutions.[4] Appendix D gives brief details of the solution of (8.23) and the calculation of the soliton energy. More detail is provided by Ref. 111. The solution is

$$a_n(t) = \exp\left\{-\frac{iE(v)t}{h}\right\}\exp\left[\frac{i\hbar v}{2M}(n - vt)\right]$$

$$\times \alpha \operatorname{sech}[2\alpha^2(n - vt - \phi_0)] \qquad (8.24)$$

where

$$\alpha = \frac{3}{\sqrt{2}}\left[\frac{M}{m(c^2 - v^2)}\right]^{1/2} \qquad (8.25)$$

ϕ_0 is an arbitrary origin, and the soliton energy $E(v)$ is given by

$$E(v) = \Delta\varepsilon - 2M + \frac{(\hbar v)^2}{4M}$$

$$-\frac{1}{M}\left[\frac{9M^2}{m(c^2 - v^2)}\right]^2\left\{1 - \frac{2}{3}\left[\frac{c^2 + v^2}{c^2 - v^2}\right]\cos^2\left(\frac{\hbar v}{2M}\right)\right\} \qquad (8.26)$$

Equation (8.24) shows that the soliton is a localized "hump" that corresponds to a localization of the internal (electronic in this case) excitation of the particles to within approximately $(2\alpha^2)^{-1}$ lattice sites. This distribution propagates with constant velocity along the lattice, accompanied by a local distortion of the bond lengths [via (8.21)].

Normally we tend to think of resonance interactions as a delocalizing influence, spreading excitation throughout a lattice. However, even the crudest linear coupling [Eq. (8.18a)] of the resonance interaction and lattice vibrations is sufficient to induce soliton phenomena whereby the resonance serves to localize the excitation. Localization of a_n contracts the bond lengths via the resonant force in (8.12). Smaller bond lengths enhance the resonance interaction via (8.17) or (8.18a) and amplify the localization of a_n via (8.10). Overall the energy of (8.11) is lowered. A few remarks about the properties of these resonance-induced solitons is in order. The soliton velocity is limited above by the velocity of sound c, which characterizes the time scale on which the lattice particles can alter their position. From (8.25), the width of the soliton is seen to decrease as the velocity increases towards c, while the amplitude increases. Clearly, there must be some velocity above which the soliton becomes so sharp that the continuum approximation is not valid. It is not clear whether a soliton survives in this regime or not. Numerical tests are necessary to establish this.

The energy of the soliton, $E(v)$, is an increasing function of its velocity for the low velocities and weak resonance for which (8.24) is valid. From (8.26) it is clear that the variation of E with v depends on both v^2/c^2 and the strength of the resonance energy M in comparison with the lattice energy mc^2. It is of course possible to associate a velocity-dependent mass[110] or momentum with the soliton and consider it as some form of "quasiparticle."

The development shown above is an intentionally simple one. It must be in order to finish with such a simple equation as the nonlinear Schrödinger equation (NSE). This cubic nonlinearity in an otherwise linear evolution equation is almost synonymous with a theory containing words like "weak coupling," "not too far from equilibrium," and so forth. The NSE is a completely integrable equation, and the soliton solution of (8.24) is a strict soliton rather than a solitary wave. However, the approximation of very weak coupling can be dispensed with,[111] so that another nonlinear evolution equation replaces the NSE, which then has solitary-wave, "near-soliton" solutions. The physical consequences of the resonance interaction is not substantially changed; only the mathematics is more difficult to handle. Numerical analysis is useful now that the broad features of the phenomena are understood.

In both Refs. 110 and 111 the potential-energy term $D(r)$ in (8.7) is retained and taken to be a change in the dispersion interaction between the atoms of the form

$$D(r) = -D\left(\frac{r}{r_{eq}}\right)^6 \tag{8.27}$$

The retention of this interaction has the effect, at the "weak coupling" level, of localizing the excitation to a greater degree: the parameter α in (8.24) and (8.25) becomes

$$\alpha = \frac{3}{\sqrt{2}} \left[\frac{4D^2 + 4DM + M^2}{Mm(c^2 - v^2)} \right]^{1/2}$$

This sharpening of the distribution, which also occurs for increasing velocity, is expected to cause a breakdown of the continuum approximation. In Ref. 111 this effect was studied by solving the equations of motion (8.10) and (8.12) numerically to find the stationary state of the lattice: the $v = 0$ soliton.

It is true to say that, when the resonance coupling is weak, the numerical calculations agree very well with the distribution of $\{a_n\}$ and the energy $E(v = 0)$ derived from the NSE. It is more interesting to consider what happens in the case of strong coupling when the NSE gives a poor representation of the dynamics.

Figures 18 and 19 show the results of two such calculations for the stationary states. Consider Fig. 18 first. There the resonance coupling $M(r)$ is sufficiently strong to make the NSE results quantitatively wrong. However, the continuum approximation is still valid. Two distributions are shown in Fig. 18: one is centered at a lattice site, one is centered between two sites.

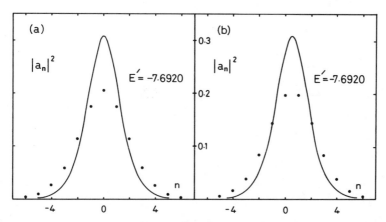

Fig. 18.   The electronic distribution, $|a_n|^2$, for the stationary soliton ($v = 0$) as a function of lattice position n, for the ground-state potential of Ref. 111 with $M/V(r_{eq}) = 3.6864$ and $D/V(r_{eq}) = 3.6 \times 10^{-5}$. The solid curves are the continuum approximation via the NSE [see (8.24) and (8.28)]; the points are exact numerical results as described in Ref. 111. The origin of the distribution is (a) at a lattice site; (b) midway between two lattice sites. The energies refer to the numerically determined distributions, $E' = [E(v = 0) - \Delta\varepsilon]/(V(r_{eq})$.

While the distributions are different, remarkably they have the same energy. In fact it is possible to center the distribution anywhere on the lattice without changing the energy. This electronically excited state is *translationally degenerate*. The arbitrary origin ϕ_0 in (8.24), which arises from the continuum approximation, is still a valid concept even though the NSE is no longer accurate. This degeneracy means that the distribution $\{a_n\}$ can move from site to site without an energy barrier. Free soliton propagation will occur. Strictly speaking, without the completely integrable NSE, the propagating wave is not a strict soliton but "solitonlike". Clearly, the notion of resonance interactions creating a localized propagating excitation is not restricted to the "weak coupling" regime of the NSE.

Figure 19 shows that the soliton picture does break down when the interaction governed by $D(r)$ becomes large and the state becomes highly localized. The translational degeneracy, seen in Fig. 18, has been broken. Free soliton propagation is no longer possible, and transport can proceed from site to site only by crossing an energy barrier. This breakdown in the soliton picture is of interest because of the energy trapping it implies and the consequences of this, in molecular crystals, for excimer formation and photochemical reaction. The value of the soliton concept in the understanding of excited molecular crystals is that by accounting for the nonrigidity of the

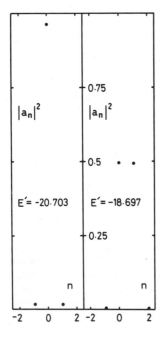

Fig. 19.   As in Fig. 18, with $M/V(r_{eq}) = 0.9216$ and $D/V(r_{eq}) = 3.6864$. The approximation derived from the NSE is in gross error, and is not shown here.

lattice, one has a theory capable of covering the whole range of energy transfer from coherent transport via soliton propagation through to incoherent transport in trapped (or at least hindered) soliton motion.

The simple NSE theory of solitons in a molecular lattice, described above, has been pursued and developed by Davydov and co-workers. The soliton solution of (8.24) depicts the localized excitation moving on an otherwise quiescent lattice. Davydov has considered the effect of soliton–acoustic-phonon collisions via an approximate quantum-statistical treatment.[112,113] As the lattice temperature rises, the perturbation of the soliton by random phonon collisions destroys, in part, the phase coherence of the soliton. As we noted above, the relative phase of a_n between sites determines whether the resonance interaction is bonding or antibonding. The loss of coherence disrupts the bonding, thereby broadening the soliton and raising its energy. A related treatment[114] considers a resonantly transferred electron in a molecular lattice and the effect that optical (rather than acoustic) phonons have on the soliton motion. Davydov and co-workers have also been concerned with the application of the simple model to energy transport in biological systems, particularly energy transport in α-helical protein molecules (see Refs. 115, 116 and references therein). In these systems the "internal" degree of freedom that is excited is an amide-group vibration. Again the resonance coupling of neighboring amide groups is via a dipole–dipole interaction.

So it would seem that the soliton phenomenon is inherent in processes involving resonance with some form of coupling to other degrees of freedom. This should not be restricted to the simple NSE, or even to cases of low excitation density. High density of excitation leads to processes involving simultaneous excitation of neighboring sites and modified soliton behavior.[117, 118] The next section considers a form of resonance well known to chemists.

IX.   RESONANCE AND SOLITONLIKE PHENOMENA IN POLYACETYLENE

Polyacetylene and related compounds have attracted considerable interest in recent years due to their semiconductor properties when doped with impurities.[119] The theoretical interpretation of these conduction properties has centered in the last few years on models involving localized electronic states that propagate almost freely along these long unsaturated chains in a solitonlike manner.

The current model stems from the rediscovery and development by Su, Schrieffer, and Heeger (SSH)[120, 121] of the Hückel-theory approach of Longuet-Higgins and Salem,[122, 123] Pople and Walmsley,[124] and Halm, McLaughlan, Dearman, and McConnell[125] of some twenty years ago. This early theory is well described in Salem's book.[126] Of course, by modern quantum-chemistry standards, this approach is a very crude one. However, even though we might put very little reliance on the quantitative results of this Hückel theory, it may be worthwhile from a qualitative viewpoint.

As Longuet-Higgins and Salem showed, in the simple Hückel model, the bonding in long linear or cyclic polyenes is alternating single and double bonds, rather than the symmetrical structure due to π-electron delocalization in smaller systems like benzene. In Fig. 20 the structure of the stable *trans* isomer of polyacetylene is shown schematically, for which there are two degenerate configurations of bonds. Polyacetylene has a lower energy in either of these two resonance structures than in a delocalized intermediate configuration. Depending on your point of view, this may be seen as a pseudo Jahn–Teller effect[126] or a Peierls instability.[127]

The resonant transfer of the double bonds is the resonance effect, which is important for soliton generation, together with a coupling to the C—C bond lengths due to the dependence of the resonance integral β on the bond length.[126]

SSH write the Hamiltonian for polyacetylene as that of a *one*-dimensional chain of CH groups whose positions are determined by their projection on the chain axis. This position coordinate U_n for the nth carbon atom is zero when all carbon atoms are equidistant (the equilibrium in a hypothetical delocalized π-electron configuration) and $U_n > 0$ or < 0 when the shorter double bond is to the right or left of this atom, respectively, as shown in Figure 20. SSH write the Hamiltonian as

$$H = \sum_n \tfrac{1}{2}M\dot{U}_n^2 + \tfrac{1}{2}K(U_{n+1} - U_n)^2$$

$$+ \sum_{n,s} t_{n+1,n}\left[C_{n+1,s}^\dagger C_{n,s} + C_{n,s}^\dagger C_{n+1,s}\right] \qquad (9.1)$$

where M is the CH mass, K is an elastic constant arising from the sp^2 hydrized σ-bonding, and $t_{n+1,n}$ (better known as β) is the resonance integral between the nth and $n+1$th carbon atoms.[121] Here $C_{n,s}^\dagger$ ($C_{n,s}$) is the operator that creates (destroys) a π-electron of spin s on the nth carbon atom. In this notation, the usual LCAO molecular-orbital states would be written as

$$|\psi\rangle = \sum_{n,s} c_{n,s} C_{n,s}^\dagger |0\rangle \qquad (9.2)$$

where $|0\rangle$ is the vacuum state and the $\{c_{n,s}\}$ are the usual coefficients determined by the secular equations.[126] We see that the resonance structure in (9.1) is of the same form as in Section VIII. A fundamental difference be-

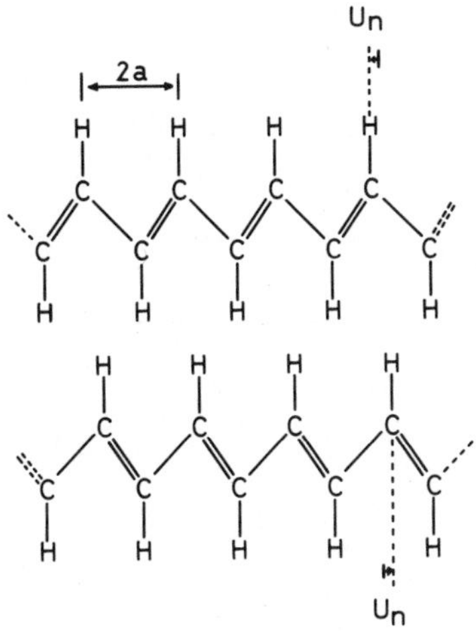

Fig. 20. A schematic representation of the two degenerate ground-state configurations in a segment of *trans*-polyacetylene. The position of the nth CH group is defined by the scalar coordinate U_n as shown. As double bonds are shorter than single bonds, the periodicity length is the axial displacement $2a$ between alternate CH groups.

tween the two systems is the presence of one π-electron for each carbon atom here, rather than one "excitation" for the whole chain as in Section VIII. Here also, the resonance integral is dependent on the bond length. In the notation of SSH,

$$t_{n+1,n} = t_0 - \alpha(U_{n+1} - U_n) + \cdots \qquad (9.3)$$

It is straightforward to show that the minimum energy is obtained when the bond lengths alternate, $U_n = (-1)^n U_0$ [or in the degenerate structure, $U_n = (-1)^{n+1} U_0$] in the limit of an infinite chain[121] or when the chain is long enough.[126] In this alternating structure a gap is created between the highest bonding molecular orbital and the lowest antibonding level. In the infinite chain these levels form two bands, the lower one being filled. This is shown in Fig. 21.

As we have seen before, a degenerate ground state of this nature is very likely associated with domains and solitonlike domain walls. It was realized very early on[124, 126] that a "mismatch" of the bonding alternation might occur in a long polyene, as shown in Fig. 22. There we represent, schematically, two widely separated defects in a long polyene. At one defect a transition occurs from (say) every even-numbered double bond to every odd double bond. In this simple picture the transition region (bonds 5 and 6 of Fig. 22) consists of long single bonds, and the middle carbon atom has an unpaired electron. Farther down the chain the reverse transition occurs and the original "phase" is restored. In a long polyene there may be many such defects, but in (say) a cyclic molecule with an even number of carbon atoms

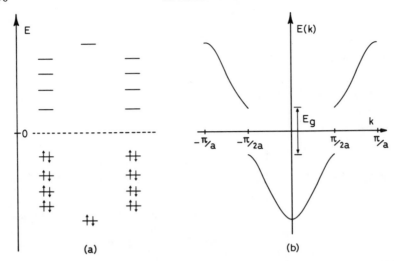

Fig. 21. (a) A schematic representation of the π-electron energy (E) levels in an even cyclic polyene, $C_{4n+2}H_{4n+2}$, for alternating bond lengths. A gap appears between the highest filled (bonding) orbital energy and the lowest unfilled (antibonding) orbital energy. (b) The energy $E(k)$ of the π-electron band states in the first Brillouin zone as a function of the wave number k. Only the lower [$E(k) < 0$] band is filled in a long chain with one π-electron per site. The band states and energy gap E_g are realized in the limit in which the polyene of (a) becomes infinitely long.

there must be an even number of these mismatches. If end effects are not invoked, then these defects will be created in pairs in a long polyene.

Pople and Walmsley investigated this type of defect under the simple assumption that there are only two single bonds surrounding the defect and only two different bond lengths and two different values of β allowed, corresponding to single and double bonds. They showed that this defect corresponds to a *nonbonding* π-orbital at the center of the energy gap in Figure 21. The molecular-orbital coefficients of (9.2) are localized about the defect site in a symmetric fashion. If the central atom is labeled $n = 0$, then in (9.2)

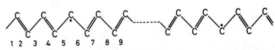

Fig. 22. A schematic representation of the carbon skeleton of *trans*-polyacetylene in which the bonds alternate for $n < 5$ and alternate for $n > 5$, but "mismatch" so that both bands 5 and 6 are single as shown. The dot represents an unpaired electron. To the right of the chain, the reverse mismatch occurs, so that even-numbered bonds are again double bonds, as they are for $n < 5$.

a nonbonding orbital has

$$c_{n,s} = \begin{cases} \left(\dfrac{-\beta_1}{\beta_2} \right)^{|n|} \left(\dfrac{\beta_2^2 - \beta_1^2}{\beta_2^2 + \beta_1^2} \right), & n \text{ even} \\ 0, & n \text{ odd} \end{cases} \tag{9.4}$$

where β_1(single bond) $< \beta_2$(double bond). This singly occupied nonbonding orbital gives rise to an unpaired electron localized near the bonding "mismatch." Pople and Walmsley estimate by difference the energy to create a pair of defects as about 5 kcal/mole,[124] which leads to roughly one defect per 70 carbon atoms at room temperature.

SSH improve this approach by allowing the variation in bond length around the defect to extend over many bonds. In the ground state, the CH group has a position coordinate of $(-1)^n U_0$ or $(-1)^{n+1} U_0$ for the two phases. An "order" parameter ψ_n is defined by

$$\psi_n = (-1)^n U_n \tag{9.5}$$

so that $\psi_n = U_0$ in the A-phase and $\psi_n = -U_0$ in the B-phase. In the presence of the defect, SSH chose a trial function for ψ_n:

$$\psi_n = \begin{cases} U_0, & n < -\nu \\ -U_0 \tanh(n/l), & -\nu < n < \nu \\ -U_0, & n > \nu \end{cases} \tag{9.6}$$

where ν is large enough to have no significant effect on the energy. The corresponding eigenvalue of (9.1) is evaluated numerically using a Green's-function technique, and the length l can be determined variationally.[121] For reasonable values of the resonance parameters t_0 and α of (9.3), the minimum energy for the defect corresponds to $l \approx 5$–7 rather than $l \approx 1$ as in Pople and Walmsley's treatment.

This "spreading out" of the defect or excitation lowers its energy (because the bond lengths are in equilibrium with respect to the electronic state) and, importantly, lowers the barrier for the movement of the defect along the chain as in Figure 23. The bond lengths change only slowly, varying from that for a single bond to that for a double bond across 10 to 14 CH units. SSH estimate the barrier to movement as of the order of 0.002 eV (≈ 5 cal/mol), which implies a very high mobility for the unpaired electron. These highly mobile spins are evident in magnetic-resonance studies of trans-polyacetylene[128, 129] and EPR studies of the cis-trans isomerization.[130]

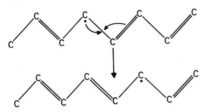

Fig. 23. The arrows represent a donation of π-electrons to form a bond whereby the unpaired electron, denoted by a dot, is resonantly transfered two places to the right in the carbon skeleton of *trans*-polyacetylene.

It is not surprising, in view of the very diffuse nature of the domain wall, that a continuum treatment of the Hamiltonian (9.1) reproduces the results of SSH,[131] except that no barrier to propagation can be obtained in this approach. In the continuum model the analytic result for the energy to create a soliton (defect) at rest is $2\Delta/\pi$,[131] where 2Δ is the band gap of Fig. 21.

The fact that the soliton state lies below the conduction (antibonding) band gives rise to interesting semiconductor properties. In the primitive model discussed above, the energy of the localized defect state does not depend on its occupancy. The neutral state has occupancy 1 and spin $\pm\frac{1}{2}$; a charged soliton has zero or two electrons, zero spin, and charge $+e$ or $-e$, respectively. Thus, doping polyacetylene with electron acceptors or donors could lead to the formation of spinless charged solitons. These might act as highly mobile charge carriers.

However, it would be difficult to place much reliance on the energy levels of these charged species, considering the limitations of a Hückel model. Moreover, recalling the last section, it is not surprising that the velocity of charged mobile solitons is limited by the velocity of sound in the chain. This is apparent from simulations of these charged species by Su and Schrieffer.[132] They employ the same classical approach to the motion of the CH groups as was used for the atoms in Section VIII.

In addition to the immediate effects that neutral and charged solitons appear to have on the magnetic and conduction properties of polyacetylene, new vibrational modes, localized at the defect, appear in the infrared absorption, which are quantitatively described in terms of charged solitons by Mele and Rice[133] and Etemad et al.[134] The solitons are also used to account for the optical absorption spectra of lightly doped *trans*-polyacetylene.[135,136] We note that in Ref. 135 Maki and Nakahara invoke the same type of correlation effect (disorder being related to soliton density) as that seen in antiferromagnets (see Section V).

It is interesting to note that the soliton–soliton interaction is attractive[137] (see also Pople and Walmsley[124]) except for the neutral-triplet case and like-charge cases. It may be that condensation occurs.

Finally, related soliton phenomena are known in charge-density-wave condensates, a "phase soliton,"[138] and SG solitons.[139-141] The phase soliton

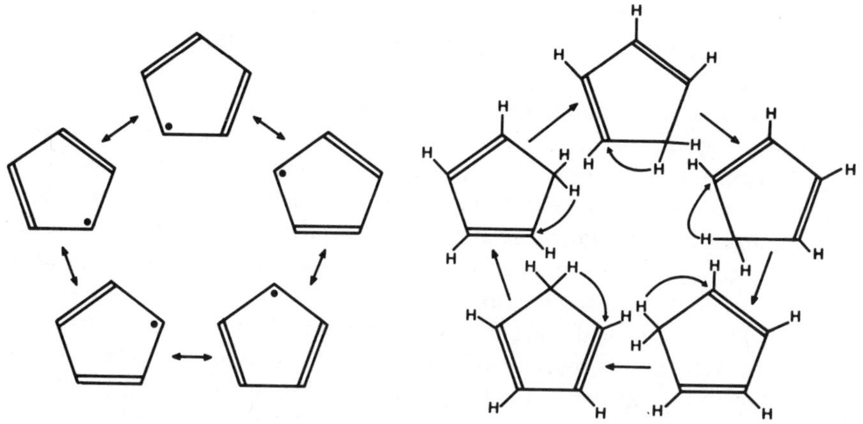

Fig. 24. The resonance structures of (a) cyclopentadienyl radical and (b) cyclopentadiene with proton transfers denoted by an arrow. Only the carbon skeleton is shown in (a).

has been used to explain semiconductor properties in TaS_3.[142] However, we shall not consider these related topics in any detail here.

We have seen how the resonance in long conjugated chains leads naturally to soliton, or rather solitonlike, phenomena. From a chemist's viewpoint it is perhaps more interesting to speculate on what effect the soliton concept could have on our understanding of smaller conjugated molecules. Cyclopentadiene (C_5H_6) and the associated radical (C_5H_5) provide well-known examples[126] of behavior similar to that discussed above. The resonance structures are illustrated in Figure 24. While a resonance or simple molecular-orbital argument suggests that the bonds in cyclopentadienyl radical are all equivalent, there is a Jahn–Teller distortion that ensures that all five structures in Figure 9.5(a) are local energy minima. The barrier to interconversion of these structures is quite small, however: only about 1 cal/mol.[126] In cyclopentadiene, the resonance is accompanied by successive 1,2 hydrogen transfers, so that a much larger barrier is expected. Similar behavior would be expected in larger molecules of this type. It may or may not be the case that the soliton concept is useful in understanding these familiar effects.

X. LATTICE VIBRATIONS

This section deals primarily with the longitudinal vibrations of a 1-D lattice. That doesn't sound very exciting, but in many ways by dealing with this subject at the end of the review we have kept the best wine till last. To my mind, this topic is the most appealing. The system studied is easily the simplest, least cluttered one, but also the subtlest, conceptually most surprising topic of all.

The direct motivation for considering lattice vibrations has been the obvious one of understanding the thermal properties of crystals. Any introductory text on the solid state[143] shows how, by making the approximation of harmonic interatomic forces, the heat capacity can be evaluated, for example. Again, within the harmonic approximation, the lattice vibrations can be used to understand the dynamical response or power spectrum $S(q, w)$, which can be measured by neutron scattering.

The 1-D lattice provides a simple model within which a clear understanding of the dynamics is more readily attainable. This type of model may be applicable to crystals with a filamentary structure, as in Section II. Moreover, the strictly longitudinal motion of crystals, in which whole planes move in unison, may be reduced to this mathematical problem in one dimension.[143]

The motivation for considering anharmonic interactions in crystals is provided by the obvious shortcomings of the harmonic model: a harmonic lattice has no significant thermal expansion, for example. More generally, the early computer experiments of Fermi, Pasta, and Ulam[144] on an anharmonic 1-D lattice have been responsible for much of the theoretical interest in nonlinear phenomena at the molecular level. They found that the anharmonicity did not result in an equipartition of energy between the harmonic normal modes as expected, but led to recurrence phenomena whereby the lattice returned intermittently to the originally excited mode. The general questions of ergodicity and energy transfer within molecules[145] will not be discussed herein.

Here we are concerned with exploring the role that solitons and related phenomena play in lattice vibrations when the harmonic approximation is discarded. Much of the early work on these nonlinear chains is due to Zabusky and co-workers[146,147] and Toda.[148-154] It is only in recent years that chemists and physicists interested in intramolecular and lattice dynamics have become generally aware of these striking nonlinear effects.

The system of interest is a 1-D lattice of identical atoms, which interact in pairs via a potential energy V that is a function of interatomic distance. The classical Hamiltonian for a chain of N atoms is given by

$$H = \sum_{n=1}^{N} \left\{ \tfrac{1}{2} m \dot{u}_n^2 + \frac{1}{2} \sum_{m \neq n} V(|u_m - u_n|) \right\}, \tag{10.1}$$

where each atom is labeled by an integer n, u_n is the position along the axis of the nth atom with respect to a fixed origin, $\dot{u}_n$ is the associated velocity, and V is the interatomic potential-energy function. Henceforth we consider only nearest-neighbor interactions, so that

$$H = \sum_n \{\tfrac{1}{2} m \dot{u}_n^2 + V(r_n)\}, \qquad (10.2)$$

where

$$r_n = u_n - u_{n-1} \qquad (10.3)$$

is the nth bond length. The pair potential $V(r)$ is assumed to have the typical atom–atom form of (say) a Lennard–Jones $(12,6)$ or a Morse potential, as sketched in Fig. 25. Such a potential has a single minimum, vanishes as $r \to \infty$, and diverges to $+\infty$ as $r \to 0$. This last attribute ensures that two atoms cannot pass through one another, so that the order of atoms on the axis cannot change. The system is complete once boundary conditions are specified. To avoid the need to account for end effects in a finite chain with free or constrained terminal atoms, we impose *periodic boundary conditions*, setting

$$u_{n+N} = u_n \qquad (10.4)$$

so that

$$r_{n+N} = r_n \qquad (10.5)$$

The limit of an infinite lattice, for which end effects are irrelevant in any case, is realized by letting $N \to \infty$ in the periodic chain.

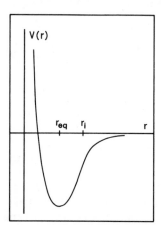

Fig. 25. A sketch of a typical atom—atom pair potential $V(r)$, for the ground electronic state, vs. interatomic separation r. $V(r) \to \infty$ as $r \to 0$, and $V(r) \to 0$ as $r \to \infty$. A single minimum occurs at $r = r_{eq}$, and r_i denotes the inflection point where $d^2 V(r)/dr^2 = 0$.

As we noted in Section VIII, since the potential energy is expressed in terms of the r_n, it is convenient to follow Toda[107] and take the $\{r_n\}$ as the coordinates. The conjugate momentum to r_n is given by p_n:

$$p_n = \frac{\partial L}{\partial \dot{r}_n} = \frac{\partial K}{\partial \dot{r}_n} \tag{10.6}$$

where L is the Lagrangian and K is the kinetic energy. Using

$$\dot{u}_n = \dot{u}_N + \sum_{j=1}^{n} \dot{r}_j, \tag{10.7}$$

$$p_k = \frac{\partial}{\partial \dot{r}_k} \sum_{n=1}^{N} \tfrac{1}{2} m \left[\dot{u}_N + \sum_{j=1}^{n} \dot{r}_j \right]^2$$

$$= \sum_{n=k}^{N} m \left[\dot{u}_N + \sum_{j=1}^{n} \dot{r}_j \right] \tag{10.8}$$

and noting that

$$\dot{u}_n = \frac{1}{m}(p_n - p_{n+1}) \tag{10.9}$$

the Hamiltonian becomes

$$H = \sum_{n=1}^{N} \frac{1}{2m}(p_n - p_{n+1})^2 + V(r_n) \tag{10.10}$$

The classical dynamics of this system is determined by Hamilton's equations of motion:

$$\frac{dr_n}{dt} = \frac{1}{m}[2p_n - p_{n-1} - p_{n+1}] \tag{10.11a}$$

$$\frac{dp_n}{dt} = -V'(r_n) \tag{10.11b}$$

where

$$V'(r_n) = \frac{dV(r_n)}{dr_n} \tag{10.12}$$

This defines the classical problem to be solved. It is worth noting in passing that Eqs. (10.10) and (10.11) also define the quantum mechanics of the chain. If we view $\{r_n\}$ and $\{p_n\}$ as dynamical variables or operators, then (10.11) are just the Heisenberg equations of motion for these operators.

The classical dynamics given by (10.11) is a seemingly intractable problem of $2N$ first-order coupled *nonlinear* differential–difference equations. There is, however, one known choice of $V'(r)$, due to Toda,[148] which allows a complete solution of the nonlinear dynamics. Before considering this Toda lattice, the harmonic case is reviewed.

In the harmonic approximation,

$$V(r) = V(r_{eq}) + \tfrac{1}{2} V''(r_{eq})(r - r_{eq})^2 \qquad (10.13)$$

The equations of motion are then linear:

$$\frac{dr_n}{dt} = \frac{1}{m}[2p_n - p_{n-1} - p_{n+1}] \qquad (10.14a)$$

$$\frac{dp_n}{dt} = -V''(r_{eq})(r_n - r_{eq}) \qquad (10.14b)$$

It is easy to see that (10.14) has fundamental solutions given by

$$r_n(t) = e^{i(\omega t - kn)} \qquad (10.15)$$

where

$$\omega = \omega(k) = \left[\frac{4V''(r_{eq})}{m} \right]^{1/2} \sin\left(\frac{k}{2}\right)$$

$$= \omega_L \sin\left(\frac{k}{2}\right) \qquad (10.16)$$

and k takes only the N discrete values in the first Brillouin zone,

$$k = \pm j\left[\frac{2\pi}{N}\right], \qquad j < \frac{N}{2} \quad \text{(integer)} \qquad (10.17)$$

and also $k = \pi/2$ if N is even. The general solution for $r_n(t)$ is constructed from a linear combination of the fundamental modes (10.15). The fact that there are just N discrete values of k arises from the *symmetry* of the chain under the periodic boundary conditions. These values of k ensure that the wavelengths of the traveling waves (10.15) fit exactly into the N-atom ring. It is interesting to compare these harmonic results with an exactly soluble nonlinear case.

A. The Toda Lattice

Although the equations of motion (10.11) are generally intractable, Toda recognized in 1967 that a particular choice of $V(r)$ allowed some analytic solutions. He proposed setting[148]

$$V(r) = \frac{a}{b} e^{-b(r - r_{eq})} + a(r - r_{eq}) \qquad (10.18)$$

where $ab > 0$ and we restrict ourselves to the case of $a, b > 0$. This potential has an exponential repulsive wall and a long-range linear attraction. Of course, this potential is not a realistic atom–atom potential-energy function, as it is unbounded as $r \to \infty$ and does not diverge at $r = 0$. This last objection is not very significant, since the exponential divergence can be made sufficiently large to preclude $r = 0$, as is done when using a Morse potential. However, the Toda potential does have the proper characteristic of a sharp repulsion and much weaker attraction. The Toda potential is quite flexible in form; it reduces to the harmonic limit as $b \to 0$ (ab finite) and to the hard-sphere limit as $b \to \infty$.

I do not propose to treat the Toda lattice in exhaustive detail, as there are fairly recent reviews available[152-154] that contain extensive derivations and references. It is important, however, to state the essential results, since this case provides a benchmark for investigations of other nonlinear lattices.

When $V(r)$ contains no inflection point, as in (10.18), the force is a monotonic function of r and Eq. (10.11b) can be inverted. Thus[153]

$$\frac{dp_n}{dt} = - V'(r_n) \qquad (10.19)$$

is equivalent to

$$r_n - r_{eq} = - \frac{1}{m} \chi \left[\frac{dp_n}{dt} \right] \qquad (10.20)$$

Then, (10.11a) gives

$$\frac{d}{dt} \chi \left[\frac{dp_n}{dt} \right] = p_{n+1} + p_{n-1} - 2 p_n \qquad (10.21)$$

Setting

$$P_n = \int^t p_n \, dt \qquad (10.22)$$

the equation of motion takes the nice form

$$\chi \left[\frac{d^2 P_n}{dt^2} \right] = P_{n+1} + P_{n-1} - 2 P_n \qquad (10.23)$$

For the Toda potential,

$$\frac{m}{b}\ln\left[1+\frac{1}{a}\frac{d^2P_n}{dt^2}\right] = P_{n+1}+P_{n-1}-2P_n \tag{10.24}$$

This is not the only convenient form for the Toda lattice equations.[153] It is often useful to take the force, $-V'(r_n)$, as the dependent variable. Setting

$$f_n = -V'(r_n) = a\{\exp[-b(r_n-r_{eq})]-1\} \tag{10.25}$$

(10.11) gives

$$\frac{m}{b}\frac{d^2}{dt^2}\ln\left[1+\frac{f_n}{a}\right] = f_{n-1}+f_{n+1}-2f_n \tag{10.26}$$

Setting

$$y_n = b(r_n-r_{eq}) \tag{10.27a}$$

$$g_n = \frac{f_n}{a} = \exp(-y_n)-1 \tag{10.27b}$$

and

$$\tau = \left[\frac{ab}{m}\right]^{1/2}t \tag{10.27c}$$

Eq. (10.26) is written as

$$\frac{d^2}{d\tau^2}\ln(1+g_n) = g_{n+1}+g_{n-1}-2g_n \tag{10.28}$$

This nonlinear lattice supports a solitary wave in the limit $N \to \infty$. Setting

$$g_n = \beta^2\mathrm{sech}^2(\alpha n \mp \beta t) \tag{10.29}$$

one can see, by substitution, that g_n of (10.29) is a solution of the Toda lattice equation (10.28) when[153]

$$\beta = \sinh\alpha. \tag{10.30}$$

g_n (essentially the interatomic force) is a localized traveling wave, which moves with uniform shape and reduced velocity

$$v = \frac{\sinh\alpha}{\alpha} \tag{10.31}$$

along the chain. The force is positive (r_n is compressed below r_{eq}) in the pulse, and decreases to zero ($r \to r_{eq}$) as $n \to \pm\infty$. The velocity is an increasing function of the pulse height β^2. Moreover, as $\beta \to 0$, so $\alpha \to 0$ and $v \to 1$; that is, as the pulse height tends to vanish, the width, roughly $2\alpha^{-1}$, diverges and the reduced velocity approaches the finite limit $v = 1$. The form of these solitary pulses is sketched in Fig. 26. From (10.27c)

$$
\begin{aligned}
\tau &= \left[\frac{V''(r_{eq})}{m} \right]^{1/2} t \\
&= \frac{c}{r_{eq}} t
\end{aligned}
\tag{10.32}
$$

where c is the velocity of low-amplitude sound waves,

$$
\frac{c}{r_{eq}} = \frac{d\omega}{dk}(k = 0) \qquad [\text{see } (10.16)].
\tag{10.33}
$$

Thus, the pulse velocity in nonreduced terms is greater than the velocity of sound. No such pulses can be found in the harmonic chain. A harmonic chain is a linear dispersive medium. A pulse could always be described at some instant as a linear combination of the plane waves (10.15). But each wave moves with a different velocity, so the pulse *must* change shape as time evolves. The compressional solitary waves of the Toda lattice maintain their form.

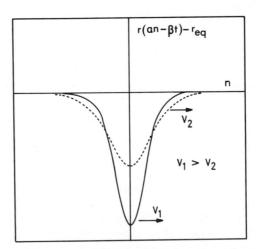

Fig. 26.   A sketch of the nearest-neighbor bond length r versus n for solitons in the Toda lattice. Two solitons are depicted for velocities v_1 and v_2 with $v_1 > v_2 > c$. Note that the solitons involve only bond compressions.

In fact, these waves are strict solitons. The collision of two solitons can be described by setting[153]

$$P_n = \frac{m}{b} \ln\left[1 + \exp\left[-2(\alpha_1 n - \beta_1 t + \delta_1)\right]\right.$$
$$+ \exp\left[-2(\alpha_2 n - \beta_2 t + \delta_2)\right]$$
$$\left. + A \exp\left\{-2\left[(\alpha_1 + \alpha_2)n - (\beta_1 + \beta_2)t + \delta_1 + \delta_2\right]\right\}\right] \quad (10.34)$$

and determining β_1, β_2, and A in terms of α_1 and α_2 by substitution in (10.24). The arbitrary constants α_1 and α_2 define the separated soliton velocities as in (10.31). The solitons collide and reemerge, suffering only a shift in position, described by δ_1 and δ_2.[153] Each soliton has a mass M given by the excess mass in the compressed lattice over the normal lattice mass:

$$M = \lim_{N \to \infty} \left[\frac{m\left[Nr_{eq} - (u_{N/2} - u_{-N/2})\right]}{r_{eq}} \right]$$
$$= \frac{m\alpha}{br_{eq}} \quad \text{(Ref. 153)} \quad (10.35)$$

During the two-soliton collision, the center of mass of the two solitons moves with constant velocity along the chain.

The general expressions for any number of solitons with arbitrary velocities on the lattice have been derived by Hirota[155] using intuitive transformations like (10.34), by Hirota and Satsuma[156] using a Bäcklund transformation, and by Flaschka[157, 158] using the inverse scattering transformation.

It is not surprising, in view of the fact that the solitary waves are strict solitons, that the Toda lattice is a completely integrable system. Henon was the first to find N integrals of the motion (conserved quantities) for the periodic and free-end chains of N atoms.[159] Flaschka produced a general scheme for finding such constants.[157, 158] Finally, the general initial-value problem has been solved using a discrete version of the inverse scattering transformation and Floquet's method by Kac and Moerbeke,[160] Flaschka and McLaughlin,[161] and Flaschka.[158] Thus it would seem to be possible to construct N action and N cyclic angle variables with which to quantize the Toda lattice. A semiclassical (WKB) approximation for the quantum energy levels associated with a single soliton in a periodic Toda lattice has been investigated by Shirafuji.[162]

The Toda lattice is apparently a very special case, since the lattice equations (10.11) are generally expected to be nonintegrable. The harmonic lattice is also integrable, but differs from the Toda lattice in not supporting solitons. Both lattices do have periodic traveling-wave solutions, given by (10.15) in the harmonic case and derived in Ref. 148 for the Toda lattice. These latter waves reduce to the sinusoidal harmonic form in the low-energy limit. At higher energy the Toda-lattice periodic waves resemble a succession of solitonlike compressions on an expanded lattice [the Toda lattice expands with increasing energy, as one would expect, since $V(r)$ is softer for $r > r_{eq}$ than for $r < r_{eq}$].

The questions we must address are these: are these solitons an isolated occurrence due to the very special form of the Toda potential? Do any aspects of the classical motion seen here carry over to lattices with more typical atom–atom forces? What is the physical interpretation of these solitons; how do we understand why they arise? The rest of this section attempts to answer these questions or at least throw some light on the matter.

B.   Solitons in Atomic Lattices

If the pair potential $V(r)$ is a typical atom–atom potential such as a Morse or Lennard–Jones (12, 6) function, the lattice equations are generally intractable by analytic methods. The search for solitons in these systems was begun by numerical simulations by Rolfe, Rice, and Dancz,[163] Valkering and De Lange,[164] Ali and Somerjai,[165] and Batteh and Powell.[166] This last study considered the longitudinal motion of a 3-D lattice, which is equivalent to the 1-D lattice when no transverse motion is allowed. We have thought about the motion in such systems in harmonic terms for so long that the numerical results come as some surprise.

Figure 27 shows the behavior of a Morse lattice in which Rolfe, Rice, and Dancz[163] set up an initial stationary Gaussian compression distribution in a chain of 64 atoms and then set the atoms free to evolve according to Hamilton's equations of motion. The localized compression splits into two pulses traveling in opposite directions. Each pulse has two peaks, which become more separated and better resolved as time goes on; each peak is a localized solitary wave. The larger wave travels faster and leaves the smaller one behind. Since the boundary conditions are periodic, the original pulses meet head on, collide, and separate to continue around the ring of atoms again and again. Clearly, the behavior seen in Fig. 27 is very solitonlike. However, the solitary-wave collisions are not perfectly elastic,[164, 166] so that some "noise" or motion is created in the lattice, as in the last snapshot of Fig. 27. Presumably, the solitonlike pulses will eventually degenerate into seemingly random lattice vibrations, but the waves are quite long lived. The pulses of Fig. 27 were followed[163] for about 300 periods of the Morse oscillator without any significant loss of integrity of the waveforms. The loss of energy from the solitary waves to the lattice following collisions seems to be more severe in the crystal simulations.[166] If the chain of oscillators were

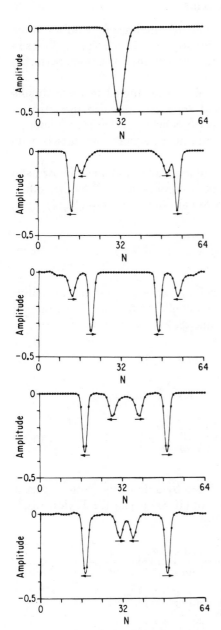

Fig. 27. Reproduced from Fig. 1 of the article by Rolfe, Rice, and Dancz, *J. Chem. Phys.* **70** 26 (1979) (with permission). The behavior, as a function of time, of a large-amplitude compressive displacement on a Morse lattice of 64 atoms under cyclic boundary conditions. Each point represents one of the 64 bond lengths, $r_n - r_{eq}$; the solid curve is a six-point interpolation to aid the eye. Each frame gives the bond-length variation at an instant, with time increasing down the figure from $t = 0$ at the top.

completely integrable, as for the Toda chain, the lifetime of the pulses would be infinite. The pulses of the Morse lattice are very like the Toda solitons in behavior and even in waveform; they are well described as a small perturbation of the Toda soliton form.[167]

A Lennard–Jones pair potential has also been studied in these simulations,[163, 164] with similar results. Ali and Somerjai[165] studied the effect of varying the potential form and came to the conclusion that soliton motion was endemic to potentials with asymmetry about equilibrium, that is, with hard repulsion and soft attraction.

The mechanism controlling these solitary waves and their consequent properties can in fact be understood from a simple analysis[168] by the author. The lattice equations of motion in Newtonian form are, from (10.11),

$$m\frac{d^2r_n}{dt^2} = V'(r_{n+1}) + V'(r_{n-1}) - 2V'(r_n) \tag{10.36}$$

This must be solved for a traveling solitary wave without approximating $V'(r)$, since it is the consequences of the nonlinear form of $V'(r)$ that we wish to understand. Let us instead assume that the solitary-wave form

$$r_n(t) = r(nr_{eq} - vt) = r(\phi) \tag{10.37}$$

does not change very rapidly with n. For this traveling wave,

$$\frac{mv^2}{r_{eq}^2}\frac{d^2r(\phi)}{dn^2} = \left[2\sinh\left(\frac{1}{2}\frac{d}{dn}\right)\right]^2 V'[r(\phi)] \tag{10.38}$$

where the time derivative has been replaced by a derivative with respect to n, from (10.37), and the right-hand side of (10.38) is just the Taylor expansion of the difference operator in (10.36):

$$f(n+1) + f(n-1) - 2f(n) = \left[\exp\left(\frac{d}{dn}\right) + \exp\left(-\frac{d}{dn}\right) - 2\right]f(n)$$

$$= 2\left[\cosh\left(\frac{d}{dn}\right) - 1\right]f(n)$$

$$= \left[2\sinh\left(\frac{1}{2}\frac{d}{dn}\right)\right]^2 f(n) \tag{10.39}$$

To avoid approximating $V'(r)$, the difference operator is eliminated by operating with its inverse, to obtain

$$\frac{mv^2}{r_{eq}^2}\frac{d^2}{dn^2}\left[\frac{1}{2}\mathrm{csch}\left(\frac{1}{2}\frac{d}{dn}\right)\right]^2 r(\phi) = V'(r) \tag{10.40}$$

That is,

$$\frac{mv^2}{r_{eq}^2}\left\{(r - r_{eq}) - \frac{1}{12}\frac{d^2r}{dn^2} + \frac{1}{240}\frac{d^4r}{dn^4} - \frac{1}{6048}\frac{d^6r}{dn^6} + \cdots\right\} = V'(r) \tag{10.41}$$

where the boundary condition $r(\phi) \to r_{eq}$ as $\phi \to \pm\infty$ has been used. The assumption that $r(\phi)$ does not change rapidly with n means that we neglect the terms of order d^4r/dn^4 and higher, to write

$$\frac{mv^2}{r_{eq}^2}\left[(r - r_{eq}) - \frac{1}{12}\frac{d^2r}{dn^2}\right] = V'(r) \tag{10.42}$$

This has been called a "quasicontinuum" approximation.[168] The continuum approximation of earlier sections neglects even the term in d^2r/dn^2, giving

$$\frac{mv^2}{r_{eq}^2}\left[r(\phi) - r_{eq}\right] = V'[r(\phi)] \tag{10.43}$$

This equation has a solution for one or two values of r, not a continuous function of n (or ϕ). Thus there are no solitons, $r(\phi)$, for a vibrating atomic lattice in the continuum limit. These more subtle solitons must require some account of discrete-lattice effects.

The equation of motion (10.42) is equivalent to that of a particle moving in a potential well $U(r)$:

$$\frac{mv^2}{24r_{eq}^2}\left(\frac{dr}{dn}\right)^2 = E - U(r)$$

$$= \frac{mv^2}{2r_{eq}^2}(r - r_{eq})^2 - \left[V(r) - V(r_{eq})\right] \tag{10.44}$$

The effective potential is the original $V(r)$ less a harmonic term that arises from the collective kinetic energy. Figure 28 sketches the form of $(dr/dn)^2$ in (10.44) when $V(r)$ has the atom–atom form of figure 10.1. The potential

$V(r)$, near equilibrium, can be written as

$$V(r)-V(r_{eq}) = \frac{1}{2}V''(r_{eq})(r-r_{eq})^2 + \frac{1}{3!}V'''(r_{eq})(r-r_{eq})^3 + \cdots$$

$$= \frac{mc^2}{2r_{eq}^2}(r-r_{eq})^2 + \frac{1}{3!}V'''(r_{eq})(r-r_{eq})^3 + \cdots \quad (10.45)$$

where c is the velocity of sound. A solitary wave like that of the Toda lattice must have zero gradient at $r = r_{eq}$ and at one shorter bond length where the compressional pulse has its turning point. Moreover, $(dr/dn)^2$ must be positive everywhere else. From Fig. 28 it is clear that this can only occur when the soliton velocity $v > c$. Then the harmonic part of $V(r)$ is weaker than the effective harmonic term from the collective kinetic energy near equilibrium, so that $(mv^2/2r_{eq}^2)(r-r_{eq})^2 - V(r)$ is positive. However, any realistic atom–atom potential has a repulsive wall, which rises faster than any harmonic. Thus $(mv^2/2r_{eq}^2)(r-r_{eq})^2 - V(r)$ decreases to zero at some value of $r < r_{eq}$, as required for a solitonlike wave.

Clearly, so long as the atom–atom potential is *more repulsive than a harmonic*, there are solitary compressional waves that *travel faster than sound*. For practical purposes this condition is always fulfilled. This simple analysis and the computer experiments[163] both imply that only compressional soli-

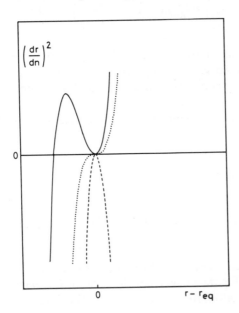

Fig. 28.   A sketch of the squared gradient of the soliton waveform, $(dr/dn)^2$, vs. $r - r_{eq}$ from Eq. (10.44) for a typical atom–atom potential as shown in Fig. 25, with $v < c$ (dashed line), $v = c$ (dotted line), and $v > c$ (full line). Since $r(n)$ is real, a physically significant solution must have $(dr/dn)^2 \geq 0$.

tary waves are possible. Dilatational pulses, seen in the reflection of a compressional pulse from a free chain end,[163] are unstable and degenerate into lattice vibrations. It is also clear that the potential form for $r > r_{eq}$ is irrelevant to these compressional waves (unless the lattice is stretched by external forces[168]). The practice of using a harmonic approximation to study lattice dynamics is not as innocuous as one might have thought. By doing so, soliton phenomena are completely excluded from the dynamics, whereas these waves are in fact found in the longitudinal motion of all atom lattices.

The simple approximation outlined above can be used to evaluate the solitary-wave form and its properties.[168] From (10.44)

$$\left(n - \frac{vt}{r_{eq}}\right) - n_0 = \int_{r_0}^{r} \left[\frac{mv^2}{24r_{eq}^2}\right]^{1/2} \frac{dr}{\left[\frac{mv^2}{2r_{eq}^2}(r - r_{eq})^2 - V(r) + V(r_{eq})\right]^{1/2}}$$

(10.46)

yields $r(\phi)$, which can then be used to evaluate properties such as the amount of kinetic and potential energy and momentum carried by a soliton. The approximate waveform from (10.46) agrees very well[168] with the numerical calculations of Rolfe, Rice, and Dancz[163] for a chain of Morse oscillators, so that the properties evaluated in this approximation should be reasonably accurate. Equation (10.46) is also useful in providing explicit expressions for the solitary waves in those cases where $V(r)$ allows the integral in (10.46) to be evaluated analytically. Two examples are provided by the cubic[169] and quartic[168] potentials:

$$V_C(r) = \begin{cases} V(r_{eq}) + \frac{1}{2}V'''(r_{eq})\left[(r - r_{eq})^2 - \frac{\alpha}{3}(r - r_{eq})^3\right], & r \leqslant 2/\alpha \\ 0, & r > 2/\alpha \end{cases}$$

(10.47)

where $\alpha = |V'''(r_{eq})|/V''(r_{eq})$, and

$$V_Q(r) = \begin{cases} \frac{1}{4}B(r - r^*)^4 - \frac{1}{2}A(r - r^*)^2, & r \leqslant r^* \\ 0, & r > r^* \end{cases}$$

(10.48)

The solitons are given by (10.46) as

$$\frac{r_C - r_{eq}}{r_{eq}} = -\frac{3}{\alpha}\left[\frac{v^2}{c^2} - 1\right]\text{sech}^2\left\{\left[\frac{3c^2}{v^2}\left(\frac{v^2}{c^2} - 1\right)\right]^{1/2}\left(\frac{vt}{r_{eq}} - n\right)\right\}$$

(10.49)

and

$$\frac{r_Q - r_{eq}}{r^* - r_{eq}} = -2\left(\frac{v^2}{c^2} - 1\right)\left\{1 + \frac{v}{c}\cosh\left[2\sqrt{3}\left(1 - \frac{c^2}{v^2}\right)^{1/2}\left(\frac{vt}{r_{eq}} - n\right)\right]\right\}^{-1}$$

$$(10.50)$$

As for the Toda solitons, these solitary waves have amplitudes that increase with velocity: $-\frac{3}{2}(v^2/c^2 - 1)$ for the cubic potential, and $-2(v/c - 1)$ for the quartic potential. The pulse widths also become smaller as v/c increases. For the quartic potential, calculations show[168] that the kinetic energy and potential energy carried by the pulse increase slowly with v/c at low velocity, while the momentum carried by the solitary wave increases rapidly at low velocity. It is interesting to note that the momentum carried by these solitary waves (the sum over all atomic momenta) is in the *opposite* direction to the wave velocity: the wave pulls the lattice towards itself as it advances.

I like to call these waves "the great railway-car solitons." The mechanism whereby these solitons arise is the same as that operating in a line of railway cars when they are shunted by a locomotive. There is a hard-wall collision between each pair of cars in turn as the impact shock is propagated down the line. The same hard-wall repulsion propagates the bond compression along a lattice of atoms at supersonic velocities.

It is not surprising that these solitons, or solitary waves, are intimately connected with the dynamics of shock fronts in solids. Recent molecular-dynamics simulations of shock waves in simple one-, two-, and three-dimensional lattices provide direct evidence for the role of solitons. A shock wave is generated when a "piston" of effectively infinite mass, traveling with constant velocity, strikes a lattice and continuously compresses it. The simplest simulations consider a 1-D lattice originally at zero temperature.[170-172] The compression produces a train of solitons, more being created as time goes on, so that a "shocked" length is created, which grows linearly with time. These are known as nonsteady shocks. Straub, Holian, and Petschek[171] have shown that the behavior of shocked 1-D Morse and Toda lattices at finite initial temperatures is essentially the same as that at zero initial temperature. Behind the shock front, soliton motion persists and the velocity distribution of the atoms is non-Maxwellian. The soliton nature of these shock waves is demonstrated by Straub, Holian, and Petschek[171] by hitting a Toda lattice from both sides with a "piston" for a finite period and then stopping the piston. Figure 29 shows the two finite trains of solitons so created advancing head on, colliding, and emerging. Similar behavior is expected for a Morse, a Lennard–Jones (12,6), or any typical atomic lattice.

SOLITON EXPERIMENT
TODA CHAIN, $a\nu$=0.525 T_0=0

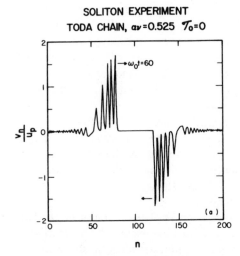

SOLITON EXPERIMENT
TODA CHAIN, $a\nu$=0.525, T_0=0

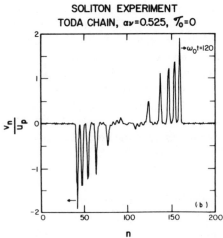

Fig. 29. Reproduced from Fig. 4 of the article by Straub, Holian, and Petschek, *Phys. Rev. B* **19** 4049 (1979) (with permission). The velocity of the atoms (as a fraction of the piston velocity u_p) in a 1-D cyclic Toda lattice of 200 atoms is presented as a function of atom number n. A piston compresses each "side" of the lattice from $\omega_0 t = 0$ to $\omega_0 t = 10$ (ω_0 is the harmonic frequency[171]). In (a) the two trains of compressions so created propagate towards each other ($\omega_0 t = 60$). In (b), at a later time ($\omega_0 t = 120$), the solitonlike compressions have collided and passed through one another. The quantity $a\nu$ (here $2\alpha u_p/c$) measures the shock strength,[171] while $T_0 = 0$ specifies the initial lattice temperature before compression.

Interestingly, the situation changes for 3-D lattices, as shown by Holian and Straub[173] and Powell and Batteh.[174] In these computer experiments the compression is applied along the z-axis to an xy plane of atoms in an fcc lattice. If the lattice is initially at rest at equilibrium ($T = 0$), the planes move as the atoms do in a 1-D chain. This 1-D behavior persists when the lattice is initially at a finite temperature $T > 0$, if the shock is weak. Holian and Straub[173] find that the quantity $2\alpha u_p/c$ (where α is the cubic anharmonicity parameter of the pair potential [see (10.47)], u_p is the piston velocity, and c is the sound velocity) is a measure of the shock strength for many different lattices. For strong shocks ($2\alpha u_p /c \geqslant 1$), the soliton motion becomes strongly coupled to transverse motions present in the thermal fluctuations of the lattice. Energy is lost to these transverse modes, and a type of transverse solitary-wave motion is observed.[174] This redistribution of energy appears to destroy the train of solitons behind the shock front, and leads to equilibration of the lattice and a Maxwellian velocity distribution[173] in the planes behind the shock front. The shock wave itself becomes a steady shock of constant width. While this transition to steady shock structures is clearly due to the coupling with transverse modes, which destroys the solitons, the situation is a little confused by the fact that the strong-shock calculations allow up to third-nearest-neighbor interactions, while the weak-shock simulations allow only nearest-neighbor forces.[173] These strong shocks produce permanent deformations in the region behind the shock.

Shock-induced transformations have been applied to chemical problems such as violent exothermic reactions in solids: explosions. Karo, Hardy and Walker[175-177] have reported molecular-dynamics studies of shock-induced detonations in 2-D lattices in which the atoms reside in a local minimum of an exothermic pair potential. A shock is applied, usually by impact of a small segment of the lattice on one end of the structure. A solitary-wave disturbance propagates through the crystal and encounters the other end of the lattice with sufficient force to spall (throw off) a small fragment of several atoms. When standard pair potentials are used (Morse, etc.), the bulk of the lattice becomes quiescent and the spalled fragment is highly vibrationally excited. When an exothermic pair potential is employed, energy is reflected back into the lattice when spalling occurs, so that the entire lattice rapidly fragments in a violent explosion. Studies of other shock-induced transformations and polymerization processes can be found among the references in Ref. 176.

Some very interesting computer experiments on lattice solitons and shocks have considered the "collision" of the waves with mass defects in the lattice. Some information is available for shock-wave scattering from defects[170] in 1-D chains where the shock is partially reflected by either a heavy or a light impurity. Rolfe and Rice[178] have conducted extensive studies of solitary waves in periodic 1-D chains of Morse oscillators where the defect-to-homogeneous atomic-mass ratio is 13:12. The interest here is to see whether the solitons can be destroyed and their energy redistributed to the lattice vibrations. The results, as yet unexplained, are quite provocative. Generally the solitons prove remarkably stable. An initial plateau is observed, during which the soliton repeatedly travels around the ring, scattering from the defect and exchanging energy back and forth with the lattice vibrations created by the scattering. However, the soliton preserves its basic integrity. Quite suddenly thereafter, the soliton is destroyed and the energy given up to vibrations. However, the length of the plateau region can vary from a few to 10^3 circuits of the chain, as the initial energy and chain size is varied, without identifiable regularity. Readers who pride themselves on a well-developed "physical intuition" for nonlinear phenomena might learn a salutory lesson from these calculations.

C. Periodic Vibrations

So far the discussion of solitons and shocks produced by massive "pistons" hitting a defenseless lattice would lead us to believe that solitons are exclusively high-energy phenomena, only relevant under somewhat special conditions; crystals in thermal equilibrium would be well described in terms of the familiar harmonic phonons. I suggest that this is not true. Classically, the fundamental "phonon" solutions for the 1-D harmonic lattice are described by real combinations of the periodic traveling waves of (10.15). Let us consider the form of these periodic vibrations when the harmonic approximation is not invoked.

It was noted in Section X.A that the Toda lattice supports traveling periodic waves that reduce to the harmonic sinusoidal form in the limit of zero amplitude. Valkering[179] has shown analytically that an atom–atom potential like that of Fig. 25 does support periodic traveling waves that are similar in form to those of the Toda lattice at high energy. The nature of these waves can be understood using a quasicontinuum treatment reported by Collins and Rice.[169]

We wish to find solutions of the equations of motion of the form

$$r_n(t) = r(\omega t - kn), \qquad p_n(t) = p(\omega t - kn), \qquad (10.51)$$

where the wave number k must take one of the N discrete values given in (10.17), because of the symmetry of the Hamiltonian (10.10) under periodic boundary conditions. The quasicontinuum form for the equation governing this motion, (10.41), can be rederived for periodic waves to give[169]

$$\nu^2 \left\{ (y - \bar{y}) - \frac{1}{12} \frac{d^2 y}{dx^2} + \frac{1}{240} \frac{d^4 y}{dx^4} - \frac{1}{6048} \frac{d^6 y}{dx^6} + \cdots \right\} = v'(y) \quad (10.52)$$

where

$$x = \frac{\omega t}{k} - n \qquad (10.53)$$

$$y(x) = \frac{r(x)}{r_{eq}} \qquad (10.54)$$

$$v'(y) = \frac{V'(r)}{r_{eq} V''(r_{eq})} \qquad (10.55)$$

$$\nu = \frac{\omega/\omega_L}{k/2} \qquad (=1 \text{ for sound}) \qquad (10.56)$$

and

$$\bar{y} = \frac{1}{N} \int_0^N y(x)\, dx \tag{10.57}$$

is the average reduced bond length for the chain. Since the pair potential is anharmonic, $\bar{y} > 1$. As in (10.44), the simplest approximation reduces to the motion of a particle in a well,

$$\frac{1}{2} \left(\frac{dy}{dx} \right)^2 + U(y) = U(y_0) \tag{10.58}$$

where

$$U(y) = \frac{12}{\nu^2} \left\{ v(y) - \frac{\nu^2}{2}(y - \bar{y})^2 \right\} \tag{10.59}$$

and y_0 is a classical turning point:

$$\frac{dy}{dx}(y = y_0) = 0.$$

To find the form of the periodic waves, one simply assumes values for $\bar{y}$ and ν (a dimensionless phase velocity) and solves (10.58) in a similar fashion to (10.46). The classical turning point y_0 is adjusted until $(1/N)\int_0^N y(x)\, dx$ equals the assumed value of $\bar{y}$. The wavenumber of the solution is found from the period $T(\nu, \bar{y})$ of the wave by $k = 2\pi/T(\nu, \bar{y})$; the lattice energy E can be computed from a quasicontinuum approximation to H in (10.10).[169] One then has the waveform and dispersion relations $\nu(k, E)$ and $\bar{y}(k, E)$, which are the nonlinear analogues of (10.15) and (10.16). Figure 30 displays the effective potential well $U(y)$ and effective energy $U(y_0)$ for two waves on a cubic lattice.[169] The well is quite similar to that for solitons, except that the energy lies below the maximum of the local potential barrier, so that periodic motion occurs. A long-wavelength vibration (small k) is one with long period of vibration. From Fig. 30 we see that this is possible when the right-hand turning point is very near the top of the energy barrier. The restoring force is very low here, so that the particle takes a long time to slow down, stop, and turn around, resulting in a long period of vibration overall. However, this also implies that the waveform is strikingly solitonlike. Figure 31 shows a snapshot of two waveforms for $k = 2\pi/50$ at different energies, in a chain of 50 xenon atoms.[169] At very low energy (the harmonic limit), this waveform is a sine function, $\sin(\omega t - kn + \text{constant})$, with $k = 2\pi/50$. At

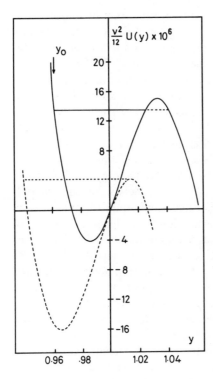

Fig. 30. The effective potential well of Eq. (10.59), in which the collective periodic motion of the lattice is represented by the motion of a single particle. Two wells are presented for the cubic pair potential of Eq. (10.47) for $\alpha = 2$ with $\bar{y} = \bar{r}/r_{eq} = 1.0005$ and $\nu = 0.99$ (full curve), $\nu = 1.01$ (dashed curve). y_0 denotes the classical turning point determined by the method discussed in the text. Each horizontal line represents the total energy of the 1-D oscillation. Note that when the phase velocity exceeds that of sound ($\nu = 1.01$), the right-hand turning point is very near the well maximum, so that long-period, long-wavelength, solitonlike motion occurs.

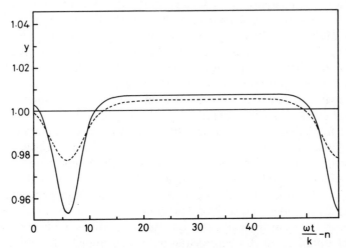

Fig. 31. The reduced bond length y vs. the traveling-wave coordinate $\omega t/k - n$, for the smallest wave number, $k = 2\pi/50$, of a cyclic lattice of 50 xenon atoms interacting via a cubic potential fitted to the Xe_2 spectroscopic data.[169] Waves at two energies are shown: $E/E_{diss} = 0.01985$, $\nu = 1.00445$, $\bar{y} - 1 = 0.6558 \times 10^{-4}$ (dashed line), and $E/E_{diss} = 0.06050$, $\nu = 1.0114$, $\bar{y} - 1 = 0.1969 \times 10^{-3}$ (solid line), where E_{diss} is the dissociation energy of the pair potential.[169]

313

higher energy the wave becomes distorted into a wave train of solitons with a very broad region of expanded bonds ($r > r_{eq}$) and short sharp compressions ($r < r_{eq}$) repeated every 50 lattice sites. The soliton nature of the vibration is reflected in a supersonic phase velocity ($\nu > 1$).

Long-wavelength modes of vibration are naturally solitonlike. This can be understood in qualitative terms, in addition to the derivation of the actual waves given above. Clearly, the anharmonicity of $V(r)$ will only be important if $|r_n - r_{eq}|$ is large for some n. But at fixed chain energy, $|r_n - r_{eq}|$ can be large at some n only if it is small elsewhere. For a wave with wave number k, $r_{n+2\pi/k} = r_n$, and if $|r_n - r_{eq}|$ is large there must be $N/(2\pi/k) = j$ bonds with the same large magnitude. Thus, for fixed energy, the larger j is, the smaller must be each magnitude $|r_n - r_{eq}|$. This means that the possibility of some bonds being significantly distorted from equilibrium at a given energy is greater if k is small, which means long wavelength. So it is not surprising that the long-wavelength modes are anharmonic solitonlike forms as shown in Fig. 31. As the qualitative argument suggests, the waves of shorter wavelength do not show this type of soliton form.[169] Detailed description of the dispersion relations for these waves in cubic and Lennard–Jones lattices in the simple quasicontinuum and higher-order approximations can be found in Ref. 169. Numerical demonstration of the stability of the waves can also be found there.

We have seen that, at high energy, all the long-wavelength vibrations of an atomic lattice with a realistic pair potential are wave trains of solitons. The soliton itself should not be seen as an isolated phenomenon, but merely as the *infinite-wavelength limit* of a more general characteristic of lattice motion. Hence, at sufficiently high energy (or temperature) a nonnegligible part of the lattice dynamics might be better described in terms of solitonlike motions than harmonic phonons. The catch phrase "sufficiently high energy" is still hanging over our heads. How high is "sufficiently high"? The only convincing way to answer this is to quantize the lattice motion so that a natural energy scale is imposed.

Some formal progress towards a quantum-mechanical description of an atomic lattice has been made by Dancz and Rice[167] and B. C. Eu[180] using a coherent-state (minimum-uncertainty wave packet) representation rather than a phonon basis set. Dancz and Rice derive the form of the quantum equations of motion for a general potential and some particular cases, whereas Eu considers the simpler equations that can be obtained for Morse oscillators. Not surprisingly, the quantum equations look very like the classical equations in this representation, which implies that the quantum equations have soliton-related solutions as the classical equations do. At present I am not aware of a complete solution of the quantum anharmonic chain. The Toda lattice could be soluble quantum-mechanically. If one can construct, for this completely integrable system, the canonical transformation from the dynamical variables $\{r_n, p_n\}$ to a set of action-angle variables and express the Hamiltonian in terms of the action variables, it ought to be possible to construct the *corresponding* unitary transformation that gives a set of N good quantum numbers and their associated stationary-state representation. Again, I am not aware at present that this has been done.

It is possible to quantize the system semiclassically. Collins and Rice[169] have quantized the motion of a cubic chain using a primitive semiclassical approach due to Miller.[181] In this approach, the eigenstate energies E_n are identified with the singularities in the quantum density of states $p(E)$:

$$p(E) = \text{trace}[\delta(E - H)]$$
$$= \sum_n \delta(E - E_n) \tag{10.60}$$

If the trace is evaluated in the coordinate representation using a simple stationary phase approximation,[181(b)] the poles in $p(E)$ are found from the energies of perfectly periodic classical trajectories if there is only *one* such trajectory associated with each discrete system symmetry. The periodic lattice of N atoms does in fact have N discrete symmetry elements, each labeled by the wavenumber k.[182] Collins and Rice find only one perfectly periodic trajectory (like those in Fig. 31) for each allowed value of k at given energy. The energy levels are then found from the simple quantization condition[169]

$$A_k(E) = 2\pi\hbar\left(n + \tfrac{1}{2}\right), \qquad n \text{ an integer} \tag{10.61}$$

where $A_k(E)$ is the abbreviated classical action

$$A_k(E) = 2\oint(\text{total kinetic energy}) \, dt \tag{10.62}$$

and the integral is taken around the closed, perfectly periodic trajectory with energy E and wave number k. This is reminiscent of the WKB result for one degree of freedom. The perfectly periodic trajectories for a cubic lattice can be found analytically in the low-order quasicontinuum approximation,[169] and the energy adjusted until (10.61) is satisfied. This approach should be reasonable for the lowest energy eigenstates: $n = 0, 1$. This semiclassical quantization has been performed for a cubic chain with atomic mass and potential parameters appropriate to xenon atoms.[169] The lowest energy levels for long-wavelength modes (small k) are found to be very close to the harmonic result,

$$E_k(n_k) = \hbar\omega_L\sin\left(\tfrac{1}{2}k\right)\left(n_k + \tfrac{1}{2}\right), \qquad n_k = 0, 1, 2, \ldots \tag{10.63}$$

with total energy $E = \sum_k E_k(n_k)$. This comes as no surprise to those who have always assumed that the harmonic approximation is valid at such low energy. For example, in the xenon chain of 50 atoms, the smallest wavenumber, $k = 2\pi/50$, has a discrete set of eigenenergies beginning with $E_k(n=0) \simeq 1$ cm^{-1} and $E_k(n=1) \simeq 3$ cm^{-1}. The classical periodic motion with these energies (spread over 50 bonds) might be expected to be sinusoidal, as the energy levels are near harmonic. However, the two periodic waves of Fig. 31 are just those that correspond to these two eigenstates via (10.62). The chain motion, even at these very small energies, is very solitonlike. To obtain harmonic sinusoidal trajectories, one must lower the energy to a small fraction of the zero-point energy $E_k(n=0)$. The harmonic approximation is only valid at unphysically low energies for the long-wavelength vibrations of this xenon chain. The xenon–xenon bond is quite anharmonic, as one would expect for a Van der Waals dimer, although the large mass would have favored low-amplitude, harmonic motion in a naive picture.

There are a number of important lessons suggested by this calculation. Firstly, there is a natural coupling between the symmetry of these chains and their anharmonicity. The long-wavelength symmetry modes of an atomic chain do not spread their small quanta of energy evenly over their wavelength. The correct classical picture is not that of a uniform-amplitude sinusoidal waveform, but a waveform like a train of solitons in which the lattice has localized a significant portion of the wave energy in a relatively small segment of the wavelength. The fact that the waveform can be as anharmonic as those in Fig. 31 while the energy is within a few percent of the harmonic value[169] suggests that eigenenergies are a poor measure of the anharmonicity of the dynamics.

It is reasonable to suppose that the anharmonicity of the classical periodic motion is more clearly reflected in the quantum eigenfunctions for these states. Thus, while the eigenenergies may be near their harmonic values, the transition amplitudes between states should differ from the harmonic values. Similarly, the anharmonicity of the classical motion should be apparent in the intensity of phonon peaks in the classical dynamical structure factor. Schneider and Stoll[183] have performed a molecular-dynamics study of the 1-D Toda lattice and reported the dynamical structure factors

$$S_{AA}(q,\omega) = 2\int_0^\infty S_{AA}(q,t)e^{i\omega t}\,dt \tag{10.64}$$

where

$$S_{AA}(q,t) = \langle A(q,t)A(-q,0)\rangle \tag{10.65}$$

$$A(q,t) = N^{-1/2}\sum_n \left[A_n(t) - \langle A\rangle\right]e^{iqnr_{eq}} \tag{10.66}$$

and A_n is a local property of each lattice site, such as the force or kinetic energy for example.

They observed a number of anharmonic effects, including peaks in these power spectra, which they identify with solitons and second sound. In addition, they observed peaks near the harmonic phonon frequencies $\omega(q)$. However, even though the peak frequencies are correctly given by the harmonic model, the peak heights, which provide a measure of the strength of the resonances, are not accounted for by the harmonic model, as we might expect from our arguments above. While this is suggestive, what is now required is the development of a phenomenological theory that evaluates the dynamical structure factor in terms of the soliton-train periodic vibrations rather than the harmonic phonons.

The calculation of the contribution of solitons and soliton-train vibrations to the thermodynamic properties of these 1-D atomic lattices is also worthwhile in developing our understanding of the fundamental excitations in these systems. The exact classical free energy can be evaluated for a 1-D lattice,[184] so that comparison with exact results is readily available. Büttner and Mertens[185] have reported a phenomenological model in the spirit of that in Section IV. For the Toda lattice they compare the exact free energy at low temperature with that arising from an ideal gas of solitons. Assuming a Boltzmann distribution of soliton energies, they obtain a free energy that has the correct temperature dependence at low T but is quantitatively incorrect. It may be possible to effect agreement with exact results by including all allowed wavelength modes rather than just solitons in the "gas."

Before concluding our discussion of lattice vibrations, we note that Dash and Patnaik have recently studied soliton and periodic motion in 1-D lattices with two atoms per unit cell. Using long-wavelength approximations, they have found various types of solitons in lattices with power-law pair potentials.[186] Solitary waves in a diatomic Toda lattice have been derived using a Fourier-expansion technique,[187] however these waves are complex solutions and do not appear to be physically relevant.

D. Molecular Vibrations

While the study of lattice vibrations is worthwhile in its own right, it is also useful in stimulating our thoughts about molecular vibrations. Since we are drawing towards the end of the review, it doesn't seem unreasonable to indulge in a little speculation and make some suggestions for interesting avenues of research without attempting to provide solutions.

Molecular vibrations have been extensively studied via the harmonic approximation, with anharmonic effects included by way of perturbation theory. Clearly, if a new insight is to be obtained, we must reject this approach entirely. The rejection is easy; replacing the traditional approach with a truly nonlinear theory is not so easy. The benzene molecule provides some stimulus for a new approach.

Figure 32 displays a schematic representation of one component of the doubly degenerate ν_{12} mode in benzene.[188] This normal mode is predominately a C—H bond-stretching vibration and is active in the infrared. There are six C—H bonds, and the motion of this normal coordinate involves a traveling wave of C—H stretch of period 6, the ortho C—H bonds being out of phase by π radians. The other component of this mode simply propagates around the ring in the opposite direction. The interesting feature of this mode is that it appears to form, in part, the fundamental of a remarkably long overtone progression.[189] The overtones are forbidden in the harmonic approximation, and so by implication the C—H stretching motion in benzene is highly anharmonic. This overtone progression, along with similar observations in other molecules with a number of C—H bonds, has led to the development of local-mode theories.[190-192]

In the light of the lattice vibrations of this section, it is tempting to *speculate* that the "high"-energy form of the classical traveling wave, ν_{12}, with period 6 is not a sinusoidal variation of the C—H bond with ring position but a soliton waveform of period 6 arising from the repulsive walls of the multidimensional potential-energy surface. This classical anharmonicity

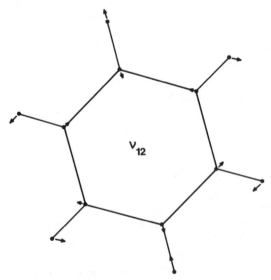

Fig. 32. A schematic representation of the normal coordinate ν_{12} of benzene. The direction and magnitude of the arrows represents the direction and magnitude of the atomic displacements in the normal-coordinate vector.[188]

should be reflected in the dipole transition moments determining the infrared absorption intensities.

However, there is another *possibility*, perhaps more in the spirit of local-mode theories. One could consider each C—H vibration as an "internal" degree of freedom associated with each ring position as in Section VIII. The C—H groups are equivalent, so that a resonance coupling exists between neighboring groups via both the carbon skeleton and direct dipole–dipole interaction. An excitation in the ν_{12} mode could be seen as a soliton like that in a molecular crystal, where the vibration is predominantly localized in one C—H bond at a time as that excitation propagates around the ring.

Raising the subject of resonance brings to mind the ubiquitous phenomenon of Fermi resonance in polyatomics. Normally the splitting of two nearly degenerate energy levels is evaluated by considering the perturbation produced by cubic terms in the potential energy on the harmonic eigenenergies. Fermi resonance is most frequently encountered in the higher energy levels, since the anharmonicity and combination effects combine to produce a greater density of levels and hence more accidental degeneracies. It is quite reasonable to suggest that only a truly anharmonic theory of molecular vibrations can provide a clear picture of this phenomenon.

On a more general note, the question of when anharmonic effects are really important (not just small perturbations) is bound up with the question of symmetry. In the lattice problem we saw how there is a natural coupling between anharmonicity and some symmetry elements: those whose representatives are labeled by small k-values. Is there a similar selective coupling in molecules whereby some modes are well described harmonically while others are unexpectedly anharmonic because of their symmetry?

To answer this question we need a convincing treatment of the symmetry of molecular motion that does not make dynamical approximations beyond, say, the Born–Oppenheimer approximation. Clearly the symmetry of the molecular Born–Oppenheimer Hamiltonian is not the symmetry of the equilibrium configuration, which the molecule may assume for an instant from time to time (see Ref. 193 and references therein). Nor should the symmetry be restricted by excluding certain configurations as improbable on dynamical grounds. For example, it makes no sense to say that ammonia belongs either to the symmetry group C_{3v} or to D_{3h}, because its instantaneous configuration might be either or neither at any one time. The use of a symmetry classification based on the equilibrium geometry makes sense when one is proceeding towards a harmonic theory of small-amplitude dynamics. Once the small-amplitude notion is discarded, so is the symmetry of the equilibrium geometry. We need a symmetry classification based on the nuclear permutation symmetry group of the Hamiltonian and the functional symmetry of the Born–Oppenheimer potential-energy hypersurface. All of which adds up to a lot of food for thought.

XI. CONCLUSION

These few examples have shown how solitons arise in the 1-D dynamics of a number of systems via different mechanisms. The SG and ϕ^4 solitons are examples of so-called "topological" solitons, which describe transitions between degenerate equilibrium configurations. The "exciton soliton" of the electronically excited molecular lattice arises from a feedback mechanism whereby the localization of the excitation is stabilized by a coupling to the lattice motion. The solitons in polyacetylene combine both these mechanisms. Finally, the solitons of longitudinal lattice vibrations arise simply from the hard-core repulsion of atomic interactions.

These systems have the common attributes of being extended, nonlinear, dispersive media. It is perhaps more difficult to imagine an extended physical system that is not both nonlinear and dispersive. In performing the mathematical analysis, it has been convenient to assume that the systems were infinite in extent. However, it is clear from the analytic results and numerical simulations that the physical requirement is only that the system be much longer than the soliton width. The soliton width is determined by the interparticle forces and the soliton velocity. In the study of the SG, ϕ^4, nonlinear-Schrödinger-equation, and polyacetylene systems, the continuum approximation is essential to the soliton concept. This provides the translational invariance necessary for free soliton propagation. A continuum treatment implies that the soliton must have a width of at least several lattice sites, so that the system itself must be correspondingly long. The longitudinal vibrations of Section X were studied in chains of 50 or more atoms. However, the soliton nature of the dynamics was found to be important for all the long-wavelength modes of these lattices, even down to a wavelength as short as 4 lattice sites at high energy.[169] It seems reasonable to suggest that solitonlike phenomena might be suspected, if not expected, in the 1-D dynamics of almost every physical system that has at least a few similar particles spatially arrayed.

This introduction to the soliton concept has concentrated mainly on the properties of individual solitons: their form, stability, and propagation. The free propagation of the solitons can obviously play an important role in the transport of energy, spin, or charge as well as governing relaxation processes in spin or dielectric systems. The interaction of solitons with low-amplitude waves was studied and shown to be essential to the understanding of the statistical mechanics in SG chains. Moreover, these interactions led to the idea of soliton diffusion and the incoherent transport associated with that behavior.

We have not dealt at length with the interactions between solitons. This is partly due to the greater complexity of the mathematics required, which would make this introduction to solitons unreasonably demanding. It is also a reflection of the physical-sciences literature, which is dominated by the application of solitons and solitary waves to phenomenological models like those considered here. The strict soliton behavior of the SG, NSE, and Toda systems raises very important questions about the nature of physical systems.

These models are completely integrable. There are as many constants of the motion as there are particles in the system, though not because the forces are linear or even because the Hamiltonian is separable. In such nonlinear systems, we do not have the principle of linear superposition to build up a general description of the classical motion from a combination of fundamental harmonic modes. Nevertheless, the motion can be decomposed into components associated with constants of the motion. The N-soliton formulas do this. For example, in a Toda chain of N particles, an expression for N coexisting solitons requires $2N$ conditions for the velocities and arbitrary positions of the solitons. This is equivalent to specifying the $2N$ initial conditions for the positions and momenta of all the particles, and gives a decomposition of the motion in terms of N solitons that retain their integrity as time evolves. It seems that systems that support strict solitons would be integrable if all possible initial conditions could be specified in this way.

However, solitonlike dynamics is not restricted to perfectly integrable systems. The atomic chains of Section X display strikingly solitonlike behavior without being perfectly integrable. So it may be that the solitonlike dynamics of these systems is related to the fact that in some energy range, or in some part of their phase space, these systems behave very like integrable systems. This has important consequences for the energy-level spectrum. Clearly, soliton research is closely bound up with the study of ergodicity, energy redistribution, and invariants, which dominates the modern work in classical mechanics and intramolecular dynamics in particular.[145]

In this more general concern with nonlinear dynamics, the solitons stand out as exciting new types of behavior that extend our understanding of many systems. To paraphrase John Scott-Russell, the soliton is not at all a singular phenomenon, but still a strikingly beautiful one.

Acknowledgments

I would like to thank my colleagues at A. N. U., Prof. D. P. Craig and Drs. J. Epstein and C. Carroll, for helpful discussions, and Dr. C. P. Mallett for assistance with a translation from the German. I gratefully acknowledge the support of a Commonwealth Travelling Fellowship from the Nuffield Foundation and the hospitality of Prof. A. D. Buckingham and the Cambridge University Chemical Laboratory where this review was completed.

APPENDIX A.   DERIVATION OF THE SG MODEL

Heisenberg's equations of motion for the spins are

$$i\frac{dS_l^\alpha}{dt} = [S_l^\alpha, H_F], \qquad \alpha = x, y, z \tag{A.1}$$

Using H_F from (2.4) and the commutation relations for angular momenta,[194] (A.1) gives

$$\frac{dS_l^x}{dt} = -2J\{S_l^z(S_{l-1}^y + S_{l+1}^y) - S_l^y(S_{l-1}^z + S_{l+1}^z)\} - A[S_l^y S_l^z + S_l^z S_l^y] \tag{A.2}$$

$$\frac{dS_l^y}{dt} = -2J\{S_l^x(S_{l-1}^z + S_{l+1}^z) - S_l^z(S_{l-1}^x + S_{l+1}^x)\}$$
$$+ A[S_l^x S_l^z + S_l^z S_l^x] + \mu_B g B^x S_l^x \tag{A.3}$$

$$\frac{dS_l^z}{dt} = -2J\{S_l^y(S_{l-1}^x + S_{l+1}^x) - S_l^x(S_{l-1}^y + S_{l+1}^y)\} - \mu_B g B^x S_l^y \tag{A.4}$$

Now, substituting the classical spin vector given in (2.7) for $\mathbf{S}$ into Eqs. (A.2) to (A.4) gives equations of motion for θ_l and ϕ_l. Note that there are only two independent spin components since $(S_l^x)^2 + (S_l^y)^2 + (S_l^z)^2 = S^2$. With this substitution (A.4) gives

$$\frac{d\theta_l}{dt} = 2JS\{\sin\phi_l[\sin\theta_{l-1}\cos\phi_{l-1} + \sin\theta_{l+1}\cos\phi_{l+1}]$$
$$- \cos\phi_l[\sin\theta_{l-1}\sin\phi_{l-1} + \sin\theta_{l+1}\sin\phi_{l+1}]\} + \mu_B g B^x \sin\phi_l \tag{A.5}$$

Then, multiplying (A.3) by $\cos\phi_l$, (A.2) by $\sin\phi_l$, and subtracting, gives

$$\sin\theta_l\frac{d\phi_l}{dt} = 2JS\{\cos\theta_l\cos\phi_l[\sin\theta_{l-1}\cos\phi_{l-1} + \sin\theta_{l+1}\cos\phi_{l+1}]$$
$$- \sin\theta_l(\cos\theta_{l-1} + \cos\theta_{l+1})$$
$$+ \cos\theta_l\sin\phi_l[\sin\theta_{l-1}\sin\phi_{l-1} + \sin\theta_{l+1}\sin\phi_{l+1}]\}$$
$$+ 2AS\cos\theta_l\sin\theta_l + \mu_B g B^x\cos\theta_l\cos\phi_l \tag{A.6}$$

That is,

$$\frac{d\theta_l}{dt} = 2JS\{\sin\theta_{l-1}\sin(\phi_l - \phi_{l-1}) - \sin\theta_{l+1}\sin(\phi_{l+1} - \phi_l)\} + \mu_B g B^x \sin\phi_l$$

$$(A.7)$$

$$\frac{d\phi_l}{dt} = 2JS\{\cot\theta_l[\sin\theta_{l-1}\cos(\phi_l - \phi_{l-1}) + \sin\theta_{l+1}\cos(\phi_{l+1} - \phi_l)]$$

$$- [\cos\theta_{l-1} + \cos\theta_{l+1}]\} + 2AS\cos\theta_l + \mu_B g B^x \cot\theta_l \cos\phi_l \qquad (A.8)$$

If $\theta_{l-1} \simeq \theta_{l+1} \simeq \theta_l (\simeq \frac{1}{2}\pi)$, then

$$\cot\theta_l[\sin\theta_{l-1}\cos(\phi_l - \phi_{l-1}) + \sin\theta_{l+1}\cos(\phi_{l+1} - \phi_l)] - \cos(\theta_{l-1} + \theta_{l+1})$$

$$\simeq (\cos\theta_l)a^2\left(\frac{\partial\phi}{\partial z}\right)^2 \ll 1 \qquad \text{and is neglected}$$

Moreover, we choose conditions under which $\mu_B g B^x \ll 2AS$, so that (A.8) gives

$$\frac{1}{2AS}\frac{\partial\phi_l}{\partial t} = \cos\theta_l + O\left(\frac{\mu_B g B^x}{2AS}\right) \qquad (A.9)$$

as in (2.11). In (A.7) we use the continuum limit

$$\sin(\phi_{l+1} - \phi_l) \simeq \sin a\left(\frac{\partial\phi_{l+\frac{1}{2}}}{\partial z}\right) \simeq a\frac{\partial\phi_{l+\frac{1}{2}}}{\partial z} \qquad (A.10)$$

where a is the lattice spacing, and

$$\sin(\phi_{l+1} - \phi_l) - \sin(\phi_l - \phi_{l-1}) \simeq a^2\frac{\partial^2\phi_l}{\partial z^2} \qquad (A.11)$$

Thus (A.7) becomes, with $\sin\theta_{l-1} = \sin\theta_{l+1} = 1$,

$$\frac{\partial\theta(z,t)}{\partial t} = -2JSa^2\frac{\partial^2\phi(z,t)}{\partial z^2} + \mu_B g B^x \sin\phi \qquad (A.12)$$

Finally, differentiating (A.9) with respect to time gives

$$\frac{1}{2AS}\frac{\partial^2\phi}{\partial t^2} = -\frac{\partial\theta}{\partial t} = 2JSa^2\frac{\partial^2\phi}{\partial z^2} - \mu_B g B^x \sin\phi \qquad (A.13)$$

with $\sin \theta_l = 1$. That is,

$$\frac{\partial^2 \phi}{\partial z^2} - [4AJS^2a^2]^{-1}\frac{\partial^2 \phi}{\partial t^2} = \frac{\mu_B g B^x}{2JSa^2}\sin \phi \qquad (A.14)$$

which is (2.10).

The assumptions used to obtain (A.14) are not very subtle but are perhaps adequate.

The continuum Hamiltonian can now be derived from H_F. Restating (2.8)

$$H_F = \sum_l -2JS^2\left[\sin \theta_l \sin \theta_{l+1}\cos(\theta_{l+1} - \phi_l) + \cos \theta_l \cos \theta_{l+1}\right]$$

$$+ AS^2\cos^2\theta_l - g\mu_B SB^x\sin \theta_l \cos \phi_l, \qquad (A.15)$$

$$\simeq 2JS^2a^2\int \frac{dz}{a}\left\{\frac{-1}{a^2}\left[\sin \theta_l \sin \theta_{l+1}\cos(\phi_{l+1} - \phi_l) + \cos \theta_l \cos \theta_{l+1}\right]\right.$$

$$\left. + \frac{A}{2Ja^2}\cos^2\theta_l - \frac{g\mu_B B^x}{2JSa^2}\sin \theta_l \cos \phi_l\right\}. \qquad (A.16)$$

Now from (A.9)

$$\frac{A}{2Ja^2}\cos^2\theta_l = \frac{1}{2}\frac{1}{C^2}\left(\frac{\partial \phi}{\partial t}\right)^2 \qquad (A.17)$$

and

$$\frac{-g\mu_B B^x}{2JSa^2}\sin \theta_l \cos \phi_l \simeq -m^2\cos \phi \qquad (A.18)$$

if $\sin \theta_l = 1$. Moreover,

$$\sin \theta_l \sin \theta_{l+1}\cos(\phi_{l+1} - \phi_l) + \cos \theta_l \cos \theta_{l+1}$$

$$\simeq \cos(\theta_{l+1} - \theta_l) - \sin \theta_l \sin \theta_{l+1}\frac{1}{2!}\left(\frac{\partial \phi}{\partial l}\right)^2 \simeq 1 - \frac{1}{2}\left(\frac{\partial \phi}{\partial l}\right)^2 \quad (A.19)$$

Thus, H_F becomes

$$H_F = 2JS^2a^2\int \frac{dz}{a}\left\{\frac{1}{2}\left(\frac{\partial \phi}{\partial z}\right)^2 - \frac{1}{a^2} + \frac{1}{2C^2}\left(\frac{\partial \phi}{\partial t}\right)^2 - m^2\cos \phi\right\} \quad (A.20)$$

We then add a constant to the Hamiltonian to give the ground state, $\phi(z, t)$

= 0, as the state of zero energy; thus,

$$H_F = 2JS^2a^2 \int \frac{dz}{a} \left\{ \frac{1}{2}\left(\frac{\partial\phi}{\partial z}\right)^2 + \frac{1}{2C^2}\left(\frac{\partial\phi}{\partial t}\right)^2 + m^2(1-\cos\phi) \right\} \quad (A.21)$$

APPENDIX B. STABILITY OF SG SOLITONS

From (3.17)

$$\frac{\partial^2 \delta u}{\partial z^2} - \frac{1}{c^2}\frac{\partial^2 \delta u}{\partial t^2} = m^2\cos[u_s(x)]\,\delta u \quad (B.1)$$

Using the kink soliton solution of (3.12) gives

$$\frac{\partial^2 \delta u}{\partial z^2} - \frac{1}{c^2}\frac{\partial^2 \delta u}{\partial t^2} = m^2\{1 - 2\operatorname{sech}^2[m\gamma(z - z_0 - vt)]\}\,\delta u \quad (B.2)$$

Fogel et al. have solved (B.2) for stationary kinks ($v = 0$).[24] In the general case, we proceed to a separable equation via a Lorentz transformation. We define

$$\xi = m\gamma[z - z_0 - vt]$$
$$\eta = m\gamma t \quad (B.3)$$

In terms of these variables, (B.2) becomes

$$\frac{\partial^2 \delta u}{\partial \xi^2} - \nu^2\frac{\partial^2 \delta u}{\partial \eta^2} + 2v\nu^2\frac{\partial^2 \delta u}{\partial \xi\,\partial \eta} = (1 - 2\operatorname{sech}^2\xi)\,\delta u(\xi, \eta) \quad (B.4)$$

where $\nu^2 = (c^2 - v^2)^{-1}$.

Now the only nonconstant coefficient in (B.4) is a function of ξ alone. Thus we can quite generally write $\delta u(\xi, \eta)$ as some linear combination of products of the form

$$\delta u(\xi, \eta) = e^{i\omega\eta}f(\xi, \omega) \quad (B.5)$$

Substituting (B.5) in (B.4) gives

$$\frac{\partial^2 f}{\partial \xi^2} + \omega^2\nu^2 f + 2i\omega v\nu^2\frac{\partial f}{\partial \xi} = (1 - 2\operatorname{sech}^2\xi)f \quad (B.6)$$

Setting

$$f(\xi) = \exp\left[-i\omega v \nu^2 \xi\right] g(\xi) \tag{B.7}$$

gives the soluble Schrödinger equation[24]

$$-\frac{\partial^2 g}{\partial \xi^2} + \left(1 - 2\operatorname{sech}^2\xi\right)g = \Omega^2 g \tag{B.8}$$

where $\Omega^2 = (\omega c \nu^2)^2$. $\tag{B.9}$
The potential, $1 - 2\operatorname{sech}^2\xi$, which arises from the kink shape, has only one bound state given by

$$\Omega = 0 \tag{B.10}$$

$$g(\xi, \Omega = 0) = 2\operatorname{sech}\xi \tag{B.11}$$

and a continuum of solutions labeled by wavenumber k, with

$$\Omega^2 = 1 + k^2 \tag{B.12}$$

$$g(\xi, k) = \frac{1}{(2\pi)^{1/2}} \frac{e^{ik\xi}}{\Omega(k)}(k + i\tanh\xi) \tag{B.13}$$

The bound state at $\Omega = 0$ is a Goldstone mode[87] associated with the translational invariance of the soliton. If

$$\delta u(\xi, \eta) = \alpha g(\xi, \omega = 0), \qquad \alpha \text{ small} \tag{B.14}$$

then

$$\begin{aligned} u(z, t) &= u_s(\xi) + \alpha g(\xi, \omega = 0) \\ &= u_s(\xi) + \alpha \frac{du_s(\xi)}{d\xi} \\ &= u_s(\xi + \alpha) + O(\alpha^2) \end{aligned} \tag{B.15}$$

Thus small perturbations associated with this zero-frequency mode are small translations of the soliton. The eigenfunctions $g(\xi, k)$ form a complete orthogonal set for functions $f(\xi)$, with inner product

$$\begin{aligned} N(k_1, k_2) &= \lim_{L \to \infty} \frac{1}{2L} \int_{-L}^{L} g(\xi, k_1) g^*(\xi, k_2)\, d\xi. \\ &= \delta(k_1 - k_2) \end{aligned} \tag{B.16}$$

This inner product vanishes if $g(\xi, k)$ is replaced by the bound-state eigen-function, $g(\xi, \omega = 0)$, and

$$\int_{-\infty}^{\infty} f(\xi, \omega = 0) f^*(\xi, \omega = 0) \, d\xi = 8 \tag{B.17}$$

The set is complete,[24]

$$\int_{\infty}^{\infty} dk \, g(k, \xi) g^*(k, \xi') + \tfrac{1}{8} g(\xi, \Omega = 0) g^*(\xi', \Omega = 0) = \delta(\xi - \xi') \tag{B.18}$$

Thus any small perturbation in an SG chain containing a kink (or antikink) can be expressed as a linear combination of terms of the form

$$\delta u(z, t) = e^{i\omega \eta} e^{-i\omega v \nu^2 \xi} g(\xi, k)$$
$$= \exp\{im\gamma[(\omega c^2 \nu^2 - kv)t + (k - \omega v \nu^2)(z - z_0)]\}$$
$$\times [2\pi(1 + k^2)]^{-1/2} \{k + i \tanh[m\gamma(z - z_0 - vt)]\} \tag{B.19}$$

Defining

$$q = m\gamma(k - \omega v \nu^2) \tag{B.20}$$

and

$$\omega(q) = m\gamma(kv - \omega c^2 \nu^2) \tag{B.21}$$
$$\delta u(z, t) = \exp\{i[\omega(q)t - q(z - z_0)]\}$$
$$\times [2\pi(1 + k^2)]^{-1/2} \{k + i \tanh[m\gamma(z - z_0 - vt)]\} \tag{B.22}$$

where (B.12) gives the same dispersion relation

$$\omega^2(q) = c^2(m^2 + q^2) \tag{B.23}$$

as in the absence of kinks. From (B.20) and (B.21)

$$k = [c^2 q - v\omega(q)][m\gamma(c^2 - v^2)]^{-1} \tag{B.24}$$

The perturbation, $\delta u(z, t)$, in (B.22) is simply a perturbed form of the low-amplitude periodic waves found in the absence of solitons. The soliton does not affect the dispersion relation of (3.8). However, the wave is perturbed *in the region* of the soliton by the tanh term in (B.22). Moreover, the wave suffers a phase shift. For as $z \to \infty$,

$$\delta u(z, t) \; \alpha \; \exp\left\{ i\left[\omega(q)t - q(z - z_0) + \arctan\left(\frac{1}{k}\right)\right]\right\},$$

and as $z \to -\infty$,

$$\delta u(z, t) \; \alpha \; \exp\left\{ i\left[\omega(q)t - q(z - z_0) + \arctan\left(\frac{-1}{k}\right)\right]\right\}$$

This corresponds to a phase shift of $\Delta(k)$:

$$\Delta(k) = \Delta(q, v) = \pi\frac{k}{|k|} - 2\arctan k \tag{B.25}$$

where $k = k(q, v)$ as given in (B.24).

Since the amplitude of the periodic mode does not increase with time, the phonon–soliton collision is elastic and the soliton is stable.

APPENDIX C. BÄCKLUND TRANSFORMATION FOR SG SOLITONS

Lamb[195] and Barnard[196], among others, have found the multiple-soliton solutions of the SG equation using Bäcklund transformations. We draw heavily here on the original derivation of the two soliton collision formulas by Seeger, Donth, and Kochendorfer.[25]

It is convenient to change variables in (3.1), setting

$$p = \frac{m}{2}(z - ct) \quad \text{and} \quad q = \frac{m}{2}(z + ct) \tag{C.1}$$

Equation (3.1) then becomes

$$\frac{\partial^2 u}{\partial p\, \partial q} = \sin u \tag{C.2}$$

or

$$\frac{\partial^2\left(\frac{1}{2}u\right)}{\partial p\, \partial q} = \sin\left(\frac{u}{2}\right)\cos\left(\frac{u}{2}\right) \tag{C.3}$$

Equation (C.3) suggests that we might consider some variation along the lines of

$$\frac{\partial^{\frac{1}{2}} u}{\partial p} = C_1 \sin\tfrac{1}{2} u \quad \text{and} \quad \frac{\partial^{\frac{1}{2}} u}{\partial q} = C_2 \sin\tfrac{1}{2} u \tag{C.4}$$

which would satisfy (C.3) if $C_1 C_2 = 1$. Forsyth has shown that if $u_1(p, q)$ and $u_2(p, q)$ are solutions of the SG equation (C.3), then[27]

$$\frac{\partial}{\partial p} \tfrac{1}{2} (u_2 - u_1) = C \sin\left[\tfrac{1}{2}(u_2 + u_1)\right]$$
$$\frac{\partial}{\partial q} \tfrac{1}{2} (u_2 + u_1) = \frac{1}{C} \sin\left[\tfrac{1}{2}(u_2 - u_1)\right] \tag{C.5}$$

or more symmetrically[25]

$$\frac{\partial}{\partial p} \tfrac{1}{2} (u_2 - u_1) = \frac{1 + \sin\sigma}{\cos\sigma} \sin\left[\tfrac{1}{2}(u_2 + u_1)\right]$$
$$\frac{\partial}{\partial q} \tfrac{1}{2} (u_2 + u_1) = \frac{1 - \sin\sigma}{\cos\sigma} \sin\left[\tfrac{1}{2}(u_2 - u_1)\right] \tag{C.6}$$

This relation between two solutions of the SG equation can be verified by cross differentiation. The essence of the Bäcklund transformation is: Given some known solution $u_1(p, q)$, a new solution $u_2(p, q)$ can be constructed from the two first-order equations of (C.6). $u_2(p, q)$ is a Bäcklund transformation of $u_1(p, q)$ for some chosen σ, or

$$u_2 = B_\sigma u_1 \tag{C.7}$$

The Bäcklund transformation may be repeated to obtain further solutions. If $u_1 = B_{\sigma_1} u_0$, $u_2 = B_{\sigma_2} u_0$ and if

$$u_3 = B_{\sigma_1} u_2 \tag{C.8}$$

then

$$u_3 = B_{\sigma_1} B_{\sigma_2} u_0 = B_{\sigma_2} B_{\sigma_1} u_0 \tag{C.9}$$

One can write a direct relation between u_0, u_1, u_2, and u_3:

$$\tan\left(\frac{u_3 - u_0}{4}\right) = \frac{\cos\tfrac{1}{2}(\sigma_1 + \sigma_2)}{\sin\tfrac{1}{2}(\sigma_1 - \sigma_2)} \tan\left(\frac{u_1 - u_2}{4}\right) \tag{C.10}$$

Special attention must be paid to the case $\sigma_1 = \sigma_2$.[25]

Now let us use (C.6) to generate some solutions. Clearly $u(p, q) = 0$ is a solution of (C.1) so let us set $u_0(p, q) = 0$, then $u_1(p, q) = B_{\sigma_1} u_0(p, q)$ is

$$\frac{\partial(\frac{1}{2}u_1)}{\partial p} = \frac{1 + \sin \sigma_1}{\cos \sigma_1} \sin \tfrac{1}{2} u_1$$

$$\frac{\partial(\frac{1}{2}u_1)}{\partial q} = \frac{1 - \sin \sigma_1}{\cos \sigma_1} \sin \tfrac{1}{2} u_1$$

(C.11)

Thus,

$$\tan \tfrac{1}{4} u_1 = \exp\left\{ \frac{1 + \sin \sigma_1}{\cos \sigma_1} p + \frac{1 - \sin \sigma_1}{\cos \sigma_1} q \right\}$$

$$= \exp\left\{ \frac{m}{\cos \sigma_1} z + \frac{m \sin \sigma_1}{\cos \sigma_1} ct \right\}$$

(C.12)

Setting $\sin \sigma_1 = v/c$, $0 < \sigma_1 < \tfrac{1}{2}\pi$,

$$\tan \tfrac{1}{4} u_1 = \exp\{ m\gamma(z - vt) \}$$

(C.13)

which is just the single soliton kink of (3.12). The antikink is obtained by putting $\sigma = \pi - \sigma_1$. Setting $\sigma_2 = -\sigma_1$ gives the kinks (antikinks) propagating in the opposite direction.

We can obtain the solutions describing kink–kink and kink–antikink collisions by repeating the Bäcklund transformation and using (C.10). For a kink–kink collision we set $\sin \sigma_1 = v/c$, $\sigma_2 = \sigma_1 + \pi$, then

$$\tan \tfrac{1}{4} u_1 = \exp\{ m\gamma(z - vt) \}$$

(C.14)

$$\tan \tfrac{1}{4} u_2 = \exp\{ -m\gamma(z + vt) \}$$

(C.15)

and (C.10) gives

$$\tan \tfrac{1}{4} u_3 = \sin \sigma_1 \frac{\sinh(m\gamma z)}{\cosh(m\gamma vt)}$$

$$= \frac{v}{c} \frac{\sinh(m\gamma z)}{\cosh(m\gamma vt)}$$

(C.16)

As $t \to -\infty$,

$$\tan \tfrac{1}{4} u_3 \overset{z \to -\infty}{\to} -\frac{v}{c} \exp\{ -m\gamma(z - vt) \}$$

$$\overset{z \to \infty}{\to} \frac{v}{c} \exp\{ m\gamma(z + vt) \}$$

(C.17)

This describes two well-separated kinks, one at large negative z with a rotation from $u = -2\pi$ to $u = 0$, and one at large positive z with a rotation from $u = 0$ to $u = 2\pi$. As $t \to \infty$,

$$\tan\tfrac{1}{4}u_3 \overset{z \to -\infty}{\to} -\frac{v}{c}\exp\{-m\gamma(z + vt)\}$$

$$\overset{z \to +\infty}{\to} \frac{v}{c}\exp\{m\gamma(z - vt)\} \tag{C.18}$$

The solitons have now passed through one another as shown in Fig. 5.

The kink–antikink collision is given by (C.10) with $\sin\sigma_1 = -v/c$, $\sigma_2 = -\sigma_1$:

$$\tan\tfrac{1}{4}u_4 = \frac{c}{v}\frac{\sinh(m\gamma vt)}{\cosh(m\gamma z)} \tag{C.19}$$

This function is also presented in Fig. 5.

The *breather mode* given in (3.20) can also be obtained via the Bäcklund transformation. This is achieved simply by taking σ_1 and σ_2 as complex parameters, so that

$$\tan\tfrac{1}{4}u_1 = \exp\left\{\frac{m}{\cos\sigma_1}\left[z - (\sin\sigma_1)ct\right]\right\} \tag{C.20}$$

This is a complex solution for σ_1 complex. Moreover, taking

$$\sigma_2 = \sigma_1^* = \alpha - i\beta \tag{C.21}$$

in (C.10), gives the real solution

$$\tan\tfrac{1}{4}u_3 = \frac{\cos\alpha}{\sin i\beta}\frac{\exp\left\{\dfrac{m}{\cos\sigma_1}\left[z - (\sin\sigma_1)ct\right]\right\} - \exp\left\{\dfrac{m}{\cos\sigma_1^*}\left[z - (\sin\sigma_1^*)ct\right]\right\}}{1 + \exp\left\{\dfrac{m}{\cos\sigma_1}\left[z - (\sin\sigma_1)ct\right] + \dfrac{m}{\cos\sigma_1^*}\left[z - (\sin\sigma_1^*)ct\right]\right\}}$$

or,

$$\tan\tfrac{1}{4}u_3 = \frac{\cos\alpha}{\sinh\beta}\frac{\sin\left[\dfrac{(\sin\alpha\sinh\beta)mz - (\sinh\beta\cosh\beta)mct}{\sinh^2\beta + \cos^2\alpha}\right]}{\cosh\left[\dfrac{(\cos\alpha\cosh\beta)mz - (\sin\alpha\cos\alpha)mct}{\sinh^2\beta + \cos^2\alpha}\right]} \tag{C.22}$$

Setting

$$\tanh \beta = \frac{\omega_B}{\omega_0} \left(1 - \frac{v^2}{c^2} \right)^{1/2}$$

$$\tan \alpha = \frac{v}{c} \left(1 - \frac{v^2}{c^2} \right)^{-1/2} \left(1 - \frac{\omega_B^2}{\omega_0^2} \right)^{-1/2}, \qquad \omega_0 = mc$$

(C.23)

We have

$$\tan \tfrac{1}{4} u_3 = \left(\frac{\omega_0^2}{\omega_B^2} - 1 \right)^{1/2} \frac{\sin\left[\gamma \omega_B \left(vz/c^2 - t \right) \right]}{\cosh\left[m\gamma(z - vt)\left(1 - \omega_B^2/\omega_0^2 \right)^{1/2} \right]} \quad (C.24)$$

which is the breather mode of Eq. (3.20). The derivation of the breather solution is clearly analogous to that for soliton collisions, which implies that the breather is a cooperative phenomenon involving two solitons.

APPENDIX D. SOLITON SOLUTION OF THE NONLINEAR SCHRÖDINGER EQUATION

To solve Eq. (8.23) we eliminate the first derivative in the standard manner by setting

$$a(\phi) = \exp\left(\frac{i\hbar v}{2M} \right) b(\phi) \tag{D.1}$$

where (8.23) gives

$$\frac{M d^2 b}{d\phi^2} + \left[\hbar\omega - H_0 - \Delta\varepsilon + 2M + \frac{(\hbar v)^2}{4M} \right] b + \frac{36 M^2}{m(c^2 - v^2)} b^3 = 0 \quad (D.2)$$

Multiplying (D.2) by $db/d\phi$, integrating from $-\infty$ to ϕ, and assuming that $b(-\infty) = db(-\infty)/d\phi = 0$, gives

$$M \left(\frac{db}{d\phi} \right)^2 = \left[H_0 + \Delta\varepsilon - 2M - \frac{(\hbar v)^2}{4M} - \hbar\omega \right] b^2 (1 - \lambda^2 b^2) \quad (D.3)$$

where

$$\lambda^2 = \frac{18 M^2}{m(c^2 - v^2)} \left[H_0 + \Delta\varepsilon - 2M - \frac{(\hbar v)^2}{4M} - \hbar\omega \right]^{-1} \quad (D.4)$$

Equation (D.3) is solved by simple integration to give $b(\phi)$ of the form

$$b(\phi) = \alpha \operatorname{sech} \beta \phi \tag{D.5}$$

The normalization condition (8.6), in the continuum limit, requires that

$$\frac{2\alpha^2}{\beta} = 1 \tag{D.6}$$

which determines the eigenvalue $\hbar\omega$ as

$$\hbar\omega = H_0 + \Delta\varepsilon - 2M - \frac{(\hbar v)^2}{4M} - \left[\frac{9M^2}{m(c^2 - v^2)}\right]^2 M^{-1} \tag{D.7}$$

This gives

$$\alpha = \frac{3}{\sqrt{2}}\left[\frac{M}{m(c^2 - v^2)}\right]^{1/2} \tag{D.8}$$

$$\beta = \frac{9M}{m(c^2 - v^2)} \tag{D.9}$$

The energy of the soliton is given by (8.11). Relative to the ground-state equilibrium, the continuum approximation is

$$E(v) = \int_{-\infty}^{\infty} d\phi \left\{ \frac{1}{2M}\left(\frac{dp}{d\phi}\right)^2 + V(r) - V(r_{eq}) + \Delta\varepsilon|a|^2 \right.$$

$$\left. + M(r)a^*(\phi)a(\phi - 1) + M[r(\phi+1)]a^*(\phi)a(\phi+1) \right\}$$

$$= \hbar\omega + \frac{(\hbar v)^2}{2M} \tag{D.10}$$

$$= \Delta\varepsilon - 2M + \frac{(\hbar v)^2}{4M} - \frac{[9M^2/m(c^2 - v^2)]^2}{M}$$

$$+ \int_{-\infty}^{\infty} d\phi \left\{ \frac{1}{2m}\left(\frac{dp}{d\phi}\right)^2 + \frac{1}{2}V''(r_{eq})[r(\phi) - r_{eq}]^2 \right\}$$

$$= \Delta\varepsilon - 2M + \frac{(\hbar v)^2}{4M} - \frac{[9M^2/m(c^2 - v^2)]^2}{M}$$

$$+ \left[\frac{1}{2}mv^2 + \frac{1}{2}V''(r_{eq})\right]\int_{-\infty}^{\infty} [r(\phi) - r_{eq}]^2 d\phi \tag{D.11}$$

from (8.12a). Using (8.21), Eq. (D.5), and the lowest-order continuum approximation,

$$E(v) = \Delta\varepsilon - 2M + \frac{(\hbar v)^2}{4M} - \frac{1}{M}\left[\frac{9M^2}{m(c^2 - v^2)}\right]^2$$

$$\times \left\{1 - \frac{2}{3}\left[\frac{c^2 + v^2}{c^2 - v^2}\right]\cos^2\left(\frac{\hbar v}{2M}\right)\right\} \tag{D.12}$$

References

1. J. Scott-Russell, *Proc. R. Soc. Edinburgh*, 319 (1844).
2. L. F. Mollenauer, R. H. Stolen, and J. P. Gordon, *Phys. Rev. Lett.* **45**, 1095 (1980).
3. H. A. Haus, *Rev. Mod. Phys.* **51**, 331 (1979).
4. A. C. Scott, F. Y. F. Chu, and D. W. McLaughlin, *Proc. IEEE* **61**, 1443 (1973).
5. R. K. Bullough, *Phys. Bull.* 78 (Feb. 1978).
6. G. Eilenberger, *Solitons*, Vol. 19, Springer Series in Solid-State Sciences, Springer-Verlag, Berlin, 1981.
7. R. Bullough and P. Caudrey, Eds., *Solitons*, Vol. 17, Topics in Current Physics, Springer-Verlag, Berlin, 1980.
8. A. R. Bishop and T. Schneider, Eds. *Solitons and Condensed Matter Physics*, Vol. 8, Springer Series in Solid-State Sciences, Springer-Verlag, Berlin, 1978.
9. W. Eckhaus and A. Van Harten, *The Inverse Scattering Transformation and the Theory of Solitons*, North-Holland, Amsterdam, 1981.
10. 'H. J. Mikeska, *J. Mag. Mag. Mater.* **13**, 35 (1979).
11. H. Yoshizawa, K. Hirakawa, S. K. Satija, and G. Shirane, *Phys. Rev. B* **23**, 2298 (1981).
12. M. Steiner, J. Villain, and C. G. Windsor, *Adv. Phys.* **25**, 87 (1976).
13. J. P. Boucher, L. P. Regnault, J. Rossat-Mignod, and J. Villain, *Solid State Commun.* **31**, 311 (1979).
14. M. Steiner and J. K. Kjems, *J. Phys. C* **10**, 2665 (1977).
15. H. J. Mikeska and E. Patzak, *Z. Phys. B* **26**, 253 (1977).
16. H. J. Mikeska, *J. Phys. C* **11**, L29 (1978).
17. H. J. Mikeska, *J. Appl. Phys.* **52**, 1950 (1981).
18. A. R. Bishop, *J. Phys. C* **13**, L67 (1980).
19. J. M. Loveluck, T. Schneider, E. Stoll, and H. R. Jauslin, *Phys. Rev. Lett.* **45**, 1505 (1980).
20. A. R. Bishop, *Z. Phys. B* **37**, 357 (1980).
21. K. Maki, *J. Low Temp. Phys.* **41**, 327 (1980).
22. J. Villain and J. M. Loveluck, *J. Phys. (Paris) Lett.* **38**, L77 (1977).
23. M. Abramowitz and I. A. Stegun, Eds. *Handbook of Mathematical Functions*, Dover, New York, 1965.
24. M. B. Fogel, S. E. Trullinger, A. R. Bishop, and J. A. Krumhansl, *Phys. Rev. B* **15**, 1578 (1977).
25. A. Seeger, H. Donth, and A. Kochendorfer, *Z. Phys.* **134**, 173 (1953).

26. J. K. Perring and T. H. R. Skyrme, *Nucl. Phys.* **31**, 550 (1962).

27. A. R. Forsyth, *Theory of Differential Equations*, Vol. 6, Dover, New York, 1959, Chap. 21.

28. M. J. Ablowitz, D. J. Kaup, A. C. Newell, and H. Segur, *The Inverse Scattering Transform-Fourier Analysis for Non-Linear Problems*, Studies in Applied Mathematics, Vol. 53, No. 4, M.I.T. Press, 1974, p. 249.

29. P. D. Lax, *Commun. Pure Appl. Math.* **21**, 467 (1968).

30. J. F. Currie, J. A. Krumhansl, A. R. Bishop, and S. E. Trullinger, *Phys. Rev. B* **22**, 477 (1980).

31. R. F. Dashen, B. Hasslacher, and A. Neveu, *Phys. Rev. D* **11**, 3424 (1975).

32. A. R. Bishop, *J. Phys. A* **14**, 883 (1981).

33. K. Maki and H. Takayama, *Phys. Rev. B* **20**, 5002 (1979); **22**, 5302 (1980).

34. W. C. Kerr, D. Baeriswyl, and A. R. Bishop, *Phys. Rev. B* **24**, 6566 (1981).

35. D. J. Scalapino, M. Sears, and R. A. Ferrell, *Phys. Rev. B* **6**, 3409 (1972).

36. N. Gupta, and B. Sutherland, *Phys. Rev. A* **14**, 1790 (1976).

37. J. F. Currie, M. B. Fogel, and F. L. Palmer, *Phys. Rev. A* **16**, 796 (1977).

38. J. A. Krumhansl and J. R. Schrieffer, *Phys. Rev. B* **11**, 3535 (1975).

39. S. E. Trullinger and R. M. DeLeonardis, *Phys. Rev. A* **20**, 2225 (1979), gives a similar treatment for systems in which phonons are partially reflected by solitons.

40. A. R. Bishop, J. A. Krumhansl, and S. E. Trullinger, *Physica D* **1**, 1 (1980).

41. R. M. DeLeonardis and S. E. Trullinger, *Phys. Rev. B* **22**, 4558 (1980).

42. P. S. Riseborough and S. E. Trullinger, *Phys. Rev. B* **22**, 4389 (1980).

43. J. F. Currie, S. E. Trullinger, A. R. Bishop, and J. A. Krumhansl, *Phys. Rev. B* **15**, 5567 (1977).

44. A. R. Bishop and W. F. Lewis, *J. Phys. C* **12**, 3811 (1979).

45. J. P. Boucher, L. P. Regnault, J. Rossat-Mignod, J. P. Renard, J. Bouillot, and W. G. Stirling, *Solid State Commun.* **33**, 171 (1980).

46. J. P. Boucher and J. P. Renard, *Phys. Rev. Lett.* **45**, 486 (1980).

47. J. P. Boucher, *Solid State Commun.* **33**, 1025 (1980).

48. A. T. Abdalian, P. Moch, and J. Cibert, *J. Appl. Phys.* **52**, 1980 (1981).

49. J. Cibert, Y. Merle d'Aubigne, J. Ferre, and M. Regis, *J. Phys. C* **13**, 2781 (1980).

50. W. Marshall and S. W. Lovesey, *Theory of Thermal Neutron Scattering*, Oxford University Press, 1971.

51. G. Reiter, preprint presented at the Neutron Scattering Conference at Argonne National Laboratory, Argonne, Ill., August 1981.

52. H. J. Mikeska, *J. Phys. C* **13**, 2913 (1980).

53. J. K. Kjems and M. Steiner, *Phys. Rev. Lett.* **41**, 1137 (1978).

54. M. Steiner, J. Kjems, K. Kakurai, and H. Dachs, *J. Mag. Mag. Mater.* **15–18**, 1057 (1980).

55. M. Steiner, *J. Mag. Mag. Mater.* **14**, 142 (1979).

56. M. Steiner, *J. Appl. Phys.* **50**, 7395 (1979).

57. G. Reiter, *Phys. Rev. Lett.* **46**, 202, 518 (1981).

58. K. M. Leung, D. W. Hone, D. L. Mills, P. S. Riseborough, and S. E. Trullinger, *Phys. Rev. B* **21**, 4017 (1980).

59. D. Hone and K. M. Leung, *Phys. Rev. B* **22**, 5308 (1980).

60. M. Steiner, R. Pynn, W. Knop, K. Kakurai, and J. Kjems, preprint (see Ref. 51).

61. J. P. Boucher, L. P. Regnault, J. Rossat-Mignod, J. P. Renard, J. Bouillot, and W. G. Stirling, *J. Appl. Phys.* **52**, 1956 (1981).

62. Y. Tanabe, T. Moriya, and S. Sugano, *Phys. Rev. Lett.* **15**, 1023 (1965).

63. K. Ebara and Y. Tanabe, *J. Phys. Soc. Jpn.* **36**, 93 (1974).

64. D. C. Mattis, *The Theory of Magnetism*, Harper and Row, New York, 1965.

65. M. Lakshmanan, Th. W. Ruijgrok, and C. J. Thompson, *Physica A* **84**, 572 (1976).

66. K. Nakamura and T. Sasada, *Phys. Lett. A* **48**, 321 (1974).

67. J. Tjon and J. Wright, *Phys. Rev. B* **15**, 3470 (1977).

68. K. Nakamura and T. Sasada, *Solid State Commun.* **21**, 891 (1977).

69. K. Nakamura and T. Sasada, *J. Phys. C* **11**, 331 (1978).

70. A. R. Bishop, K. Nakamura, and T. Sasada, *J. Phys. C* **13**, L515 (1980).

71. J. F. Currie, S. Sarker, A. R. Bishop, and S. E. Trullinger, *Phys. Rev. A* **20**, 2213 (1979).

72. J. F. Currie and A. R. Bishop, *Can. J. Phys.* **56**, 890 (1979).

73. C. Kawabata and A. R. Bishop, *Solid State Commun.* **33**, 453 (1980).

74. B. A. Ivanov, A. M. Kosevich, and I. V. Manzhos, *Solid State Commun.* **34**, 417 (1980).

75. I. V. Bar'yakhtar and B. A. Ivanov, *Solid State Commun.* **34**, 545 (1980).

76. V. I. Ozhogin and V. L. Preobrazhenskii, *Sov. Phys.-JETP* **46**, 523 (1977).

77. V. I. Ozhogin and A. Yu. Lebedev, *J. Mag. Mag. Mater.* **15–18**, 617 (1980).

78. V. I. Ozhogin and A. Yu. Lebedev, and A. Yu. Yakubovskii, *Sov. Phys.-JETP* **27**, 313 (1978).

79. S. Aubry, *J. Chem. Phys.* **62**, 3217 (1975); **64**, 3392 (1976).

80. M. A. Collins, A. Blumen, J. F. Currie, and J. Ross, *Phys. Rev. B* **19**, 3630 (1979).

81. J. F. Currie, A. Blumen, M. A. Collins, and J. Ross, *Phys. Rev. B* **19**, 3645 (1979).

82. A. R. Bishop, E. Domany, and J. A. Krumhansl, *Phys. Rev. B* **14**, 2966 (1976).

83. S. K. Sarker and J. A. Krumhansl, *Phys. Rev. B* **23**, 2374 (1981).

84. V. H. Schmidt, *Phys. Rev. B* **20**, 4397 (1979).

85. T. Sasada, *Phys. Lett. A* **75**, 5 (1979).

86. Y. Wada and J. R. Schrieffer, *Phys. Rev. B* **18**, 3897 (1978).

87. D. Forster, *Hydrodynamic Fluctuations, Broken Symmetry and Correlation Functions*, Benjamin, Reading, 1975.

88. T. Schneider, E. P. Stoll, and R. Morf, *Phys. Rev. B* **18**, 1417 (1978).

89. J. D. Murray, *J. Theor. Biol.* **52**, 459 (1975); **56**, 329 (1976).

90. (a) M. A. Collins and J. Ross, *J. Chem. Phys.* **68**, 3774 (1978); (b) A. Nitzan, P. Ortoleva, and J. Ross, *Faraday Symp. Chem. Soc.* **9**, 241 (1974); (c) H. Metiu, K. Kitahara, and J. Ross, *J. Chem. Phys.* **64**, 292 (1976).

91. E. Fatuzzo, *Phys. Rev.* **127**, 1999 (1962); also J. C. Burfoot, *Ferroelectrics*, Van Nostrand, London, 1967, p. 218; R. C. Miller and G. Weinreich, *Phys. Rev.* **117**, 1460 (1960).

92. J. F. Ryan and K. Hisano, *J. Phys. C* **6**, 566 (1973).

93. C. H. Perry and D. K. Agrawal, *Solid State Commun.* **8**, 225 (1970).

94. R. A. Cowley, J. D. Axe, and M. Iizumi, *Phys. Rev. Lett.* **36**, 806 (1976).

95. J. L. Skinner and P. G. Wolynes, *J. Chem. Phys.* **73**, 4015 (1980).

96. J. L. Skinner and P. G. Wolynes, *J. Chem. Phys.* **73**, 4022 (1980).

97. M. Mansfield and R. H. Boyd, *J. Polym. Sci. Polym. Phys. Ed.* **16**, 1227 (1978).

98. M. L. Mansfield, *Chem. Phys. Lett.* **69**, 383 (1980).

99. M. Büttiker and R. Landauer, *Phys. Rev. Lett.* **43**, 1453 (1979), and preprint.

100. C. H. Bennett, M. Büttiker, R. Landauer, and H. Thomas, *J. Stat. Phys.*, to be published.

101. M. Büttiker and R. Landauer, *J. Phys. C* **13**, L325 (1980); *Phys. Rev. Lett.* **46**, 75 (1981), and preprint.

102. M. Büttiker and H. Thomas, Periodic and Solitary States of the Driven Sine–Gordon Chain, preprint.

103. L. Gunther and Y. Imry, *Phys. Rev. Lett.* **46**, 76 (1981); **44**, 1225 (1980).

104. K. Fesser, *Z. Phys. B* **39**, 47 (1980).

105. J. M. Thomas, S. E. Morin, and J. P. Desvergne, in *Advances in Physical Organic Chemistry*, Vol. 15, V. Gold and D. Bethell, Eds., Academic Press, New York, 1977, p. 63.

106. J. A. Barker, in *Rare Gas Solids*, M. C. Klein and J. A. Venerable, Eds., Academic Press, New York, 1975, Chap. 4.

107. M. Toda, *Prog. Theor. Phys. Suppl.* **36**, 113 (1966).

108. A. S. Davydov, *Theory of Molecular Excitons*, Plenum Press, New York, 1971.

109. A. S. Davydov and N. I. Kislukha, *Phys. Status Solidi B* **59**, 465 (1973).

110. A. S. Davydov and N. I. Kislukha, *Sov. Phys.-JETP* **44**, 571 (1976).

111. M. A. Collins and D. P. Craig, to be published in *Chemical Physics* (1983).

112. A. S. Davydov, *Int. J. Quantum Chem.* **16**, 5 (1979).

113. A. S. Davydov, *Sov. Phys.-JETP* **51**, 397 (1980).

114. A. S. Davydov and V. Z. Enol'skii, *Sov. Phys.-JETP* **52**, 954 (1980).

115. A. S. Davydov, A. A. Eremenko and A. I. Sergienko, *Ukr. Fiz. Zh.* **23** (6), (1978).

116. A. S. Davydov, *Phys. Scr.* **20**, 387 (1979).

117. V. K. Fedyanin, V. G. Makhankov and L. V. Yakushevich, *Phys. Lett. A* **61**, 256 (1977).

118. V. K. Fedyanin and V. G. Makhankov, *Phys. Scr.* **20**, 552 (1979).

119. *Physics Today*, p. 19, (September 1979).

120. W. P. Su, J. R. Schrieffer and A. J. Heeger, *Phys. Rev. Lett.* **42**, 1698 (1979).

121. W. P. Su, J. R. Schrieffer and A. J. Heeger, *Phys. Rev. B* **22**, 2099 (1980).

122. H. C. Longuet-Higgins and L. Salem, (a) *Proc. R. Soc. Lond. Ser. A* **251**, 172 (1959); (b) *Proc. R. Soc. Lond. Ser. A* **257**, 445 (1960).

123. L. Salem and H. C. Longuet-Higgins, *Proc. R. Soc. Lond. Ser. A* **255**, 435 (1960).

124. J. A. Pople and S. H. Walmsley, *Mol. Phys.* **5**, 15 (1962).

125. M. W. Hahn, A. D. McLachlan, H. H. Dearman, and H. M. McConnell, *J. Chem. Phys.* **37**, 361, 3008 (1962).

126. L. Salem, *The Molecular Orbital Theory of Conjugated Systems*, Benjamin, New York, 1966.

127. R. E. Peierls, *Quantum Theory of Solids*, Oxford University Press, London, 1955, p. 108.

128. B. R. Weinberger, E. Ehrenfreund, A. Pron, A. J. Heeger, and A. G. MacDiarmid, *J. Chem. Phys.* **72**, 4749 (1980).

129. S. Ikehata, J. Kaufer, T. Woerner, A. Pron, M. A. Druy, A. Sivak, A. J. Heeger, and A. G. MacDiarmid, *Phys. Rev. Lett.* **45**, 1123 (1980).

130. J. C. W. Chien, F. E. Karasz, and G. E. Wnek, *Nature* **285**, 390 (1980).

131. H. Takayama, Y. R. Lin-Liu, and K. Maki, *Phys. Rev. B* **21**, 2388 (1980).

132. W. P. Su and J. R. Schrieffer, *Proc. Natl. Acad. Sci. USA* **77**, 5626 (1980).

133. E. J. Mele and M. J. Rice, *Phys. Rev. Lett.* **23**, 926 (1980).

134. S. Etemad, A. Pron, A. J. Heeger, A. G. Macdiarmid, E. J. Mele, and M. J. Rice, *Phys. Rev. B* **23**, 5137 (1981).

135. K. Maki and M. Nakahara, *Phys. Rev. B* **23**, 5005 (1981).

136. N. Suzuki, M. Ozahi, S. Etemad, A. J. Heeger, and A. G. MacDiarmid, *Phys. Rev. Lett.* **45**, 1209 (1980).

137. Y. R. Lin-Liu and K. Maki, *Phys. Rev. B* **22**, 5754 (1980).

138. B. Horowitz, *Phys. Rev. B* **22**, 1101 (1980).

139. B. Horowitz, *Phys. Rev. Lett.* **46**, 742 (1981).

140. J. A. Krumhansl, B. Horowitz, and A. J. Heeger, *Solid State Commun.* **34**, 945 (1980).

141. T. Nakano and H. Fukuyama, *J. Phys. Soc. Jpn.* **49**, 1679 (1980).

142. T. Takoshima, M. Ido, K. Tsutsumi, T. Sambongi, S. Honma, K. Yamaya, and Y. Abe, *Solid State Commun.* **35**, 911 (1980).

143. C. Kittel, *Introduction to Solid State Physics*, Wiley, New York, 1976.

144. E. Fermi, J. R. Pasta, and S. M. Ulam, Los Alamos Sci. Lab. Rep. LA-1940 (1955) and in *Collected Works of Enrico Fermi*, Vol. II, University of Chicago Press, Chicago, 1965 p. 978.

145. See, for example, S. A. Rice, in *Quantum Dynamics of Molecules*, R. G. Woolley, Ed., Plenum Press, New York, 1980, p. 257.

146. N. J. Zabusky, *Comput. Phys. Commun.* **5**, 1 (1973).

147. N. J. Zabusky, in *Nonlinear Partial Differential Equations*, W. Ames, Ed., Academic Press, New York, 1967, p. 223.

148. M. Toda, *J. Phys. Soc. Jpn.* **22**, 431 (1967).

149. M. Toda, *J. Phys. Soc. Jpn.* **23**, 501 (1967).

150. M. Toda, *J. Phys. Soc. Jpn. Suppl.* **26**, 235 (1969).

151. M. Toda, *Prog. Theor. Phys. Suppl.* **45**, 179 (1970).

152. M. Toda, *Phys. Rep.* **18**, 1 (1975).

153. M. Toda, *Prog. Theor. Phys. Suppl.* **59**, 1 (1976).

154. M. Toda, *Rocky Mtn. J. Math.* **8**, 197 (1978).

155. R. Hirota, *J. Phys. Soc. Jpn.* **35**, 286 (1973).

156. R. Hirota and J. Satsuma, *Prog. Theor. Phys. Suppl.* **59**, 64 (1976).

157. H. Flaschka, *Phys. Rev. B* **9**, 1924 (1974).

158. H. Flaschka, *Prog. Theor. Phys.* **51**, 703 (1974).

159. M. Henon, *Phys. Rev. B* **9**, 1921 (1974).

160. M. Kac and P. Van Moerbeke, *Proc. Natl. Acad. Sci. USA* **72**, 1627 (1975); **72**, 2879 (1975).

161. H. Flaschka and D. W. McLaughlin, *Prog. Theor. Phys.* **55**, 438 (1976).

162. T. Shirafuji, *Prog. Theor. Phys. Suppl.* **59**, 126 (1976).

163. T. J. Rolfe, S. A. Rice, and J. Dancz, *J. Chem. Phys.* **70**, 26 (1979).

164. T. P. Valkering and C. de Lange, *J. Phys. A* **13**, 1607 (1980).

165. M. K. Ali and R. L. Somorjai, *J. Phys. A* **12**, 2291 (1979).

166. J. H. Batteh and J. D. Powell, *Phys. Rev. B* **20**, 1398 (1979).

167. J. Dancz and S. A. Rice, *J. Chem. Phys.* **67**, 1418 (1977).

168. M. A. Collins, *Chem. Phys. Lett.* **77**, 342 (1981).

169. M. A. Collins and S. A. Rice, *J. Chem. Phys.*, **77**, 2607 (1982).

170. B. L. Holian and G. K. Straub, *Phys. Rev. B* **18**, 1593 (1978).

171. G. K. Straub, B. L. Holian and R. G. Petschek, *Phys. Rev. B* **19**, 4049 (1979).

172. J. R. Hardy and A. M. Karo, in *Proceedings of the International Conference on Lattice Dynamics, Paris 1977*, Flammarion, Paris, 1978, p. 163.

173. B. L. Holian and G. K. Straub, *Phys. Rev. Lett.* **43**, 1598 (1979).

174. J. D. Powell and J. H. Batteh, *J. Appl. Phys.* **51**, 2050 (1980).

175. A. M. Karo and J. R. Hardy, *Int. J. Quant. Chem.* **12**, Suppl. 1, 333 (1977).

176. A. M. Karo and J. R. Hardy, in *Fast Reactions in Energetic Systems*, C. Capellos and R. F. Walker, Eds., Reidel, London, 1981.

177. A. M. Karo, J. R. Hardy and F. E. Walker, *Acta Astronautica* **5**, 1041 (1978).

178. T. J. Rolfe and S. A. Rice, *Physica D* **1**, 375 (1980).

179. T. P. Valkering, *J Phys. A* **11**, 1885 (1978).

180. B. C. Eu, *J. Chem. Phys.* **73**, 2405 (1980).

181. W. H. Miller, (a) *J. Chem. Phys.* **56**, 38 (1972); (b) *Adv. Chem. Phys.* **25**, 69 (1974).

182. J. M. Ziman, *Elements of Advanced Quantum Theory*, Cambridge University Press, Cambridge, 1969, Section 7.7.

183. T. Schneider and E. Stoll, *Phys. Rev. Lett.* **45**, 997 (1980).

184. H. Takahashi, *Proc. Phys.-Math. Soc. Jpn.* **24**, 60 (1942); and in *Mathematical Physics in One Dimension*, E. H. Lieb and D. C. Mattis, Eds., Academic Press, New York, 1966, p. 25.

185. H. Büttner and F. G. Mertens, *Solid State Commun.* **29**, 663 (1979).

186. P. C. Dash and K. Patnaik, *Prog. Theor. Phys.* **65**, 1526 (1981).

187. P. C. Dash and K. Patnaik, *Phys. Rev. A* **23**, 959 (1981).

188. G. Herzberg, *Infrared and Raman Spectra*, Van Nostrand, New York, 1945, pp. 118, 362ff

189. J. W. Ellis, *Trans. Faraday Soc.* **25**, 888 (1929).

190. B. R. Henry and W. Siebrand, *J. Chem. Phys.* **49**, 5369 (1968).

191. B. R. Henry, *Vibr. Spectra Struct.* **10** (1980).

192. M. L. Sage and J. Jortner, *Adv. Chem. Phys.* **47**, 293 (1981).

193. P. R. Bunker, *Molecular Symmetry and Spectroscopy*, Academic Press, New York, 1979.

194. A. Messiah, *Quantum Mechanics*, Vol. 1, North-Holland, Amsterdam, 1970, p. 208.

195. G. L. Lamb, *Rev. Mod. Phys.* **43**, 99 (1971).

196. T. Barnard, *Phys. Rev. A* **7**, 373 (1973).

A DEDUCTIVE APPROACH
TO REDUCED DYNAMICS AND ENTROPY
IN FINITE QUANTUM SYSTEMS

PETER PFEIFER*

The Fritz Haber Research Center for Molecular Dynamics
The Hebrew University
Jerusalem 91904, Israel

CONTENTS

*Present address: Fakultät für Chemie, Universität Bielefeld, D-4800 Bielefeld, F.R.G.

I. INTRODUCTION

Most successful attempts to understand irreversibility, approach to equilibrium, and related phenomena on the basis of first-principles quantum theory are of the constructive type. That is, for some specific model (e.g., a system coupled to a heatbath), one analyzes the dynamics of well-selected quantities, typically by setting up some master equation (ME) for these quantities, which then is studied in various limits (weak coupling, low density, long time, etc.). While this approach has beautifully elucidated some basic mechanisms of irreversible behavior for systems with infinitely many degrees of freedom (see Refs. 1–3 for some complementary reviews), it has failed so far to give comparable insight to similar phenomena in finite closed quantum systems. One of the difficulties in the finite case is that the customary MEs either do not meet the actual situation (such as correlated initial states), or else lack the simplifying limits characteristic of infinite systems.

It is the purpose of the following two subsections to prepare for an appreciation of MEs as a necessary concept for whatever theory of irreversibility, rather than as a conditionally sufficient tool for case studies. This shift in logical order leads to a framework for reduced dynamics (Section II) that provides a natural coordinate system for the locus of any one (generalized) ME. Equipped with a general notion of entropy, this paves the road for a systematic search for successful descriptions of irreversibility and the like, even for finite systems. Although the latter application will be illustrated explicitly in Section IV.B, the primary aim of the examples (Sections III, IV) is to demonstrate the unifying versatility of this reduction scheme, that is, that there is much more to reduced quantum dynamics than what the two or three standard examples[4] seem to suggest.

A. Irreversibility from Incomplete Resolution of States

The goal of "a theory of irreversible (elementary) processes from which all statistical elements foreign to pure quantum dynamics have disappeared"[5] is essentially this: to explain why, in a reactive collision of two molecules say, the system—quite irrespectively of its initial state—appears to evolve into one of a rather restricted class of persistent states; and why the population of these final states can often be rationalized by surprisal analysis.[6,7] Thus above all, such a theory should establish that

(i)   for sufficiently many (ideally all) initial states $\rho(0)$ of the system, the state $\rho(t)$ at time t approaches some limit state $\rho(\infty)$ as $t \to \infty$; and that

(ii) for many different initial states $\rho(0)$, $\rho(t)$ evolves into one and the same final $\rho(\infty)$ as $t \to \infty$.

But if we assume (as we shall) the dynamics to be Hamiltonian, then (i) may fail due to Poincaré recurrence cycles, and (ii) cannot be because unitary time evolution preserves distance between states. So the program, as formulated, is too narrow a transcription of the indicated experimental results. Indeed, since the state of a system may be identified with the list of expectation values of all observables of the system, statements of type (i) or (ii) cannot be inferred from experiment unless enough observables have been measured—which is very rare (except perhaps for pure spin systems). In practice one rather measures a few distinguished observables, and it is for the expectation values of these alone that one knows the existence of infinite-time limits and the like. If we denote this actualized set of observables by $\mathcal{Q}$, then a realistic version of (i) and (ii) reads:

(i') For sufficiently many initial states $\rho(0)$, the expectation-value limits $\lim_{t \to \infty} \mathrm{tr}(\rho(t)A)$ exist for all $A \in \mathcal{Q}$ (existence of $\mathcal{Q}$-effective equilibrium states).

(ii') For many initial states $\rho(0)$ with different expectation values on $\mathcal{Q}$, the limits $\lim_{t \to \infty} \mathrm{tr}(\rho(t)A)$ coincide for all $A \in \mathcal{Q}$ (asymptotic contraction of relevant observables).

Now by the hypothesis that only observables from $\mathcal{Q}$ are available, we can no longer distinguish between states that give the same expectation values on $\mathcal{Q}$. Call any such two states $\mathcal{Q}$-equivalent. The question then arises whether, for a given set $\mathcal{S}$ of initial states, the time evolution maps $\mathcal{Q}$-equivalent states into $\mathcal{Q}$-equivalent states again. For if so, a reduced description in terms of time-evolving $\mathcal{Q}$-equivalence classes ($\mathcal{Q}$-effective states) is possible, in which case (i) and (ii) may well be realizable by appropriate reference to effective states (instead of ordinary states) to yield (i') and (ii').[8]

Of course, it is by no means necessary that time evolution preserve $\mathcal{C}$-equivalence for all times (Sections III, IV.A) in order to recover (i′) and (ii′). Considerably larger sets $\mathcal{S}$ —often even the set of *all* initial states—have the property that the time evolution respects $\mathcal{C}$-equivalence just in the infinite-time limit (Section IV.B) to give (i′) and (ii′). This reflects the circumstance that the dynamics at intermediate times depends also on initial-state details hidden with respect to $\mathcal{C}$, but that such memory effects tend to decay as $t \to \infty$.[‡] So for initial states not particularly prepared, still an asymptotic reduced description may be possible.

B. Master Equations Revisited

Quantum-mechanical MEs are usually introduced as a shortcut to determine selected properties of a system at time t from these very properties at time 0 without having to solve for the full dynamics. But by construction,[4] any such ME propagates the selected properties only for certain well-picked initial states. So the propagator (like the time-ordered exponential of the probability matrix in Ref. 9), solving an initial-value problem, does not depend on initial data—and yet it does depend (implicitly) on an underlying specific class of initial states. While this dichotomy may be viewed as analogue of the moving vs. space-fixed frames of reference for the total vs. partial time derivative of the density of an incompressible fluid, say, its origin remains obscure in the customary inductive theory of MEs. For example, it is not immediate why only a particular choice of initial states yields the desired "local universality" of the propagator; or what the propagator, provided that one exists, would be for some other class of initial states (recall the beginning of this section).

Such problems find straightforward answers once we realize that all known quantum MEs can be interpreted as examples for dynamics of $\mathcal{C}$-equivalence classes as introduced in Section I.A (see examples 2, 5, 6 in Section III, henceforth called standard MEs, for the pertinent $\mathcal{C}$'s and $\mathcal{S}$'s). Indeed, preservation of $\mathcal{C}$-equivalence of initial states in the course of time is the backbone of any ME and hence of "locally universal" propagators—just as for fluids the incompressibility condition $\operatorname{div} \mathbf{v} = 0$ (together with the general continuity equation) implies $dn/dt = 0$ (cf. Table I).

More specifically, some merits of this amplification of the ME concept are as follows:

1. The usual construction of MEs proceeds via projection of states onto some subspace of state space (Nakajima–Zwanzig scheme[4]). But it is

---

[‡]It remains to be seen if this type of situation may also be formalized as a ("hidden-variable") stochastic process in the space of linear functionals on $\mathcal{C}$ that is nondeterministic for finite t, and asymptotically deterministic for $t \to \infty$.

TABLE I

Analogy between Incompressible Fluids and Master Equations

Fluid dynamics	Quantum dynamics
General: Time t	General: Initial state $\rho(0)$
Density n (in velocity field $\mathbf{v}$) obeying the continuity equation $\partial n / \partial t + \operatorname{div}(n\mathbf{v}) = 0$	Propagator taking $\operatorname{tr}(\rho(0)A)$ into $\operatorname{tr}(\rho(t)A)$ for all $A \in \mathcal{C}$
Incompressible: Conservation of density in moving reference frame: $dn/dt = 0$	Reduced: Propagator does not depend on $\rho(0) \in \mathcal{S}$
Nonconservation of density in space-fixed frame: $\partial n / \partial t \neq 0$	Propagator depends on $\mathcal{S}$
Incompressibility condition: $\operatorname{div} \mathbf{v} = 0$	ME condition: $\rho(t), \sigma(t)$ are $\mathcal{C}$-equivalent if $\rho(0), \sigma(0) \in \mathcal{S}$ are so

only by means of examples, and therefore in an *ad hoc* way, that this projection acquires physical meaning. Here, the physics (kinematics) enters naturally through $\mathcal{C}$ and $\mathcal{S}$, and this conceptual universality allows implementation of whatever theoretical abstractions (e.g., tailored to some experiment) one may wish to consider for the system. The above projection then turns out to be a universal function of both $\mathcal{C}$ and $\mathcal{S}$ (see Section III.A).

2. The required specification of relevant observables, $\mathcal{C}$, replaces the customary (and often rather dubious) procedure of averaging, coarse-graining, or the like over what is considered as irrelevant. This is nontrivial in that, due to the non-Boolean structure of quantum theory, such complementary views are not equivalent in general: Here a ME which is meager because of conservative choice of $\mathcal{C}$ or $\mathcal{S}$ may be improved (by enlarging $\mathcal{C}$ or $\mathcal{S}$) without loss of results previously obtained. Conversely, the fact that the partially ordered "coordinate axes" $\mathcal{C}, \mathcal{S}$ generate a partial order among MEs (Fig. 1) suggests a

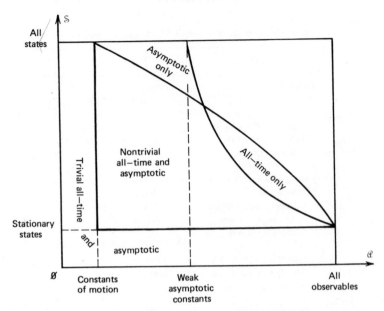

Fig. 1.   Schematic locus of various types of master equations (for fixed time evolution generated by a time-independent Hamiltonian) in the natural $\mathcal{C}$, S "coordinate system." The order among the $\mathcal{C}$'s and S's, as implied by the two axes (each directed by successive inclusion) is only partial, of course. Similarly to a phase diagram, the five fields indicate for what $\mathcal{C}$'s and S's a ME of the type shown exists. See Section II.A for all-time vs. asymptotic ME's, Section IV.A for the trivial ME, and Section IV.B for weak asymptotic constants.

perspective of theory reduction (in the sense of Ref. 10, Chaps. 5.5 and 6) that includes the dynamics from the very outset.

3.   Once $\mathcal{C}$ is fixed, we can distinguish between *purely kinematical* MEs based on sets S such that an arbitrary time evolution preserves $\mathcal{C}$-equivalence (such S's are characterized[‡] by $\mathcal{C}$-inequivalence of any two states in S and underlie the standard MEs, see Section III); and *purely dynamical* ones based on time evolutions that preserve $\mathcal{C}$-equivalence for arbitrary S (this occurs if, e.g., the time evolution Heisenberg-maps $\mathcal{C}$ into itself; see Section IV). So the present framework embraces a spectrum of reduced descriptions of which the standard MEs are just the most meager ones.

[‡]The exceptional case where $\mathcal{C}$ contains only multiples of the unit operator 1 is classed as purely dynamical.

4. Identifying the Nakajima–Zwanzig-projected part of a state as $\mathcal{C}$-effective state, we shall recover well-known entropy expressions for standard MEs (Section III) from a definition of the entropy of $\mathcal{C}$-effective states which carries a well-defined thermodynamic meaning and offers a natural "renormalization", again by putting all the physics into $\mathcal{C}$, of alternative prescriptions for entropy in incompletely resolved systems (Section II.C). For some simple MEs (examples 5, 10, 11 in Sections III.C and IV.B) one can even show that this effective-state entropy may, but need not, increase when effective states coalesce in the sense of (ii′), Section I.A. (Another case of entropy increase occurs in example 4 [see Section III.B].)

5. The result that the ME condition is necessary and sufficient for the dynamics of expectation values (for $\mathcal{C}$, and with initial-state set $\mathcal{S}$) to be expressible in terms of $\mathcal{C}$-observables evolving effectively in $\mathcal{C}$ (Section II.B) puts on general grounds earlier instances of such dual reduced dynamics (see Ref. 3, p. 138–142; and Ref. 11). Typically this reduced Heisenberg picture (RHP) is just a family of maps from $\mathcal{C}$ to $\mathcal{C}$. Neither does it have a group structure, nor do the maps extend to algebraic morphisms if the linear span of $\mathcal{C}$ is an algebra. It is this second departure from the basic assumptions of algebraic quantum theory (see e.g. Ref. 10, Chaps. 4.2–3) that allows for asymptotic contraction of relevant observables and hence appears as ultimate source of irreversibility independent of initial states.

C. A Caveat

Only for examples 5 and 8–11 will explicit (but exact) solutions of MEs appear in this paper. Its main purpose is rather to expound how to check (in several ways) whether some, say experimentalist-given triple ($\mathcal{C}$, $\mathcal{S}$, Hamiltonian) lends itself to a ME; how to ensure certain continuity, range, and fixed-point properties of this ME; how to guarantee a unique RHP (indeed, the price for interesting subdynamics, which typically demand an infinite-dimensional Hilbert space for the system, is that even modest properties of this kind require qualification); how to compute the effective-state entropy for special $\mathcal{C}$'s; how to construct the Nakajima–Zwanzig projector for suitable $\mathcal{C}$ and $\mathcal{S}$ (examples 2–7); and how to use Hilbert-space methods (examples 6, 10, 11) and Lie-algebraic techniques (examples 4, 8, 9) to arrive at an analytically tractable ME when the projection scheme is inapplicable or, depending on the Hamiltonian, not the most convenient one. This confinement is with the understanding that references such as those in Table II provide ample guidance as to the final setting up of the MEs in question, and to the numerous possible modes of their subsequent analysis.

II.   GENERAL THEORY

We assume the system to live in a separable Hilbert space $\mathcal{K}$ (with inner product $\langle\ |\ \rangle$), so that[‡]

1. the observables are given by self-adjoint elements of $\mathcal{B}(\mathcal{K})$, the Banach algebra of all bounded (linear) operators on $\mathcal{K}$ with the usual operator norm $\|\ \|$;[§]

2. the (normal) states are given by unit-trace positive elements of $\mathcal{T}(\mathcal{K})$, the Banach space of all trace-class operators on $\mathcal{K}$ with the usual trace norm $\mathrm{tr}(|\ |)$; thus $\mathrm{tr}(\rho A)$ is the expectation value of observable A in state (density operator) ρ;

3. the time evolution is given by a family of propagators $\Phi_t: \mathcal{B}(\mathcal{K}) \to \mathcal{B}(\mathcal{K})$, $\Phi_t(B) \equiv U_t B U_t^{-1}$ $[B \in \mathcal{B}(\mathcal{K})]$, where the unitary operators U_t on $\mathcal{K}$ satisfy $\dot{U}_t = -iH_t U_t$, $U_0 = 1$, with self-adjoint, possibly time-dependent Hamiltonian H_t ($t \in \mathbb{R}$); thus in the Schrödinger (Heisenberg) picture, $\Phi_t(\rho)$ $[\Phi_t^{-1}(A)]$ is the state ρ [observable A] at time t.

In the sets $\mathcal{C}$ and $\mathcal{S}$ of actualized observables and initial states we conveniently include, by abuse of language, also all (complex) linear combinations of such observables and states, respectively. That is, $\mathcal{C}$ and $\mathcal{S}$ are subspaces —not necessarily closed, but spanned by self-adjoint and positive elements, respectively— of $\mathcal{B}(\mathcal{K})$ and $\mathcal{T}(\mathcal{K})$.[¶] Such $\mathcal{C}$ and $\mathcal{S}$ will be called *appropriate*.

Let $\mathcal{C}^*$ be the Banach space of all bounded linear functionals on $\mathcal{C}$ with the usual norm for functionals, and define the (contraction) mapping $\hat{\ }: \mathcal{T}(\mathcal{K}) \to \mathcal{C}^*$ by

$$\hat{\rho}(A) \equiv \mathrm{tr}(\rho A) \qquad \text{for all} \quad A \in \mathcal{C}, \tag{2.1}$$

$\rho \in \mathcal{T}(\mathcal{K})$. If ρ is a state, then $\hat{\rho}$ is called $\mathcal{C}$-*effective state* (generated by ρ). By linearity of $\hat{\ }$, the set of all $\mathcal{C}$-effective states is convex. By $\hat{\Phi}_t(\rho)$ we mean the application of the mapping $\hat{\ }$ to $\Phi_t(\rho)$, and similarly for the circumflex on other compound expressions.

A.   Master Equations

Since $\hat{\Phi}_t(\rho)$ is stripped of all redundancy of $\Phi_t(\rho)$ relative to $\mathcal{C}$, one expects a simpler time dependence (the more marked the smaller $\mathcal{C}$) for the

---

[‡]The functional-analytic apparatus used for the development is not deep and may be found for instance in Ref. 12. Its use is for more than mere convenience, however. It safeguards against ambiguities and pitfalls hardly avoidable otherwise.

[§]Observables like energy have a place in $\mathcal{B}(\mathcal{K})$ as bounded functions, particularly spectral projectors, of the corresponding unbounded self-adjoint operators.

[¶]If also correlations between (and particularly, dispersions of) observables are open to actual measurement, then $\mathcal{C}$ is likely to be even an algebra.

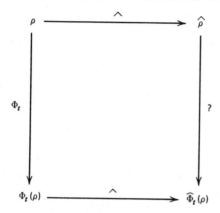

Fig. 2.   Indetermination of the $\mathcal{C}$-effective state at time t by its initial value.

$\mathcal{C}$-effective state than for the state generating it. It is the object of a ME to describe this dynamics of $\mathcal{C}$-effective states.

Now $\hat{\Phi}_t(\rho)$ is typically not a function of $\hat{\rho}$ alone; or equivalently, $\rho, \sigma \in \mathcal{T}(\mathcal{H})$ with $\rho \neq \sigma$ and $\hat{\rho} = \hat{\sigma}$ may lead to $\hat{\Phi}_t(\rho) \neq \hat{\Phi}_t(\sigma)$ (Fig. 2). Hence such effective-state dynamics, if to be formulated as an initial-value problem (e.g., as a first-order integrodifferential equation), can no longer be global, but is possible only for specific subsets of initial states.

The triple $(\mathcal{C}, \mathcal{S}, \Phi.)$ is said to admit an *all-time ME* if $\hat{\Phi}_t(\rho) = \hat{\Phi}_t(\sigma)$ for all $t \in \mathbb{R}$ whenever $\rho, \sigma \in \mathcal{S}$ satisfy $\hat{\rho} = \hat{\sigma}$, that is, if, for every $t \in \mathbb{R}$, there exists a (linear) mapping $\Theta_t: \hat{\mathcal{S}} \to \mathcal{C}*$,[‡] called the $(\mathcal{C}, \mathcal{S})$-*effective propagator*, such that

$$\hat{\Phi}_t(\rho) = \Theta_t(\hat{\rho}) \qquad \text{for all} \quad \rho \in \mathcal{S}. \qquad (2.2)$$

Any equation determining Θ. is called ME. $(\mathcal{C}, \mathcal{S}, \Phi.)$ admits an *asymptotic ME* if $w^*\text{-}\lim_{t \to \pm\infty} \hat{\Phi}_t(\rho)$ exists [this means (Ref. 12, p. 140) that $\lim_{t \to \pm\infty} \text{tr}(\Phi_t(\rho)A)$ exists for all $A \in \mathcal{C}$ and implies that the limit functional, called the $\mathcal{C}$-*effective equilibrium state* if ρ is a state, is bounded] for every $\rho \in \mathcal{S}$, and $w^*\text{-}\lim_{t \to \pm\infty} \hat{\Phi}_t(\rho) = w^*\text{-}\lim_{t \to \pm\infty} \hat{\Phi}_t(\sigma)$ whenever $\rho, \sigma \in \mathcal{S}$ satisfy $\hat{\rho} = \hat{\sigma}$; that is, if there exists a (linear) mapping $\Theta_{\pm\infty}: \hat{\mathcal{S}} \to \mathcal{C}*$ (called the asymptotic $(\mathcal{C}, \mathcal{S})$-effective propagator) such that

$$w^*\text{-}\lim_{t \to \pm\infty} \hat{\Phi}_t(\rho) = \Theta_{\pm\infty}(\hat{\rho}) \qquad \text{for all} \quad \rho \in \mathcal{S}. \qquad (2.3)$$

[‡]By the usual shorthand for a set of images of a mapping, $\hat{\mathcal{S}} \equiv \{\hat{\rho} | \rho \in \mathcal{S}\}$, $\Phi_t(\mathcal{S}) \equiv \{\Phi_t(\rho) | \rho \in \mathcal{S}\}$, and so on.

1. Functionals vs. Equivalence Classes

To see that this corresponds to the discussion in Section I.A, we first note that two states ρ, σ are $\mathcal{C}$-equivalent (as defined in Section I.A) if and only if $\hat{\rho} = \hat{\sigma}$. $\mathcal{S}_t \equiv \Phi_t(\mathcal{S})$ is the linear span of states, originating from $\mathcal{S}$, at time t. Its decomposition into mutually disjoint $\mathcal{C}$-equivalence classes is given (Ref. 12, p. 140) by the quotient space $\mathcal{S}_t / \mathcal{S}_t^{\mathcal{C}}$ with

$$\mathcal{S}_t^{\mathcal{C}} = \{ \rho \mid \rho \in \mathcal{S}_t, \hat{\rho} = \hat{0} \}.$$

Clearly, $\mathcal{S}_t / \mathcal{S}_t^{\mathcal{C}}$ and $\hat{\mathcal{S}}_t$ are vector-space isomorphic, so $\mathcal{C}$-equivalence class and $\mathcal{C}$-effective state are essentially synonymous. For example, in Eq. (2.2) we recognize the earlier condition that Φ_t should preserve $\mathcal{C}$-equivalence when applied to $\mathcal{S}$, which may usefully be written as

$$\Phi_t\left(\mathcal{S}_0^{\mathcal{C}}\right) \subseteq \mathcal{S}_t^{\mathcal{C}}. \tag{2.4}$$

In fact, (2.4) is the full-fledged ME analogue of the incompressibility condition in Table I. Trivial as its being valid for $\mathcal{S}_0^{\mathcal{C}} = \{0\}$ (irrespectively of Φ.) may seem, the latter is precisely the characteristic of purely kinematical MEs (recall item 3 in Section I.B). As illustrated by Eq. (2.3), the formalism in terms of $\hat{}$ is more appropriate, however, in that it naturally carries the physically relevant topology associated with expectation values: For $\rho \in \mathcal{S}_t$, we have

$$\sup_{\substack{A \in \mathcal{C} \\ \|A\| \leqslant 1}} |\mathrm{tr}(\rho A)| \leqslant \inf_{\substack{\sigma \in \mathcal{S}_t \\ \hat{\sigma} = \hat{\rho}}} \mathrm{tr}(|\sigma|)$$

where the two sides equal the $(\mathcal{C}^*)$ norm of $\hat{\rho}$ and the (quotient-space) norm of the corresponding $\mathcal{C}$-equivalence class, respectively (for the latter, see Ref. 12, p. 140, and note that $\mathcal{S}_t^{\mathcal{C}}$ is closed relative to $\mathcal{S}_t$). So $\hat{\mathcal{S}}_t$ and $\mathcal{S}_t / \mathcal{S}_t^{\mathcal{C}}$ are not isomorphic as naturally normed spaces.

2. Continuity Properties

Fix $-\infty \leqslant t \leqslant \infty$ and let Θ_t be effective propagator for $(\mathcal{C}, \mathcal{S}, \Phi.)$. For each $s \in \mathbb{R}$, Φ_s maps $\mathcal{T}(\mathcal{H})$ continuously (with respect to the trace norm) onto itself, so one expects some of this continuity to survive in $\Theta_t : \hat{\mathcal{S}} \to \mathcal{C}^*$. Indeed, it follows from Eq. (2.7) below that Θ_t is w^*-continuous, that is, that $\rho_1, \rho_2, \ldots \in \mathcal{S}$ and $w^*\text{-}\lim_{n \to \infty} \hat{\rho}_n = \hat{0}$ imply $w^*\text{-}\lim_{n \to \infty} \Theta_t(\hat{\rho}_n) = \hat{0}$, and (Ref. 12, Theorem III.5.28) that Θ_t is closable (with respect to the norm). So in the practically important case of norm-closed $\hat{\mathcal{S}}$ (examples 1–5, 7–11), the closed-graph theorem (Ref. 12, p. 166) tells us that Θ_t is even norm continuous, that is, that $\rho_1, \rho_2, \ldots \in \mathbb{S}$ and $\lim_{n \to \infty} \|\hat{\rho}_n\| = 0$ imply $\lim_{n \to \infty} \|\Theta_t(\hat{\rho}_n)\| = 0$. Another desideratum,

$$\Theta_t(\hat{\mathbb{S}}) \subseteq \hat{\mathfrak{T}}(\mathcal{H}) \tag{2.5}$$

is trivially satisfied for $-\infty < t < \infty$, and holds as well for $t = \pm\infty$ if $\mathcal{C}$ is a von Neumann algebra (see Ref. 13, including also examples where (2.5) fails). Finally, for an all-time ME with $\Phi.$ induced by unitaries U_t strongly operator continuous in t (such as when H_t is t-independent), $\Theta_t(\hat{\rho})$ depends norm-continuously on t, that is,

$$\|\Theta_t(\hat{\rho}) - \Theta_s(\hat{\rho})\| \to 0 \qquad \text{as} \quad |t - s| \to 0$$

for all $\rho \in \mathbb{S}$ (cf. Ref. 13).

3. Time-Translation Invariance

Let $(\mathcal{C}, \mathbb{S}, \Phi.)$ admit a ME with $\Phi.$ generated by a time-independent Hamiltonian [so $\Phi.$ is a group, $\Phi_t(\Phi_s(\ )) = \Phi_{t+s}(\ )$ for all $t, s \in \mathbb{R}$] and with $\Phi_t(\mathbb{S}) \subseteq \mathbb{S}$ for all $t \in \mathbb{R}$. Then, depending on the case, time-translation invariance shows up as

$$\Theta_{t+s}(\hat{\rho}) = \Theta_t(\hat{\Phi}_s(\rho))$$

or

$$\Theta_{\pm\infty}(\hat{\rho}) = \Theta_{\pm\infty}(\hat{\Phi}_s(\rho))$$
$$= \Theta_{\pm\infty}(\Theta_{\pm\infty}(\hat{\rho})) \qquad \text{if} \quad \Theta_{\pm\infty}(\hat{\rho}) \in \hat{\mathbb{S}} \tag{2.6}$$

for all $\rho \in \mathbb{S}$. Equation (2.6) enhances the equilibrium property of the output of $\Theta_{\pm\infty}$ by showing that the intersection of the domain and range of $\Theta_{\pm\infty}$ equals the set of all fixed points of $\Theta_{\pm\infty}$.

4. How Big Can $\mathcal{C}$ and $\mathbb{S}$ Be?

Two bounds on the possibilities of asymptotic MEs are as follows (Fig. 1): For $\Phi.$ generated by a time-independent Hamiltonian, $(\mathfrak{B}(\mathcal{H}), \mathbb{S}, \Phi.)$ admits an asymptotic ME if and only if $\mathbb{S}$ is spanned by stationary states (see Ref. 13, Theorem 5); and $(\mathcal{C}, \mathfrak{T}(\mathcal{H}), \Phi.)$ admits an asymptotic ME if and only if $\mathcal{C}$ is an (appropriate) subset of $\mathcal{C}_\pm$ with $\Gamma_\pm(\mathcal{C}) \subseteq \mathcal{C}$ (see Section IV.B, where $\mathcal{C}_\pm$ and $\Gamma_\pm$ are explained). This illustrates also the elementary but noteworthy point that if $(\mathcal{C}, \mathbb{S}, \Phi.)$ admits a ME, then so does $(\mathcal{C}, \mathbb{S}', \Phi.)$ for every (appropriate) $\mathcal{S}' \subseteq \mathcal{S}$, but that neither $\mathcal{Q}' \subseteq \mathcal{Q}$ nor $\mathcal{Q}' \supseteq \mathcal{Q}$ implies anything about $(\mathcal{Q}', \mathcal{S}, \Phi.)$ in general. Concerning extensions: If $(\mathcal{Q}, \mathcal{S}, \Phi.)$ admits an all-time ME, then so do $(\mathcal{Q}, \mathcal{S}', \Phi.)$, $(\mathcal{Q}', \mathcal{S}, \Phi.)$, $(\mathcal{Q}', \mathcal{S}', \Phi.)$ whenever $\mathcal{S}'$ and $\mathcal{Q}' \supseteq \mathcal{Q}$ are (appropriate) subsets of the trace-norm closure of $\mathcal{S}$, and of the set of limits of weakly operator-convergent sequences in $\mathcal{Q}$, respectively. For such $\mathcal{Q}'$, any two states are $\mathcal{Q}'$-equivalent if and only if they are $\mathcal{Q}$-equivalent, so that $(\mathcal{Q}, \mathcal{S}, \Phi.)$ and $(\mathcal{Q}', \mathcal{S}, \Phi.)$ have essentially the same ME.

B.   Reduced Heisenberg Picture

A conceptually different way of formalizing the idea of reduced dynamics is this: $(\mathcal{Q}, \mathcal{S}, \Phi.)$ is said to admit an *all-time RHP* if, for every $t \in \mathbb{R}$, there exists a (linear) mapping $\Gamma_t : \mathcal{Q} \to \mathcal{Q}$ such that

$$\mathrm{tr}\big(\Phi_t(\rho)A\big) = \mathrm{tr}\big(\rho\Gamma_t(A)\big) \qquad \text{for all} \quad \rho \in \mathcal{S}, \quad A \in \mathcal{Q}.$$

$(\mathcal{Q}, \mathcal{S}, \Phi.)$ admits an *asymptotic RHP* if there is a (linear) mapping $\Gamma_{\pm\infty}$: $\mathcal{Q} \to \mathcal{Q}$ such that

$$\lim_{t \to \pm\infty} \mathrm{tr}\big(\Phi_t(\rho)A\big) = \mathrm{tr}\big(\rho\Gamma_{\pm\infty}(A)\big) \qquad \text{for all} \quad \rho \in \mathcal{S}, \quad A \in \mathcal{Q}.$$

1.   Equivalence of MEs and RHPs

Not too surprisingly, the duality between state evolution and evolution of observables, as mediated by the unitary implementation of the propagators Φ_t, persists in all reduced descriptions: $(\mathcal{Q}, \mathcal{S}, \Phi.)$ admits a ME if and only if it admits a RHP (of the same type). Indeed, a variant of the ME condition (2.4) is that, for every $\rho \in \mathcal{S}$, $\mathrm{tr}(\rho\mathcal{Q}) = \{0\}$ must imply $\mathrm{tr}\big(\rho\Phi_t^{-1}(\mathcal{Q})\big) = \{0\}$. But such can be if and only if, for every $A \in \mathcal{Q}$, there is some $A' \in \mathcal{Q}$ such that $\mathrm{tr}\big(\rho\Phi_t^{-1}(A)\big) = \mathrm{tr}(\rho A')$ for all $\rho \in \mathcal{S}$. The proof for the asymptotic case is analogous. Duality manifests itself through the corollary that $\Theta_t : \hat{\mathcal{S}} \to \mathcal{Q}^*$ is adjoint to $\Gamma_t : \mathcal{Q} \to \mathcal{Q}$ ($-\infty \leqslant t \leqslant \infty$) (Ref. 12, p. 167):

$$[\Theta_t(\hat{\rho})](A) = \hat{\rho}(\Gamma_t(A)) \qquad \text{for all} \quad \rho \in \mathcal{S}, \quad A \in \mathcal{Q}. \tag{2.7}$$

2.   Is the RHP Unique? Continuous?

At variance with the unreduced dynamics, the correspondence between the Schrödinger and the Heisenberg picture is no longer one-to-one at all levels of reduction: While the ME propagator Θ_t is always unique (by construction), the same is not true for the mapping Γ_t in the associated RHP if $\mathcal{S}$ is too small (see example 6 in Section III.D for an instance of complete characterization of all possible Γ_t's for given $(\mathcal{Q}, \mathcal{S}, \Phi.)$ and t). This is why no general results on continuity and the like are available for Γ_t as for Θ_t, and why the ME a priori is more convenient than the RHP.

However, if

$$\mathcal{C} \text{ is normed-spaced isomorphic to a subspace of } \hat{\mathcal{S}}* \qquad (2.8)$$

($\hat{\mathcal{S}}* = $ Banach space of all bounded linear functionals on $\hat{\mathcal{S}}$), then Γ_t (t fixed) is unique and is closable as a linear transformation from the (norm) completion of $\mathcal{C}$ to itself; if additionally $\hat{\mathcal{S}}$ is norm closed, then Γ_t is even norm continuous. The proof is simple: By hypothesis and the fact that $\mathcal{C}$ is normed-space isomorphic to a subspace of $\mathcal{C}**$, we can regard $\Gamma_t: \mathcal{C} \to \mathcal{C}$ as a (typically nondensely defined) linear transformation from $\mathcal{C}**$ to $\hat{\mathcal{S}}*$. As such, it is adjoint to $\Theta_t: \hat{\mathcal{S}} \to \mathcal{C}*$ and hence equals the restriction to $\mathcal{C}$ (taken as a subspace of $\mathcal{C}**$) of *the* adjoint $\Theta_t^\dagger$ of Θ_t, $\Theta_t^\dagger$ being a (densely defined) closed linear transformation from $\mathcal{C}**$ to $\hat{\mathcal{S}}*$, which is bounded if Θ_t is so (Ref. 12, pp. 167–168).

This criterion [part (2.8) of which is trivially satisfied when $\hat{\mathcal{S}}$ is norm dense in $\mathcal{C}*$] can be used to establish uniqueness and continuity in norm (for fixed t) of the RHP in examples 1–5, 7–11.

C. Entropy of Effective States

The notion of entropy as measure of the mixedness of a state (or of missing information about the system in that state) generalizes naturally to our $\mathcal{C}$-effective states if we accordingly translate the usual definition of pure and mixed states (see e.g. Ref. 1, Chaps. 1.2–1.4) and the Jaynes–Uhlmann characterization[14] of the entropy of density matrices: Let $\rho \in \mathcal{T}(\mathcal{H})$ be a state. Call $\hat{\rho}$ *mixed* if there are states $\sigma, \tau \in \mathcal{T}(\mathcal{H})$ such that $\hat{\sigma} \neq \hat{\tau}$ and

$$\hat{\rho} = p\hat{\sigma} + (1-p)\hat{\tau} \qquad \text{for some} \quad 0 < p < 1.$$

Else $\hat{\rho}$ is called *pure*. Let $\mathcal{P}(\hat{\rho})$ be the set of all numerical sequences $(p_1, p_2, \dots)$ with $1 \geqslant p_1 \geqslant p_2 \geqslant \cdots \geqslant 0$ and $\sum_{n=1}^{\infty} p_n = 1$ for which there exist states $\rho_1, \rho_2, \dots \in \mathcal{T}(\mathcal{H})$ such that

$$\hat{\rho}_1, \hat{\rho}_2, \dots \text{ are pure} \qquad (2.9a)$$

$$\hat{\rho}_n \neq \hat{\rho}_m \qquad \text{whenever} \quad p_n p_m \neq 0 \quad (n, m = 1, 2, \dots) \qquad (2.9b)$$

$$\hat{\rho} = w*\text{-} \sum_{n=1}^{\infty} p_n \hat{\rho}_n. \qquad (2.9c)$$

Then the *entropy* of $\hat{\rho}$ is

$$S(\hat{\rho}) \equiv \inf_{(p_1, p_2, \dots) \in \mathcal{P}(\hat{\rho})} \left(-\sum_{n=1}^{\infty} p_n \ln p_n \right) \qquad (2.10)$$

where $0\ln 0 \equiv 0$, and $S(\hat{\rho}) \equiv \infty$ for empty $\mathcal{P}(\hat{\rho})$. Note that all this does not make use of $\mathcal{S}$ and $\Phi_{.}$.

1. Elementary Properties of $S(\hat{\rho})$

Trivially, $0 \leqslant S(\hat{\rho}) \leqslant \infty$; $S(\hat{\rho}) = 0$ if $\hat{\rho}$ is pure; and $S(\hat{\rho}) = \infty$ may occur for nonempty $\mathcal{P}(\hat{\rho})$. But if $1 \notin \mathcal{Q}$ (as may be in examples 3–5), there may be states ρ with $\hat{\rho} = \hat{0}$, in which case there are no pure effective states and $\mathcal{P}(\hat{\rho})$ is empty for every state ρ. Another cause for the absence of pure effective states is found in examples 7 and 9. Since $\hat{\rho}$ need not be pure even if ρ is (and vice versa), $S(\hat{\Phi}_t(\rho))$ may change as time t passes. In particular, a ME can describe "reduction of the wave packet" by taking a pure $\hat{\rho}$ into a mixed $\Theta_t(\hat{\rho})$ (example 10). To be sure, (2.10) yields the ordinary-state entropy if the reduction $\hat{\ }$ is trivial (example 1).

2. Interpretation of $S(\hat{\rho})$

It is typical of quantum systems that every mixed state can be decomposed into pure states in infinitely many different ways (Ref. 10, pp. 94, 167). So we cannot expect decomposition of a mixed $\mathcal{Q}$-effective state $\hat{\rho}$ into pure components, as stipulated by (2.9), to be given by a unique sequence $(p_1\hat{\rho}_1, p_2\hat{\rho}_2,\dots)$. A special decomposition of $\hat{\rho}$ may be singled out, however, by asking the probability distribution $(p_1, p_2,\dots)$ to be least mixed ("most peaked"), in which case $-\sum_{n=1}^{\infty} p_n \ln p_n$ is minimal (Ref. 15, Section II.C). Thus $S(\hat{\rho})$ can be interpreted as the minimal lack of information about $\mathcal{Q}$ in state ρ.

The thermodynamic meaning of $S(\hat{\rho})$ is as follows: Call two effective states $\hat{\rho}, \hat{\sigma}$ *orthogonal* if [16]

$$\|\hat{\rho} - \hat{\sigma}\| = \|\hat{\rho}\| + \|\hat{\sigma}\|.$$

(This generalizes orthogonality, $\rho\sigma = 0$, of ordinary states ρ, σ.) For all examples in Sections III and IV that allow pure effective states, it is easily verified that every effective state $\hat{\rho}$ has a unique decomposition into mutually orthogonal pure $\hat{\rho}_1, \hat{\rho}_2,\dots$ where the weights $p_1, p_2,\dots$ satisfy

$$S(\hat{\rho}) = - \sum_{n=1}^{\infty} p_n \ln p_n.$$

So we can parallel von Neumann's argument (Ref. 17, Chap. V.2)

1. to identify $\hat{\rho}$ with an ideal gas (at temperature T, with volume V) of boxes, each containing a copy of the system in one of the effective states $\hat{\rho}_1, \hat{\rho}_2,\dots$ with relative frequency $p_1, p_2,\dots$;

2. to check that orthogonality of the $\hat{\rho}_1, \hat{\rho}_2, \ldots$ is sufficient for construction, in terms of $\mathcal{C}$, of a semipermeable wall that separates (e.g.) $\hat{\rho}_1$-boxes from the others with zero change of thermodynamic entropy;

3. to use process 2 for thermodynamically reversible separation of the initial $\hat{\rho}$-"gas" into a $\hat{\rho}_1$-"gas" of volume $p_1 V$, a $\hat{\rho}_2$-"gas" of volume $p_2 V$, and so on, which requires $-kT\sum_{n=1}^{\infty} p_n \ln p_n$ of mechanical work (k = Boltzmann's constant).

Thus in a thermodynamically reversible process (at temperature T) governed by operations from $\mathcal{C}$, the system in state ρ can do the work $kTS(\hat{\rho})$ at most.

3. Another Effective-State Entropy

For $\mathcal{C}$ as in example 5 with finite-dimensional projectors $P_1, \ldots, P_n$ ($n \leq \infty$), von Neumann (Ref. 17, Chap. V.4) introduced an entropy of $\hat{\rho}$ which generalizes to

$$S'(\hat{\rho}) \equiv \sup_{\sigma \in \mathfrak{T}(\mathcal{H}), \, \sigma \text{ a state}, \, \hat{\sigma} = \hat{\rho}} \left[-\text{tr}(\sigma \ln \sigma) \right] \tag{2.11}$$

for arbitrary $\mathcal{C}$-effective states. That is, $S'(\hat{\rho})$ is the ordinary-state entropy of the least-biased generator of $\hat{\rho}$,[6] if such exists. In all examples with $1 \in \mathcal{C}$, one has

$$S(\hat{\rho}) \leq S'(\hat{\rho}) \qquad \text{for all states} \quad \rho \in \mathfrak{T}(\mathcal{H}),$$

equality holding only in exceptional cases. Reference 18 compares (2.10) with (2.11) for $\mathcal{C}$'s as in example 5.

Now if $\mathcal{C}$ is as in example 5 and $P_1, \ldots, P_n$ are the eigenprojectors of a discrete-spectrum Hamiltonian H, then (by definition) evaluation of (2.11) for

$$\rho = \frac{e^{-\beta H}}{\text{tr}(e^{-\beta H})} \qquad \left[\text{if} \quad e^{-\beta H} \in \mathfrak{T}(\mathcal{H}), \quad \beta \equiv (kT)^{-1} \right] \tag{2.12}$$

reproduces the (Gibbs) equilibrium entropy; whereas Eq. (2.10) gives a smaller value unless $P_1, \ldots, P_n$ are all one-dimensional. This failure of (2.10) is only an apparent one: for example, if we determine the thermodynamic entropy from heat-capacity measurements, we really probe the density of energy eigenstates, rather than that of energy eigenvalues (multiplicities not counted) as implied by the above $\mathcal{C}$. So the pertinent $\mathcal{C}$ is a larger one, spanned by mutually orthogonal one-dimensional projectors, and results in an $S(\hat{\rho})$ for (2.12) that agrees with the equilibrium entropy. We could also drop some projectors from $\mathcal{C}$ to mimic a heat-capacity experiment where these are "frozen," and would obtain [from (2.10), (2.12)] the corresponding restricted-equilibrium entropy.

Such an implementation of (2.10) seems apt to render superfluous the use of (2.11), which is plagued by the tendency of $S'(\hat{\rho})$ to be $\equiv \infty$ (even with the extra rule that states σ varying over "frozen degrees of freedom" must be excluded) when the reduction $\hat{}$ ignores sufficiently many aspects of the original description (cf. examples 2–5, 10, 11). Indeed, while $S(\hat{\rho})$ is a function of $\hat{\rho}$ alone (i.e., completely disregards the "internal structure" ρ of $\hat{\rho}$), the quantity $S'(\hat{\rho})$ requires also to know what state space the underlying ρ comes from. (A seemingly similar, implicit reference of $S(\hat{\rho})$ to $\mathcal{T}(\mathcal{H})$ just serves to locate the effective states in $\mathcal{C}^*$.)

III.   EXAMPLES FOR PURELY KINEMATICAL MASTER EQUATIONS

For purely kinematical MEs, and only for these, the effective-propagator family $\Theta.$ is always an explicit function of $\Phi.$. To see this in a constructive way, we show how every such ME fits naturally into the premises of the Nakajima–Zwanzig formalism (and vice versa). This will set the stage for Sections III.B–D.

A.   General Projection Scheme

Fix $\mathcal{C}$, and let $\mathcal{S}$ satisfy

$$\rho \neq \sigma \implies \hat{\rho} \neq \hat{\sigma} \qquad \text{for all} \quad \rho, \sigma \in \mathcal{S} \tag{3.1a}$$

(recall Sect. I.B, item 3). Without loss of generality we may assume $\mathcal{S}$ to be so large that

$$\hat{\mathcal{S}} = \hat{\mathcal{T}}(\mathcal{H}). \tag{3.1b}$$

Equations (3.1a,b) and the extension results in Section II.B imply that $\mathcal{S}$ is $\mathrm{tr}(|\ |)$-closed, and hence that the inverse $I: \hat{\mathcal{S}} \to \mathcal{S}$,

$$I(\hat{\rho}) = \rho \qquad \text{for all} \quad \rho \in \mathcal{S},$$

is well defined, is closed [as a transformation from $\mathcal{C}^*$ to $\mathcal{T}(\mathcal{H})$], and induces a linear transformation $\Pi: \mathcal{T}(\mathcal{H}) \to \mathcal{T}(\mathcal{H})$ by

$$\Pi(\rho) \equiv I(\hat{\rho}) \qquad \text{for all} \quad \rho \in \mathcal{T}(\mathcal{H}).$$

Trivially,

$$\Pi(\mathcal{T}(\mathcal{H})) = \mathcal{S} \tag{3.2a}$$

$$\Pi(\rho) = \rho \qquad \text{for all} \quad \rho \in \mathcal{S} \tag{3.2b}$$

$$\Pi(\rho) = \Pi(\sigma) \quad \Leftrightarrow \quad \hat{\rho} = \hat{\sigma} \qquad \text{for all} \quad \rho, \sigma \in \mathcal{T}(\mathcal{H}) \tag{3.2c}$$

$$\Pi \text{ is tr}(|\ |)\text{-bounded} \tag{3.2d}$$

[use Ref. 12, Problem III.5.22 for (3.2d)]. So Π is a projector (Ref. 12, p. 155) that maps $\mathcal{T}(\mathcal{H})$ onto $\mathcal{S}$ and decomposes every $\rho \in \mathcal{T}(\mathcal{H})$ into an $\mathcal{Q}$-relevant part $\Pi(\rho)$ and a maximal remainder $\mathcal{Q}$-equivalent to 0 (however, Π need not map states into states if $1 \notin \mathcal{Q}$). Conversely, existence of Π with the properties (3.2) implies (3.1).

This—the fact that we can I-represent the time-evolved effective state as Π-projected action of Φ_t on the I-representation of the initial effective state,

$$I(\Theta_t(\hat{\rho})) = \Pi(\Phi_t(I(\hat{\rho}))) \qquad \text{for all} \quad \rho \in \mathcal{S} \tag{3.3}$$

—is the starting point of the Nakajima–Zwanzig method.[‡] The significance of (3.3) is that I and Π are independent of $\Phi_{\cdot}$. Indeed, a projector $\Pi : \mathcal{T}(\mathcal{H}) \to \mathcal{T}(\mathcal{H})$ and a transformation $I : \mathcal{S} \to \Pi(\mathcal{T}(\mathcal{H}))$ such that (3.2c) and (3.3) hold, exist for any ME; but these depend on $\Phi_{\cdot}$ unless (3.1a) is satisfied.

B. The Examples (Formal)

The examples are collected in Table II as follows: columns 1–9 contain kinematical information independent of $\mathcal{S}$. The MEs can be set up in the form of Eq. (3.3) from columns 10, 11 (with more explicit forms indicated in column 12); and column 13 displays the corresponding RHPs.

Unless stated otherwise, $\rho \in \mathcal{T}(\mathcal{H})$ and $A \in \mathcal{Q}$ are arbitrary. The functional $\hat{\rho}$ is always identified with its representation introduced in columns 3, 5. In examples 3–5, $\perp$-projectors are mutually orthogonal self-adjoint projectors; $\hat{\rho}_1, \ldots, \hat{\rho}_n$ are the components of the respective $\hat{\rho}$ in column 3; and if $n = \infty$, then span$\{\cdots\}$ and $\sum_{m=1}^{n} \cdots$ are in the sense of strong operator limits for observables or trace-class limits for states (similarly in example 7). The operation tr( ) is always over the Hilbert space on which the operator argument is defined. Likewise the varying homes of the operators 1 and 0 are clear from the context.

Most entries are straightforward from the definitions. [Use Ref. 14 to infer column 8 from column 7. To prove column 9, use the positivity of the

---

[‡] The absence of the basic condition (3.2c) in the usual treatments of the subject[4] matches with their neglect of $\mathcal{Q}$, so (3.2c) may conversely be taken to reconstruct "the" $\mathcal{Q}$ of any given ME in projected form.

TABLE II
Examples for Purely Kinematical Master Equations

Example	1 Hilbert space $\mathcal{K}$	2 Actualized observables, $\mathscr{A}$	3 Effective state $\hat{\rho}$	4 Effective-state space $\hat{\mathfrak{S}}(\mathcal{K})$
1. Complete state resolution	Arbitrary	$\mathscr{B}(\mathcal{K})$	ρ	$\mathfrak{S}(\mathcal{K})$
2. Subsystem dynamics	$\mathfrak{G}\otimes\mathcal{K}$ with arbitrary Hilbert spaces $\mathfrak{G},\mathcal{K}$	$\langle B\otimes 1 \mid B\in\mathscr{B}(\mathfrak{G})\rangle$	$\mathrm{tr}_{\mathcal{K}}(\rho)$	$\mathfrak{S}(\mathfrak{G})$
3. Measurement process		$\mathrm{span}\{B\otimes P_m,\ m=1,\ldots,n\}$ with fixed $\perp$-projectors $P_1,\ldots,P_n\in\mathscr{B}(\mathcal{K})$	$\begin{pmatrix} \mathrm{tr}_{\mathcal{K}}(\rho[1\otimes P_1]) \\ \vdots \\ \mathrm{tr}_{\mathcal{K}}(\rho[1\otimes P_n]) \end{pmatrix}$	$\mathfrak{S}(\mathfrak{G})^n$ if $n<\infty$ [else space of trace-summable sequences in $\mathfrak{S}(\mathfrak{G})$]
4. Radiationless transitions	Arbitrary	$\mathrm{span}\{P_m B P_m \mid B\in\mathscr{B}(\mathcal{K}),\ m=1,\ldots,n\}$ with fixed $\perp$-projectors $P_1,\ldots,P_n\in\mathscr{B}(\mathcal{K})$	$\begin{pmatrix} P_1\rho{\restriction}P_1\mathcal{K} \\ \vdots \\ P_n\rho{\restriction}P_n\mathcal{K} \end{pmatrix}$ [${\restriction}$ = restriction to]	$\displaystyle\bigoplus_{m=1}^{n}\mathfrak{S}(P_m\mathcal{K})$
5. Level population	Arbitrary	$\mathrm{span}\{P_1,\ldots,P_n\}$ with fixed $\perp$-projectors $P_1,\ldots,P_n\in\mathscr{B}(\mathcal{K})$	$\begin{pmatrix} \mathrm{tr}(\rho P_1) \\ \vdots \\ \mathrm{tr}(\rho P_n) \end{pmatrix}$	$\mathbf{C}^n$ if $n<\infty$ [else space of summable sequences in $\mathbf{C}$]
6. Generalized suprisal analysis	Arbitrary	Appropriate $\mathscr{A}$ consisting of Hilbert–Schmidt (HS) operators on $\mathcal{K}$ and being closed in HS norm	$\mathrm{pr}_{\mathscr{A}}(\rho)$ [for HS operator B, $\mathrm{pr}_{\mathscr{A}}(B)\in\mathscr{A}$ with $\mathrm{tr}(A^\dagger\mathrm{pr}_{\mathscr{A}}(B))=\mathrm{tr}(A^\dagger B)$ for all $A\in\mathscr{A}$]	HS-dense subspace of $\mathscr{A}$
7. Quantum trajectories	$L^2(\mathbf{R}^{J+I})$ $[J,I<\infty]$	$\{$multiplication by function a of variables $1,\ldots,j\mid a\in L^\infty(\mathbf{R}^J)\}$	$\int_{\mathbf{R}^I}\tilde{\rho}(\mathbf{x},\mathbf{y};\mathbf{x},\mathbf{y})\,d(\mathbf{y})$ $(\mathbf{x}\in\mathbf{R}^J)$ [$\tilde{\rho}$ = integral kernel of ρ]	$L^1(\mathbf{R}^J)$

5 Expectation value $\mathrm{tr}(\rho A)$ in terms of $\hat{\rho}$	6 $\hat{\rho}$ is effective state if and only if	7 $\hat{\rho}$ is pure if and only if	8 Entropy of $\hat{\rho}$, $S(\hat{\rho})$	9 Alternative entropy $S'(\hat{\rho})$
$\mathrm{tr}(\hat{\rho}A)$	$\hat{\rho} \geqslant 0;\ \mathrm{tr}(\hat{\rho}) = 1$	$\hat{\rho} = \|\psi\rangle\langle\psi\|$ for some normalized $\psi \in \mathcal{H}$	$-\mathrm{tr}(\hat{\rho}\ln\hat{\rho})$	$S(\hat{\rho})$
$\mathrm{tr}(\hat{\rho}B)$ for $A = B \otimes 1$ $[B \in \mathfrak{B}(\mathcal{G})]$		$\hat{\rho} = \|\psi\rangle\langle\psi\|$ for some normalized $\psi \in \mathcal{G}$		$S(\hat{\rho}) + \ln(\dim\mathcal{K})$
$\mathrm{tr}(\hat{\rho}_m B)$ for $A = B \otimes P_m$ $[B \in \mathfrak{B}(\mathcal{G}),\ m = 1,\ldots,n]$	$\hat{\rho}_1,\ldots,\hat{\rho}_n \geqslant 0;$ $\sum\limits_{m=1}^{n} \mathrm{tr}(\hat{\rho}_m) = 1$ [a]	$\hat{\rho}_m = \delta_{km}\|\psi\rangle\langle\psi\|\ (m=1,\ldots,n)$ for some $1 \leqslant k \leqslant n$ and normalized $\psi \in \mathcal{G}$ [b]	$-\sum\limits_{m=1}^{n} \mathrm{tr}(\hat{\rho}_m \ln\hat{\rho}_m)$ [c]	$S(\hat{\rho}) + \sum\limits_{m=1}^{n} \mathrm{tr}(\hat{\rho}_m)\ln[\mathrm{tr}(P_m)]$ [d]
$\sum\limits_{m=1}^{n} \mathrm{tr}(\hat{\rho}_m A \!\restriction\! P_m\mathcal{K})$		$\hat{\rho}_m = \delta_{km}\|\psi_m\rangle\langle\psi_m\|$ with normalized $\psi_m \in P_m\mathcal{K}$ $(m = 1,\ldots,n)$ for some $1 \leqslant k \leqslant n$ [b]		$S(\hat{\rho})$ [d]
$\hat{\rho}_m$ for $A = P_m$ $(m = 1,\ldots,n)$	$\hat{\rho}_1,\ldots,\hat{\rho}_n \geqslant 0;$ $\sum\limits_{m=1}^{n} \hat{\rho}_m = 1$ [a]	$\hat{\rho}_m = \delta_{km}\ (m = 1,\ldots,n)$ for some $1 \leqslant k \leqslant n$ [b]	$-\sum\limits_{m=1}^{n} \hat{\rho}_m \ln\hat{\rho}_m$ [c]	$S(\hat{\rho}) + \sum\limits_{m=1}^{n} \hat{\rho}_m \ln[\mathrm{tr}(P_m)]$ [d]
$\mathrm{tr}(\hat{\rho}A)$	$\hat{\rho} = \hat{\rho}^\dagger;\ \mathrm{tr}(\hat{\rho}^2) \leqslant 1;\ldots?$	?	?	$-\mathrm{tr}(\hat{\rho}\ln\hat{\rho})$ if $\mathcal{A}$ is an algebra with $1 \in \mathcal{A}$
$\int_{\mathbf{R}^J} \hat{\rho}(\mathbf{x})a(\mathbf{x})d(\mathbf{x})$ for $A =$ multiplication by $a \in L^\infty(\mathbf{R}^J)$	$\hat{\rho} \geqslant 0;$ $\int_{\mathbf{R}^J} \hat{\rho}(\mathbf{x})d(\mathbf{x}) = 1$	$-$ [$\hat{\rho}$ is always mixed]	∞	∞

[a] If $1 \in \mathcal{A}$; else $\leqslant 1$. [c] If $1 \in \mathcal{A}$; else $= \infty$.

[b] If $1 \in \mathcal{A}$; else $\hat{\rho}$ is always mixed. [d] If $1 \in \mathcal{A}$.

359

TABLE II (*Continued*)

10 Initial states, $\mathbb{S}$	11 Projection $\Pi(\rho)$	12 Equation of motion for Θ_t	13 Reduced Heisenberg picture $\Gamma_t(A)$
$\mathfrak{T}(\mathcal{H})$	ρ	Liouville–von Neumann	$\Phi_t^{-1}(A)$
$\langle \tau \otimes \omega \mid \tau \in \mathfrak{T}(\mathfrak{H}) \rangle$ with fixed $\omega \in \mathfrak{T}(\mathfrak{H})$, $\omega \geq 0$, $\mathrm{tr}(\omega) = 1$	$\mathrm{tr}_{\mathfrak{H}}(\rho) \otimes \omega$	Ref. 1, Chap. 10; Ref. 2, Section V; Ref. 3, pp. 137–143; Ref. 19, Section 2	$\mathrm{tr}_{\mathfrak{H}}([1 \otimes \omega]\Phi_t^{-1}(A)) \otimes 1$
$\mathrm{span}\langle \tau \otimes \omega_m \mid \tau \in \mathfrak{T}(\mathfrak{H}), m = 1,\ldots,n \rangle$ with fixed $\omega_m \in \mathfrak{T}(\mathfrak{H})$, $\omega_m \geq 0$, $\mathrm{tr}(\omega_m) = 1$, $\mathrm{tr}(\omega_m P_k) = \delta_{mk}$ $(m,k = 1,\ldots,n)$ $\mathrm{span}\langle \tau_m \mid \tau_m \in \mathfrak{T}(P_m\mathfrak{H}); \ m = 1,\ldots,n \rangle$	$\displaystyle\sum_{m=1}^{n} \mathrm{tr}_{\mathfrak{H}}(\rho[1 \otimes P_m]) \otimes \omega_m$ $\displaystyle\sum_{m=1}^{n} P_m \rho P_m$	(Ref. 1, Chap. 10.4)	$\displaystyle\sum_{m=1}^{n} \mathrm{tr}_{\mathfrak{H}}([1 \otimes \omega_m]\Phi_t^{-1}(A)) \otimes P_m$ $\displaystyle\sum_{m=1}^{n} P_m \Phi_t^{-1}(A) P_m$ [see also Section III.D]
$\mathrm{span}\langle \omega_1,\ldots,\omega_n \rangle$ with fixed $\omega_m \in \mathfrak{T}(\mathcal{H})$, $\omega_m \geq 0$, $\mathrm{tr}(\omega_m) = 1$, $\mathrm{tr}(\omega_m P_k) = \delta_{mk}$ $(m,k = 1,\ldots,n)$	$\displaystyle\sum_{m=1}^{n} \mathrm{tr}(\rho P_m) \omega_m$	Ref. 1, Chap. 10	$\displaystyle\sum_{m=1}^{n} \mathrm{tr}(\omega_m \Phi_t^{-1}(A)) P_m$
$\mathrm{span}\langle \tau \mid \tau \in \mathcal{C}, \tau \geq 0, \mathrm{tr}(\tau) = 1 \rangle$	$\mathrm{pr}_{\mathcal{C}}(\rho)^e$	(Ref. 1, Chap. 10.4); Ref. 9; Ref. 20, Section II; Sect. III.D	$\mathrm{pr}_{\mathcal{C}}(\Phi_t^{-1}(A))$ [see also Section III.D]
$\mathrm{span}\langle\|\psi_m\rangle\langle\psi_m\| \mid m = 1,\ldots,n \rangle$ with fixed $\psi_m(\mathbf{x},\mathbf{y}) = \varphi(\mathbf{x},\mathbf{y})\chi_m(\mathbf{x})$, $\int_{\mathbf{R}^j} |\varphi(\mathbf{x},\mathbf{y})|^2 d(\mathbf{y}) = 1$ $(\mathbf{x} \in \mathbf{R}^j, \mathbf{y} \in \mathbf{R}^j, m = 1,2,\ldots)$ and $|x_1|^2, |x_2|^2, \ldots$ being a basis for $L^1(\mathbf{R}^j)$	$\displaystyle\sum_{m=1}^{\infty} c_m(\rho)\|\psi_m\rangle\langle\psi_m\|$ where, for all $\mathbf{x} \in \mathbf{R}^j$, $\displaystyle\int_{\mathbf{R}^j} \tilde\rho(\mathbf{x},\mathbf{y};\mathbf{x},\mathbf{y})d(\mathbf{y}) = \sum_{m=1}^{\infty} c_m(\rho)|\chi_m(\mathbf{x})|^2$	(Ref. 1, Chap. 10.4)	Multiplication by $\displaystyle\sum_{m=1}^{\infty} \langle \psi_m \mid \Phi_t^{-1}(A)\psi_m \rangle a_m$, where $a_m \in L^\infty(\mathbf{R}^j)$, $\displaystyle\int_{\mathbf{R}^j} a_m(\mathbf{x})\chi_k(\mathbf{x})|^2 d(\mathbf{x}) = \delta_{mk}$ $(m,k = 1,2,\ldots)$

^{e}For this Π, only (3.2b,c) hold if $\mathbb{S} \neq \mathrm{pr}_{\mathcal{C}}(\mathfrak{T}(\mathcal{H}))$ (see Section III.D). Equations (3.2a–d) do hold if (e.g.) $\mathcal{C}$ is a $\|\cdot\|$-closed algebra $\subseteq \mathfrak{T}(\mathcal{H})$.

relative entropy of ordinary states[15] to show that obvious candidates for "the least-biased" generator of $\hat{\rho}$ do give the supremum in (2.11).] As is apparent from Fig. 3, example 6 covers such a wide range of possibilities that general closed-form expressions for the ?-marked entries are unlikely to exist.

In Table II we recognize familiar items like the reduced density operator (column 3, example 3); the presence of the so-called reference state ω in column 11, example 3; state reduction upon a measurement (column 11, example 4 [see also Ref. 1, Chap. 4.3]); the coarse-grained density operator (column 11, example 5 with P_m finite-dimensional and $\omega_m = P_m / \text{tr}(P_m)$ for all m [see also Ref. 15, Section I.B.4]); the coarse-grained entropy (column 9, example 5 with all P_m finite-dimensional); Shannon-type entropies in column 8. These and many other aspects are well-treated in the literature, most of which (including the classics) can be traced back from References 4, 15, and those in column 12 (see also Section III.C below). One important point should here be recalled, however: For none of the ensuing MEs does the effective-state entropy increase generally with time. This is little surprising in view of their purely kinematical origin. Only example 4, with $1 \in \mathcal{C}$, does have

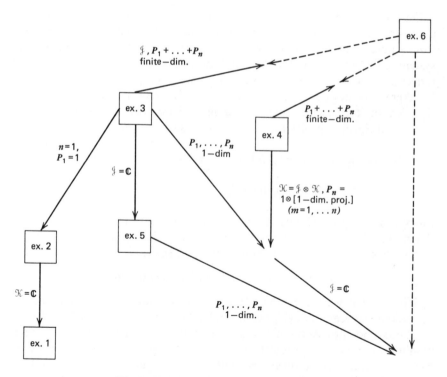

Fig. 3.   Interrelations of examples in Table II.

the property that the density operator $I(\hat{\rho})$ is not more mixed than $I(\Theta_t(\hat{\rho}))$ (use Eq. (3.3); column 11; and Ref. 15, Section II.C) and hence that $S(\hat{\rho}) \leqslant S(\Theta_t(\hat{\rho}))$ for all $\rho \in \mathcal{S}$ and t.

C. The Meat of the Examples

The $\mathcal{S}$'s in Table II (and the ensuing entries in columns 11 and 13) are far from exhaustive, of course. Others are obtained by unitary transformation of these. But the present choices are typical of many applications:

1. Example 2: Subsystem Dynamics

If $\mathcal{J}$ and $\mathcal{K}$ correspond to the observed system and ignored "reservoir," respectively, then the total system is assumed in an uncorrelated initial state $\tau \otimes \omega$. The fixed $\omega \in \mathcal{T}(\mathcal{K})$ may be a thermal-equilibrium state of a heatbath, the vacuum of the radiation field, or the like. If $\mathcal{J}$ and $\mathcal{K}$ describe electrons and nuclei of a molecule, however, then reasonable initial states are adiabatically correlated and hence cannot fall in this class (see example 7).

2. Example 3: Measurement Process

Let the $P_1,\ldots,P_n$ represent the n different pointer positions of a measuring instrument (see example 11 for extraction of such classical observables from a pure quantum description). So $\mathcal{J}$ and $\mathcal{K}$ are the Hilbert spaces of the system to be measured and of the meter. Initial states $\tau \otimes \omega_m$ of indicated type reflect again absence of correlations, and a well-defined pointer position m. Different choices of ω_m, m fixed, amount to different internal meter conditions leading to the same pointer reading. If there exists an observable $E \in \mathcal{B}(\mathcal{J})$ with eigenprojectors $E_1,\ldots,E_n \in \mathcal{B}(\mathcal{J})$ such that, for fixed ω_m we have

$$\lim_{t \to \infty} \mathrm{tr}\big((\tau_{k,t}E_l - \delta_{kl}\tau_{k,t})B\big) = 0 \qquad (B \in \mathcal{B}(\mathcal{J}), \quad k,l=1,\ldots,n)$$

$$\tau_{k,t} \equiv \mathrm{tr}_{\mathcal{K}}\big(\Phi_t(\tau \otimes \omega_m)[1 \otimes P_k]\big) \qquad (k=1,\ldots,n, \quad t \in \mathbb{R})$$

for every $\tau \in \mathcal{T}(\mathcal{J})$ (i.e., such that $\tau_{k,t}$ becomes concentrated in $E_k\mathcal{J}$ as $t \to \infty$), this means that the ω_m-prepared meter measures the observable E of the $\mathcal{J}$-system. For yet stronger measuring properties, see example 11.

3. Example 4: Radiationless Transitions

Suppose that we can model electronic energy transfer in a molecule as transitions between mutually orthogonal subspaces $\mathcal{K}_1,\ldots,\mathcal{K}_n$ of $\mathcal{K}$, where $\mathcal{K}_m$ accommodates vibrations, rotations, and so on in the mth electronic manifold. Let $P_1,\ldots,P_n$ project onto $\mathcal{K}_1,\ldots,\mathcal{K}_n$, so each observable $A \in \mathcal{A}$ can be simultaneously measured with electronic excitation (if also $1 \in \mathcal{A}$—i.e., $\sum_{m=1}^{n} P_m = 1$—then $\mathcal{A}$ contains all such "electronic A's"). Hence the suggested initial states are (mixtures of) states conditioned to well-defined electronic excitation. The latter is not conserved by $\Phi_.$: Even in the present case where $\Phi_.$ is to be the free-molecule time evolution, $\Phi_t^{-1}(P_m)$ varies with t, albeit perhaps slowly, to effect a conversion

$$\rho = P_1 \rho P_1 \mapsto P_2 \Phi_t(\rho) P_2 \tag{3.4}$$

—for example, of an electronically excited state into a vibrotational state of the electronic ground manifold. Depending on the nature of the P_m's, processes of the type (3.4) describe internal conversion or intersystem crossing.

By hypothesis (i.e., choice of $\mathcal{C}$), superpositions of state vectors belonging to different subspaces, $\mathcal{H}_1$ and $\mathcal{H}_2$ say, are not observable. So the example may also be interpreted in terms of superselection rules, however phenomenological (compare Ref. 10 or Ref. 27).

4. Example 5: Level Population

In a scattering experiment one can often measure only the energy of a particular reactant. Let $P_1, \ldots, P_n$ be the eigenprojectors of this reactant's Hamiltonian (acting in the Hilbert space of the total system). A typical initial state is $|\psi\rangle\langle\psi|$ with $\psi \in P_m \mathcal{H}$ for some m. For fixed m, different preparations of the other species lead to different $\psi \in P_m \mathcal{H}$. If accordingly $\omega_m = |\psi_m\rangle\langle\psi_m|$ with normalized $\psi_m \in P_m \mathcal{H}$ $(m = 1, \ldots, n)$, then $\mathrm{tr}(\Phi_t(\omega_m) P_k)$ is the probability of finding the reactant in level k (irrespectively of the other species) after having been ω_m-prepared in level m. And for the instance of potential scattering with one-dimensional $P_1, \ldots, P_n$, column 11 reduces to the customary diagonal-part scheme

$$\Pi(\rho) = \sum_{m=1}^{n} \langle \psi_m | \rho \psi_m \rangle |\psi_m\rangle\langle\psi_m|,$$

where $n = 1$ corresponds to decay of a single pure state.

To see an exactly solvable example (somewhat different from the above scattering situation) of how high-dimensional P_m's and a Hamiltonian with continuous spectrum tend to simplify such population dynamics, we consider a two-level molecule interacting with the radiation field in the dipole-velocity–rotating-wave approximation. The Hamiltonian, first with N discrete modes of the field, reads[21]

$$H = \begin{pmatrix} 0 & 0 \\ 0 & \delta \end{pmatrix} \otimes 1 + 1 \otimes \sum_{\nu=1}^{N} \varepsilon_\nu b_\nu^\dagger b_\nu$$

$$+ \sum_{\nu=1}^{N} \lambda_\nu \left[\begin{pmatrix} 0 & 0 \\ 1 & 0 \end{pmatrix} \otimes b_\nu + \begin{pmatrix} 0 & 1 \\ 0 & 0 \end{pmatrix} \otimes b_\nu^\dagger \right]$$

where $\delta > 0$ is the level separation; $b_\nu, b_\nu^\dagger$ are the boson annihilation and creation operators for the νth field oscillator of frequency $\varepsilon_\nu > 0$; and $\lambda_\nu \in \mathbb{R}$ are coupling constants ("Wigner–Weisskopf–Dicke model"). But H, acting in $\mathfrak{H} = \mathbb{C}^2 \otimes \mathfrak{K}$, where $\mathfrak{K}$ is the N-mode Hilbert space, may also be interpreted as coupling two electronic states

$$|1\rangle \equiv \begin{pmatrix} 1 \\ 0 \end{pmatrix}, \qquad |2\rangle \equiv \begin{pmatrix} 0 \\ 1 \end{pmatrix}$$

of a molecule to N of its vibrational modes—that is, as another model for radiationless transitions. Clearly the state $|1\rangle \otimes |0\rangle$ with the molecule down and the field in the vacuum $|0\rangle \in \mathfrak{K}$ ($b_\nu |0\rangle = 0$, $\nu = 1, \ldots, N$) is an eigenstate of H—actually[21] the ground state if (and only if) $\delta \geq \sum_{\nu=1}^{N} \lambda_\nu^2 \varepsilon_\nu^{-1}$. The nondecay probability $p(t) = |\langle \psi | e^{-iHt} \psi \rangle|^2$ of an initial state $\psi = |2\rangle \otimes |0\rangle$, with the molecule excited and no photons present, is[21]

$$p(t) = \left| \sum_{\substack{\varepsilon \in \mathbb{R} \\ f(\varepsilon) = 0}} \frac{e^{-i\varepsilon t}}{f'(\varepsilon)} \right|^2$$

$$f(\varepsilon) \equiv \varepsilon - \delta - \sum_{\nu=1}^{N} \frac{\lambda_\nu^2}{\varepsilon - \varepsilon_\nu}.$$

So if we choose the projectors P_m to be eigenprojectors $|m\rangle\langle m| \otimes 1$ ($m = 1, 2$) of the "reactant's" Hamiltonian $\begin{pmatrix} 0 & 0 \\ 0 & \delta \end{pmatrix} \otimes 1$, and take initial states to be arbitrary populations of the two levels in the absence of photons,

$$\mathbb{S} = \mathrm{span}\{ |m\rangle\langle m| \otimes |0\rangle\langle 0| \,|\, m = 1, 2 \},$$

then the corresponding ME (with representation

$$\hat{\rho} \equiv \begin{pmatrix} \hat{\rho}_1 \\ \hat{\rho}_2 \end{pmatrix} \in \mathbb{C}^2$$

according to Table II) is

$$\Theta_t \left(\begin{pmatrix} \hat{\rho}_1 \\ \hat{\rho}_2 \end{pmatrix} \right) = \begin{pmatrix} 1 & 1 - p(t) \\ 0 & p(t) \end{pmatrix} \begin{pmatrix} \hat{\rho}_1 \\ \hat{\rho}_2 \end{pmatrix} \qquad (\rho \in \mathbb{S}).$$

Since the function f has (exactly) $N + 1$ zeros, $p(t)$ so far is an almost periodic function of t. In the continuum limit, however, where $N \to \infty$ and ε_ν –

$\varepsilon_{\nu-1} \to 0$ so that $\lambda_\nu^2 (\varepsilon_\nu - \varepsilon_{\nu-1})^{-1}$ becomes a function $g(\varepsilon) > 0$ of the continuous frequency variable $0 < \varepsilon < \infty$, one can prove[21] that

$$p(t) = \left| q(t) + \int_0^\infty \frac{e^{-i\varepsilon t} g(\varepsilon)}{\pi^2 g^2(\varepsilon) + f^2(\varepsilon)} \, d\varepsilon \right|^2 \qquad (p(0) = 1)$$

$$f(\varepsilon) \equiv \varepsilon - \delta - \int_0^\infty \frac{g(\varepsilon')}{\varepsilon - \varepsilon'} \, d\varepsilon' \qquad (\varepsilon \in \mathbb{R})$$

$$q(t) \equiv \begin{cases} 0 & \text{if } \delta \geqslant \displaystyle\int_0^\infty g(\varepsilon) \varepsilon^{-1} d\varepsilon \\[2ex] \dfrac{e^{-iEt}}{f'(E)} & \text{otherwise} \quad (\text{where } f(E) = 0,\, E < 0). \end{cases}$$

Thus $\lim_{t \to \infty} p(t) = 0$ by the Riemann–Lebesgue lemma whenever $\delta \geqslant \int_0^\infty g(\varepsilon)\varepsilon^{-1} d\varepsilon$ [the case $q(t) \neq 0$ will be discussed elsewhere[21]]. The associated asymptotic ME

$$\Theta_\infty \left(\begin{pmatrix} \hat\rho_1 \\ \hat\rho_2 \end{pmatrix} \right) = \begin{pmatrix} \hat\rho_1 \dotplus \hat\rho_2 \\ 0 \end{pmatrix} \qquad (\rho \in \mathbb{S})$$

describes the ultimate decay of the excited level to the ground state, with zero entropy of the $\mathcal{C}$-effective equilibrium state. Had we chosen, for example, $\mathbb{S} = \mathrm{span}\langle |1\rangle\langle 1| \otimes |0\rangle\langle 0|, |2\rangle\langle 2| \otimes |1_\nu\rangle\langle 1_\nu|\rangle$, where $|1_\nu\rangle \equiv b_\nu^\dagger |0\rangle$, ν fixed (induced emission), then a different 2×2 matrix for Θ_t would have obtained (the ME depends on $\mathbb{S}$), but Θ_∞ would presumably coincide with the present one.

5. Example 6: Generalized Surprisal Analysis

With $P_1, \ldots, P_n$ as in example 5 (but now assumed finite-dimensional and with $n < \infty$), $\mathcal{C} = \mathrm{span}\langle P_1, \ldots, P_n\rangle$ satisfies the hypotheses of example 6. So initial states in $\mathbb{S}$ ($= \mathcal{C}$) are mixtures of *states* P_m, each being an equal-weight mixture of whatever pure states carried by $P_m \mathcal{H}$. This is Laplace's principle of assigning equal probabilities to completely unresolved states. More generally, if $\mathcal{C}$ is an (appropriate) $\| \; \|$-closed algebra contained in $\mathcal{T}(\mathcal{H})$, then each state $\rho \in \mathbb{S}$ ($= \mathcal{C}$) is of maximum entropy subject to $\mathcal{C}$, that is,

$$-\mathrm{tr}(\sigma \ln \sigma) < -\mathrm{tr}(\rho \ln \rho) < \infty$$

for all states $\sigma \in \mathcal{T}(\mathcal{H})$ with $\sigma \neq \rho$ and $\hat\sigma = \hat\rho$ (Ref. 22, Section II.B). If additionally $1 \in \mathcal{C}$ (whence $\dim \mathcal{H} < \infty$), then $\mathbb{S}$ contains all such maximum-entropy states, and the projected ME (3.3) sends each of them into the maximum-entropy generator ρ_t [$\equiv \mathrm{pr}_{\mathcal{Q}}(\Phi_t(\rho))$] of $\hat{\Phi}_t(\rho)$; so $\rho \mapsto \rho_t$ is the least-biased prediction for the time evolution of $\mathcal{Q}$-resolved states.

6.   *Example 7: Quantum Trajectories*

'Let variables $1, \ldots, j$ and $j+1, \ldots, j+l$ correspond to the nuclei and electrons of one or more reacting molecules. So $\mathcal{Q}$ focuses on the nuclear density. The initial-state ψ_m's may be understood as adiabatic wave functions with electronic ground-state function $\varphi(\mathbf{x}, \cdot)$ (depending parametrically on the nuclear position $\mathbf{x}$) and nuclear functions χ_m, where m indexes different vibrotational states. So the indicated $\mathcal{S}$ is nothing but a particular model for the subspace associated with the electronic ground manifold in example 4. The technical requirement that the "conformations" $|\chi_m|^2$ should be linearly independent and complete is likely to be satisfied automatically.

The ME computes the nuclear particle density at time t from that at time 0, whereas the RHP generates trajectories in the following sense: If A is the "operator" of formal multiplication by $\delta(\cdot - \mathbf{x}_0)$, $\mathbf{x}_0 \in \mathbb{R}^j$ fixed, then $\Gamma_t(A)$ in column 13 describes the diffusion of the initial configuration $\mathbf{x}_0$.

D.   Variants

1.   Ad Example 4

Consider the case (different from that in Section III.C) where $P_1, \ldots, P_n$ are the eigenprojectors of a discrete-spectrum Hamiltonian K. Suppose that Φ. is generated by the Hamiltonian $K + V$, where V is some perturbation. Then the RHP can be formally expressed[23] as the time average (ergodic limit[13])

$$\Gamma_t(A) = \lim_{\varepsilon \downarrow 0} \varepsilon \int_0^\infty e^{-\varepsilon s} e^{-iKs} e^{i(K+V)t} A e^{-i(K+V)t} e^{iKs}\, ds$$

$$= \lim_{\varepsilon \downarrow 0} \varepsilon \int_0^\infty e^{-\varepsilon s} e^{i(K+V(s))t} A e^{-i(K+V(s))t}\, ds$$

where

$$V(s) \equiv e^{-iKs} V e^{iKs}.$$

This is particularly useful when the operators K, V, A generate a tractable Lie commutator algebra (recall that $[K, A] = 0$ for all $A \in \mathcal{Q}$). It also shows $\Gamma_t(A)$ to be the average over s of the evolution of A under the dressed Hamiltonian $K + V(s)$, where the "driving force" $V(s)$ depends on s only to the extent that V does not commute with K and hence does couple relevant observables (i.e., $\mathcal{Q}$) to irrelevant ones (note that s-dependence of $V(s)$ is indeed the source of all t-dependence of $\Gamma_t(A)$). So here is an explicit instance of the idea[8] that the "reservoir" drives the effective (reduced) dynamics in a way which predisposes to equilibration. For instance, from Table II and the above expression for $\Gamma_t(A)$ we can obtain $\Pi(\Phi_t(\rho))$, $\rho \in S$, in a form that manifests mixing enhancement (recall Section III.B) similarly to Gaussian semigroups (Ref. 2, Section V.B).

2.   Ad Example 6

For S as in column 10, $\Pi(\ ) = \mathrm{pr}_{\mathcal{C}}(\ )$ may fail to satisfy (3.2a, d) because S may be so small (even empty, since $\mathcal{C}$ need contain neither positive nor trace-class elements) that (3.1b), necessary for Π of the type (3.2) to exist, does not hold. More natural than to enlarge S and construct Π satisfying (3.2), however, is to retain S and use $\mathrm{pr}_{\mathcal{C}}(\ )$ for a projection scheme operating in the space of HS (instead of trace-class) operators. Such is precisely the setting of the Alhassid–Levine approach:[9] Since $\mathcal{C}$ is separable with respect to the HS norm (because $\mathcal{K}$ is so), we can write

$$\mathrm{pr}_{\mathcal{C}}(B) = \sum_{k,m=1}^{n} [\mathbf{A}^{-1}]_{km} \mathrm{tr}\left(A_m^\dagger B\right) A_k \tag{3.5}$$

for every HS operator B on $\mathcal{K}$. Here $A_1,\ldots,A_n$ ($n \leqslant \infty$) is any basis for $\mathcal{C}$;[‡] $\mathbf{A}^{-1}$ is the inverse of the Gram matrix

$$\mathbf{A} \equiv \left\{\mathrm{tr}\left(A_k^\dagger A_m\right)\right\}_{k,m=1}^{n};$$

and the sum converges in HS norm. Still identifying $\hat{\rho}$ with $\mathrm{pr}_{\mathcal{C}}(\rho)$ as in column 3 [so that $\hat{\rho} = \rho$ and

$$\Theta_t(\hat{\rho}) = \mathrm{pr}_{\mathcal{C}}\left(\Phi_t(\hat{\rho})\right) \tag{3.6}$$

for all $\rho \in S$], we expand $\hat{\rho}$ in (3.6) according to (3.5) to obtain

$$\Theta_t(\hat{\rho}) = \sum_{k,m=1}^{n} \mathrm{pr}_{\mathcal{C}}\left(\Phi_t(A_k)\right)[\mathbf{A}^{-1}]_{km} \mathrm{tr}\left(A_m^\dagger \hat{\rho}\right),$$

[‡]For example, with $\mathrm{tr}(A_k^\dagger A_m) = \delta_{km}$ for all k, m. Independently, one can also pick it with $A_k^\dagger = A_k$ for all k.

or in the numerical-sequence representation,

$$\mathbf{a}_t = \mathbf{A}_t \mathbf{A}_0^{-1} \mathbf{a}_0 \tag{3.7}$$

$$\mathbf{a}_t \equiv \left\{ \mathrm{tr}\left(A_l^\dagger \Phi_t(\rho) \right) \right\}_{l=1}^n$$

$$\mathbf{A}_t \equiv \left\{ \mathrm{tr}\left(A_l^\dagger \Phi_t(A_k) \right) \right\}_{l,k=1}^n$$

for all $\rho \in \mathcal{S}$ and $t \in \mathbb{R}$. This is the integrated form of the ME in Ref. 9. The inverse of $\mathbf{A}_t$ required for the differential version

$$\dot{\mathbf{a}}_t = \dot{\mathbf{A}}_t \mathbf{A}_t^{-1} \mathbf{a}_t$$

of (3.7) exists if $\Phi_t^{-1}(\mathcal{Q}) \supseteq \mathcal{Q}$, but also simply by continuity if $|t|$ is sufficiently small. Noninvertibility of $\mathbf{A}_t$ means that part of "$\mathbf{a}$-space" becomes inaccessible ("contraction of phase space").

If $\mathcal{S}$ is not HS-dense in $\mathcal{Q}$, then the RHP in column 13 is not unique. We may replace it by

$$\Gamma_t(A) = \mathrm{pr}_{\bar{\mathcal{S}}}\left(\Phi_t^{-1}(A) \right) \qquad (A \in \mathcal{Q}) \tag{3.8}$$

where $\bar{\mathcal{S}}$ is the HS closure of $\mathcal{S}$. Reference 24 shows that, out of all possible RHPs in this case, (3.8) is the maximally reduced one. In fact, if this Γ_t is considered as an operator acting in $\mathcal{Q}$, then it is simply *the* adjoint of Θ_t where Θ_t is viewed as an operator acting in $\mathcal{Q}$ with nondense domain $\mathcal{S}$.[24]

IV. EXAMPLES FOR PURELY DYNAMICAL MASTER EQUATIONS

Almost by definition, MEs not purely kinematical can be gathered only as more or less disconnected events. The following examples, both for the all-time and for the asymptotic type, are no exception. This and the preceding section differ in that here [in Section III] we fix Φ. [$\mathcal{Q}$] and look for $\mathcal{Q}$ [$\mathcal{S}$] such that $(\mathcal{Q}, \mathcal{S}, \Phi.)$ admits a ME for arbitrary $\mathcal{S}$ [Φ.]. We focus on the maximal choice $\mathcal{S} = \mathcal{T}(\mathcal{H})$ with the understanding that the ME may simplify for · smaller $\mathcal{S}$.

A. Nonexpansion of Relevant Observables

This means that

$$\Phi_t^{-1}(\mathcal{Q}) \subseteq \mathcal{Q} \qquad \text{for all} \quad t \in \mathbb{R},$$

so $(\mathcal{Q}, \mathcal{T}(\mathcal{H}), \Phi.)$ admits a RHP (with $\Gamma_t = \Phi_t^{-1}$) and hence a ME. Trivial instances are example 1 and the constants of motion of a given $\Phi.$; the latter yields the *trivial ME* $\Theta_t(\hat{\rho}) = \hat{\rho}$ for all ρ and t. More instructive are these:

1. Example 8: Observables Closed under Commutation with the Hamiltonian

This is to say that there are operators A_k and numerical functions $a_{km,\cdot}$ such that

$$[H_t, A_k] = \sum_{m=1}^{n} a_{km,t} A_m \qquad (k = 1, \ldots, n) \qquad (4.1a)$$

and implies that the Heisenberg evolution of $A_1, \ldots, A_n$ is obtained from solving a first-order $n \times n$-matrix differential equation:

$$\Phi_t^{-1}(A_k) = \sum_{m=1}^{n} g_{km,t} A_m$$

$$\dot{g}_{km,t} = i \sum_{l=1}^{n} a_{kl,t} g_{lm,t}, \qquad g_{km,0} = \delta_{km}.$$

So if $\mathcal{Q} = \mathrm{span}\{A_1, \ldots, A_n\} = \mathrm{span}\{A_1^\dagger, \ldots, A_n^\dagger\} \subseteq \mathcal{B}(\mathcal{H})$, then the resulting ME computes final expectation values from initial ones according to

$$[\Theta_t(\hat{\rho})](A_k) = \sum_{m=1}^{n} g_{km,t} \hat{\rho}(A_m)$$

for all $\rho \in \mathcal{T}(\mathcal{H})$. Explicit examples for (4.1a) with $n < \infty$ often come[25] from the much stronger Lie property

$$[A_k, A_m] \in \mathcal{Q} \qquad (k, m = 1, \ldots, n) \qquad (4.1b)$$
$$H_t \in \mathcal{Q} \qquad (t \in \mathbb{R}). \qquad (4.1c)$$

Notable exceptions based just on (4.1a) or (4.1a, b) can be found in Refs. 22 and 26. [A special feature of (4.1b) is that any e^B with $B \in \mathcal{Q}$, such as a state of maximum entropy subject to $\mathcal{Q}$ can be expressed[25] as a product of exponentials of $A_1, \ldots, A_n$ separately; whereas (4.1b,c) imply[25] that even the evolution operators U_t are of the form $U_t = e^{-iB_t}$, $B_t = B_t^\dagger \in \mathcal{Q}.$]

2. Example 9: Simplest Continuous Version of Example 8

Let Q, P be the position and momentum operators acting in $\mathcal{H} = L^2(\mathbb{R})$; consider the time-independent Hamiltonian $H_t = P^2 + \lambda Q$ with fixed $\lambda \in \mathbb{R}$

("free fall"); and let $\mathcal{C}$ consist of all bounded functions of P. So formally we have $[H_t, A] \in \mathcal{C}$ for essentially all $A \in \mathcal{C}$, and hence a continuous analogue of (4.1a,b) if we generate $\mathcal{C}$, for example, from $A_k = e^{ikP}$, $k \in \mathbb{R}$. The point is that an infinite-dimensional $\mathcal{C}$ closed under commutation with H_t need not be intractable: Here $\Phi_t^{-1}(g(P)) = g(P - \lambda t)$ for all $g \in L^\infty(\mathbb{R})$, of course; and similarly to example 7, the ME amounts to sending the momentum density $\tilde{\rho}(p, p)$ ($p \in \mathbb{R}$) into the density $\tilde{\rho}(p + \lambda t, p + \lambda t)$ if $\tilde{\rho}$ stands for the integral kernel of $\rho \in \mathcal{T}(\mathcal{H})$ in the momentum representation.

B. Weak Asymptotic Constants

Let Φ. be generated by a time-independent Hamiltonian. Then all purely dynamical asymptotic MEs are based on the weak asymptotic constants

$$\mathcal{C}_\pm \equiv \left\{ A \mid A \in \mathcal{B}(\mathcal{H}); \Gamma_\pm(A) \equiv \underset{t \to \pm\infty}{w\text{-lim}} \Phi_t^{-1}(A) \text{ exists} \right\}$$

[see Ref. 1, Chap. 1.6 for a résumé of w(eak) and strong operator convergence to synthesize $\mathcal{C}_\pm$ partially from the smaller set of strong asymptotic constants (Ref. 27, p. 111)]. Since

$$A \in \mathcal{C}_\pm \quad \Rightarrow \quad A^\dagger \in \mathcal{C}_\pm$$

$$\Phi_t^{-1}(\mathcal{C}_\pm) = \mathcal{C}_\pm \qquad (t \in \mathbb{R})$$

$$\Gamma_\pm\left(\Phi_t^{-1}(A)\right) = \Phi_t^{-1}(\Gamma_\pm(A)) = \Gamma_\pm(A) \qquad (A \in \mathcal{C}_\pm, \ t \in \mathbb{R})$$

[so $\Gamma_\pm(\mathcal{C}_\pm)$ equals the constants of motion and hence is in $\mathcal{C}_\pm$], it follows that $\left(\mathcal{C}_\pm, \mathcal{T}(\mathcal{H}), \Phi.\right)$ admits an asymptotic RHP with associated ME given by Eq. (2.7) with $\mathcal{C} = \mathcal{C}_\pm$, $\mathcal{S} = \mathcal{T}(\mathcal{H})$, $t = \pm\infty$, $\Gamma_{\pm\infty} = \Gamma_\pm$.

A rich source of asymptotic constants is scattering theory, of course: If the generalized wave operators exist, then every constant of motion of the free system, when projected onto the subspace of scattering states, is a (strong) asymptotic constant of the interacting system (Ref. 27, p. 126).[30] The final two examples illustrate how (weak) asymptotic constants arise when the wave operator does not exist (or equals 0 if taken as the weak operator limit), and that these need not "preexist" as constants of the motion of some free Hamiltonian but rather may be created by the interaction (example 11).

1. Example 10: Maximum-Entropy Asymptotics

Let Q, P be the position and momentum operators acting in $L^2(\mathbb{R})$; in $\mathcal{H} = \mathbb{C}^2 \otimes L^2(\mathbb{R})$ consider the time evolution

$$U_t = \exp\left(-it\left[\begin{pmatrix} 1 & 0 \\ 0 & 1 \end{pmatrix} \otimes P^2 + \lambda \begin{pmatrix} 1 & 0 \\ 0 & -1 \end{pmatrix} \otimes Q\right]\right)$$

with constant $\lambda > 0$ ("Stern–Gerlach experiment"); and assume the spin operators to be the actualized observables,

$$\mathcal{A} = \{ B \otimes 1 \,|\, B \in \mathcal{B}(\mathbf{C}^2) \}.$$

Then $\mathcal{A}$ can be proved[28] to consist of weak asymptotic constants, with $\Gamma_{\pm}$ sending every $B \otimes 1$ into (diagonal part of $B) \otimes 1$. Upon identification of $\hat{\rho}$ with

$$\mathrm{tr}_{L^2(\mathbf{R})}(\rho) \equiv \begin{pmatrix} \hat{\rho}_{11} & \hat{\rho}_{12} \\ \hat{\rho}_{21} & \hat{\rho}_{22} \end{pmatrix} \in \mathcal{T}(\mathbf{C}^2)$$

(recall Table II, example 2), the asymptotic ME results in

$$\Theta_{\pm\infty}\left(\begin{pmatrix} \hat{\rho}_{11} & \hat{\rho}_{12} \\ \hat{\rho}_{21} & \hat{\rho}_{22} \end{pmatrix} \right) = \begin{pmatrix} \hat{\rho}_{11} & 0 \\ 0 & \hat{\rho}_{22} \end{pmatrix} \qquad [\rho \in \mathcal{T}(\mathcal{H})], \qquad (4.2)$$

is mixing-enhancing (Ref. 15, Section II.C), and hence entails nondecrease of the effective-state entropy. In fact, Eq. (4.2) implies that $\Theta_{\pm\infty}(\hat{\rho})$ always has entropy greater than any other effective state with the same expectation value for

$$A \equiv \begin{pmatrix} 1 & 0 \\ 0 & -1 \end{pmatrix} \otimes 1.$$

So here we have a quantum dynamical model for strongly successful surprisal analysis:[6] Suppose that, of the initial $\hat{\rho}$, we just know the expectation value $\hat{\rho}_{11} - \hat{\rho}_{22}$ for this A; that therefore we replace $\hat{\rho}$ by the effective state

$$\hat{\rho}' = \begin{pmatrix} \hat{\rho}_{11} & 0 \\ 0 & \hat{\rho}_{22} \end{pmatrix}$$

of maximum entropy subject to $\hat{\rho}'(A) = \hat{\rho}(A)$; that we propagate $\hat{\rho}'$ and use the ensuing $\Theta_\infty(\hat{\rho}')$ to predict expectation values. Since $\Theta_\infty(\hat{\rho}') = \Theta_\infty(\hat{\rho})$, we find that the two give the same results not only for the constraint A (as is the method's goal) but actually for all observables in $\mathcal{A}$.

2. Example 11: Spin Measurement

Assume the hypotheses of example 10, but choose a larger set of actualized observables ("improved resolution"):

$$\mathcal{A} = \mathrm{span}\{ B \otimes P_m \,|\, B \in \mathcal{B}(\mathbf{C}^2), \, m = 1, 2 \}$$

where the pointer positions $P_1, P_2 \in \mathcal{B}(L^2(\mathbb{R}))$ are the projectors onto the functions with support in $]-\infty,0]$ and $[0,\infty[$ respectively. Still,[28] $\mathcal{C}$ consists of weak asymptotic constants where $\Gamma_\pm$ maps

$$\begin{pmatrix} B_{11} & B_{12} \\ B_{21} & B_{22} \end{pmatrix} \otimes P_1 \quad \text{and} \quad \begin{pmatrix} B_{11} & B_{12} \\ B_{21} & B_{22} \end{pmatrix} \otimes P_2$$

into

$$\begin{pmatrix} B_{11} & 0 \\ 0 & 0 \end{pmatrix} \otimes 1 \quad \text{and} \quad \begin{pmatrix} 0 & 0 \\ 0 & B_{22} \end{pmatrix} \otimes 1$$

respectively. Identifying $\hat\rho$ with the two-component vector whose entries are the partial traces

$$\mathrm{tr}_{L^2(\mathbb{R})}(\rho[1 \otimes P_m]) \equiv \begin{pmatrix} \hat\rho_{m,11} & \hat\rho_{m,12} \\ \hat\rho_{m,21} & \hat\rho_{m,22} \end{pmatrix} \in \mathcal{T}(\mathbb{C}^2)$$

($m = 1,2$; recall Table II, example 3), we obtain

$$\Theta_{\pm\infty}\left(\begin{pmatrix} \begin{pmatrix} \hat\rho_{1,11} & \hat\rho_{1,12} \\ \hat\rho_{1,21} & \hat\rho_{1,22} \end{pmatrix} \\ \begin{pmatrix} \hat\rho_{2,11} & \hat\rho_{2,12} \\ \hat\rho_{2,21} & \hat\rho_{2,22} \end{pmatrix} \end{pmatrix} \right) = \begin{pmatrix} \begin{pmatrix} \hat\rho_{1,11} + \hat\rho_{2,11} & 0 \\ 0 & 0 \end{pmatrix} \\ \begin{pmatrix} 0 & 0 \\ 0 & \hat\rho_{1,22} + \hat\rho_{2,22} \end{pmatrix} \end{pmatrix} \qquad [\rho \in \mathcal{T}(\mathcal{H})]$$

(4.3)

for the corresponding asymptotic ME. This says that, for every initial state ρ, the resulting $\mathcal{C}$-effective equilibrium state is a mixture of the pure $\mathcal{C}$-effective states

$$\begin{pmatrix} \begin{pmatrix} 1 & 0 \\ 0 & 0 \end{pmatrix} \\ \begin{pmatrix} 0 & 0 \\ 0 & 0 \end{pmatrix} \end{pmatrix} \quad \leftrightarrow \quad \text{spin up, particle in }]-\infty,0]$$

and

$$\begin{pmatrix} \begin{pmatrix} 0 & 0 \\ 0 & 0 \end{pmatrix} \\ \begin{pmatrix} 0 & 0 \\ 0 & 1 \end{pmatrix} \end{pmatrix} \quad \leftrightarrow \quad \text{spin down, particle in } [0,\infty[$$

with weights equal to the first and second diagonal element of $\mathrm{tr}_{L^2(\mathbb{R})}(\rho)$. So the binary meter (P_1, P_2) measures the spin observable $\begin{pmatrix} 1 & 0 \\ 0 & -1 \end{pmatrix}$ irrespectively of the initial state of the total system (recall Section III.C). Typically of a measurement process, the above decomposition of $\Theta_{\pm\infty}(\hat{\rho})$ into pure effective states is unique. Note that the effective-state entropy nondecreases if (e.g.) the initial state is localized in one of the two half spaces, but nonincreases if (e.g.) $\rho = [|\psi\rangle\langle\psi|] \otimes \omega$ with ψ an eigenvector of $\begin{pmatrix} 1 & 0 \\ 0 & -1 \end{pmatrix}$ and arbitrary $\omega \in \mathfrak{I}(L^2(\mathbb{R}))$.

For the dynamics (4.3), the maximum-entropy procedure works similarly to example 10 when again based on the constant of motion

$$ A \equiv \begin{pmatrix} 1 & 0 \\ 0 & -1 \end{pmatrix} \otimes 1. $$

Knowing just the expectation value $\hat{\rho}_{1,11} + \hat{\rho}_{2,11} - (\hat{\rho}_{1,22} + \hat{\rho}_{2,22})$ of A for the initial $\hat{\rho}$, we take (instead of $\hat{\rho}$) the effective state

$$ \hat{\rho}' = \frac{1}{2} \left(\begin{pmatrix} \hat{\rho}_{1,11} + \hat{\rho}_{2,11} & 0 \\ 0 & \hat{\rho}_{1,22} + \hat{\rho}_{2,22} \end{pmatrix} \\ \begin{pmatrix} \hat{\rho}_{1,11} + \hat{\rho}_{2,11} & 0 \\ 0 & \hat{\rho}_{1,22} + \hat{\rho}_{2,22} \end{pmatrix} \right) $$

of maximum entropy subject to $\hat{\rho}'(A) = \hat{\rho}(A)$ and predict expectation values from $\Theta_\infty(\hat{\rho}')$. Again we find $\Theta_\infty(\hat{\rho}') = \Theta_\infty(\hat{\rho})$—with the difference, however, that here $\Theta_\infty(\hat{\rho}')$ is not always of maximum entropy subject to A.[‡] This corresponds to the fact that in example 10 there is no surprise in $\Theta_\infty(\hat{\rho})$ as compared to the prior $\hat{\rho}'$, while here $\Theta_\infty(\hat{\rho}) \neq \hat{\rho}'$.

V. SUMMARY

It is reviewed why a rigorous theory of irreversible quantum dynamics, typically with small range of accessible final states, requires specification of the set $\mathcal{Q}$ of actualized observables. The ensuing notion of $\mathcal{Q}$-equivalent states naturally singles out sets $\mathbb{S}$ of initial states (density operators) for which the Hamiltonian time evolution Φ_t preserves $\mathcal{Q}$-equivalence for all times t or in

[‡]This differs somewhat from the framework in Refs. 22, 26, where the maximal entropy of the final state, subject to the initial constraints, is guaranteed by unitary state evolution; and from Ref. 29, where the constraints for which the final effective state is of maximal entropy, if they are the same as the initial ones, depend on the initial state.

the limit $t \to \pm\infty$. This identifies reduced descriptions universally as instants of time-evolving $\mathcal{Q}$-equivalence classes (generalized master equations), but has also a counterpart in terms of effective dynamics within $\mathcal{Q}$ (reduced Heisenberg picture). The common quantum-mechanical master equations (subsystem and level dynamics) and some less familiar examples (measurement process, radiationless transitions, generalized surprisal analysis, quantum trajectories) are shown to hold for arbitrary Φ. because $\mathcal{S}$ is taken sufficiently small relative to $\mathcal{Q}$ (purely kinematical master equations). It is demonstrated how to implement the Nakajima–Zwanzig scheme for any other choice of $\mathcal{Q}$ and not too big $\mathcal{S}$. Unrestricted irreversibility, however, may emerge only from purely dynamical master equations where Φ. preserves $\mathcal{Q}$-equivalence for *all* initial states. Notable examples are of the asymptotic type, including two obeying a maximum-entropy principle. The latter rests on a definition of entropy of $\mathcal{Q}$-equivalence classes which, for special $\mathcal{Q}$'s, coincides with certain standard expressions for effective entropy (but not with those based on least-biased states) and turns out to be a measure of the maximal work $\mathcal{Q}$-extractable from the system in any one state of the class.

Acknowledgments

The author is indebted to R. D. Levine for very helpful and patient discussions on the subject, for critical reading of several versions of the manuscript, and for valuable suggestions for improvements. Thanks are also due to R. Kosloff, N. Tishby, and G. A. Raggio for formative discussions and correspondence. The Fritz Haber Research Center is supported by the Minerva Gesellschaft für die Forschung, mbH, München, BRD. This work was supported by the Office of Naval Research and by the Swiss National Science Foundation.

References

1. E. B. Davies, *Quantum Theory of Open Systems*, Academic Press, London, 1976.

2. H. Spohn, *Rev. Mod. Phys.* **53**, 569 (1980).

3. W. Thirring, *Lehrbuch der Mathematischen Physik*, Vol. 4, Springer, Wien, 1980.

4. See, for example, Ref. 1, Chap. 10; Ref. 19, Section 2. See also I. Oppenheim, K. E. Shuler, and G. H. Weiss, *Stochastic Processes in Chemical Physics: The Master Equation*, MIT Press, Cambridge, 1977, Chap. 4.1.

5. R. K. Nesbet, *Chem. Phys. Lett.* **42**, 197 (1976).

6. See, for example, R. D. Levine and J. L. Kinsey, "Information-theoretic approach: Application to molecular collisions," in *Atom-Molecule Collision Theory*, R. B. Bernstein, Ed., Plenum, New York, 1979.

7. For selected initial states, model and ab initio studies of this second problem can be found in G. L. Hofacker and R. D. Levine, *Chem. Phys. Lett.* **33**, 404 (1975); D. F. Heller, *Chem. Phys. Lett.* **45**, 64 (1977); D. C. Clary and R. K. Nesbet, *Chem. Phys. Lett.* **59**, 437 (1978); D. C. Clary and R. K. Nesbet, *J. Chem. Phys.* **71**, 1101 (1979); and Refs. 5, 22, 26.

8. Both the notion of $\mathcal{Q}$-effective states and the idea that the degrees of freedom not in $\mathcal{Q}$ should act as a "reservoir" allowing the $\mathcal{Q}$-effective states to "relax to equilibrium" are as old as quantum mechanics (see, e.g., Ref. 17, Chap. V). By predominant application to large

systems, $\mathcal{C}$ and $\mathcal{C}$-effective states have become known as macroobservables and macrostates (see, e.g., R. Kubo, *Statistical Mechanics*, North-Holland, Amsterdam, 1965, Chap. 1; H. Primas and U. Müller-Herold, *Adv. Chem. Phys.* **38**, 1 (1978), Chap. V; Ref. 10, Chap. 4.3). Their relevance for small systems has been emphasized in Ref. 29. See also S. Augustin and H. Rabitz, *J. Chem. Phys.* **64**, 1223 (1976); and Ref. 20.

9. Y. Alhassid and R. D. Levine, *Chem. Phys. Lett.* **72**, 401 (1980).

10. H. Primas, *Chemistry, Quantum Mechanics, and Reductionism*, Lecture Notes in Chemistry, No. 24, Springer, Berlin, 1981.

11. I. Oppenheim and R. D. Levine, *Physica* **99A**, 383 (1979).

12. T. Kato, *Perturbation Theory for Linear Operators*, 2nd ed., Springer, Berlin, 1976.

13. E. Prugovečki and A. Tip, *J. Math. Phys.* **15**, 275 (1974).

14. E. T. Jaynes, *Phys. Rev.* **108**, 171 (1957), Section 7; A. Uhlmann, *Wiss. Z. Karl-Marx-Univ. Leipzig* **22**, 139 (1973), Section 7.

15. A. Wehrl, *Rev. Mod. Phys.* **50**, 221 (1978).

16. S. Sakai, *C*-Algebras and W*-Algebras*, Springer, Berlin, 1971, Section 1.14.

17. J. von Neumann, *Mathematische Grundlagen der Quantenmechanik*, Springer, Berlin, 1932.

18. R. D. Levine, *Adv. Chem. Phys.* **47**, 239 (1981), Appendix.

19. F. Haake, *Statistical Treatment of Open Systems by Generalized Master Equations*, Springer Tracts in Modern Physics, No. 66, Berlin, 1973.

20. S. Mukamel, *Adv. Chem. Phys.* **47**, 509 (1981).

21. H. H. Chan and M. Razavy, *Chem. Phys. Lett.* **2**, 202 (1968); S. Swain, *J. Phys. A* **5**, 1587 (1972); R. Davidson and J. J. Kozak, *J. Math. Phys.* **14**, 414, 423 (1973); P. Pfeifer, *Phys. Rev. A* **26**, 701 (1982).

22. Y. Alhassid and R. D. Levine, *J. Chem. Phys.* **67**, 4321 (1977).

23. H. Primas, *Rev. Mod. Phys.* **35**, 710 (1963).

24. P. Pfeifer, *J. Math. Phys.* **22**, 1619 (1981); **23**, 1231 (1982).

25. See, for example, R. M. Wilcox, *J. Math. Phys.* **8**, 962 (1967).

26. Y. Alhassid and R. D. Levine, *Phys. Rev. A* **18**, 89 (1978).

27. W. Thirring, *Lehrbuch der Mathematischen Physik*, Vol. 3, Springer, Wien, 1979.

28. P. Pfeifer, *Helv. Phys. Acta* **53**, 410 (1980).

29. Y. Alhassid and R. D. Levine, *Phys. Rev. C* **20**, 1775 (1979).

30. See also P. Pfeifer and R. D. Levine, *J. Chem. Phys.* **78**, in press (1983).

AUTHOR INDEX

Numbers in parentheses are reference numbers and indicate that the author's work is referred to although his name is not mentioned in the text. Numbers in *italics* show the pages on which the complete references are listed.

SUBJECT INDEX

385